HOLLOW STRUCTURAL SECTION
CONNECTIONS and TRUSSES

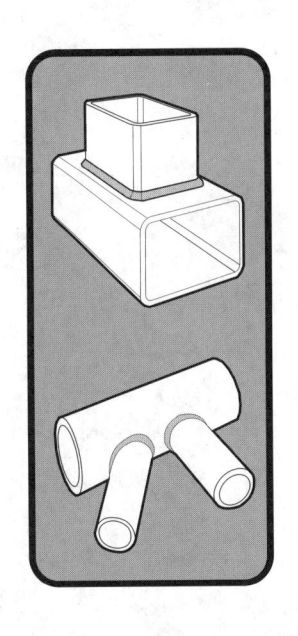

— a Design Guide

By

Jeffrey A. Packer, P.Eng.
Professor of Civil Engineering
University of Toronto
Toronto, Ontario

J.E. (Ted) Henderson, P.Eng.
Henderson Engineering Services
Milton, Ontario

Canadian Institute of Steel Construction

COPYRIGHT © 1997
by
Canadian Institute of Steel Construction

All rights reserved. This book or any part thereof must not be reproduced in any form without the written permission of the publisher.

First Edition
First Printing, July 1992
Revised Second Printing, February 1996

Second Edition
First Printing, June 1997
Revised Second Printing, November 2003

CISC is located at:

201 Consumers Road, Suite 300, Willowdale, Ontario, Canada, M2J 4G8

Telephone: +1-416-491-4552
Fax: +1-416-491-6461
Email: info@cisc-icca.ca
Web site: http://www.cisc-icca.ca

Cover photo:
Screaming Machine, Vancouver, B.C.
COMSTOCK / W. Gordon

ISBN 0-88811-086-3

Printed in Canada by
Quadratone Graphics Limited
Toronto, Ontario

"... I wish to discuss the strength of hollow solids, which are employed in art—and still oftener in nature—in a thousand operations for the purpose of greatly increasing strength without adding to weight; examples of these are seen in the bones of birds and in many kinds of reeds which are light and highly resistant both to bending and breaking."

Galileo Galilei

"Discorsi e dimostrazioni matematiche, intorno a due nuove scienze attenenti alla meccanica e i movimenti locali"

Leiden, The Netherlands, 1638

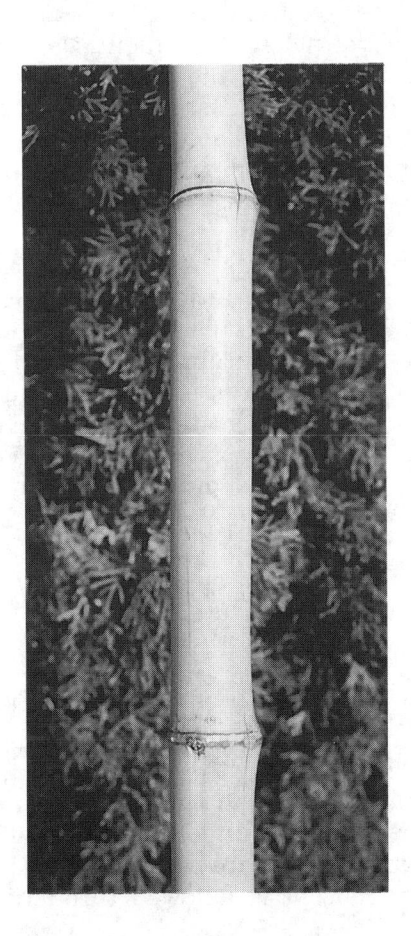

Bamboo:
> A naturally-occurring Hollow Structural Section, and still used in building construction in parts of the world today.

PREFACE

This book, a compendium of current design information on the topic of Hollow Structural Section (HSS) connections and trusses, is directed to practising structural engineers. Connection strength frequently governs the selection of HSS members in structural steelwork, and the impracticality of providing internal reinforcement can pose concerns for the design engineer. However, with some insight into the connection behaviour at the conceptual design stage, subsequent connection reinforcement can frequently be averted, and maximum economy gained from the use of HSS.

In preparing this volume, heavy reliance has been placed upon the sponsored research work of CIDECT (Comité International pour le Développement et l'Etude de la Construction Tubulaire), plus its associated monographs and guides, as well as the design recommendations of the Subcommission on Welded Joints in Tubular Structures of IIW (International Institute of Welding). The authors are also grateful to Stelco Inc. for providing some material contained in a forerunner of the first edition of this book, the Stelco "Hollow Structural Sections Design Manual for Connections" (2nd. edition, 1981). With encouragement of Canadian HSS producers, the publication of this book has been made possible by the financial support of the Steel Structures Education Foundation, the Canadian Welding Bureau and the American Iron and Steel Institute.

Because of the proximity of the American market, all dimensional properties provided in Chapter 1 are given in imperial as well as SI units. It should be noted that the properties given pertain to HSS manufactured in accordance with CAN/CSA-G40.20-M92. Hence, these engineering properties are not appropriate for HSS manufactured to ASTM A500, either in the U.S.A. or in Canada. Wherever possible, formulae that are not dimension-dependant are used, but all design examples are performed in SI units. Furthermore, the limit states design concept is followed throughout, in conformity with CAN/CSA-S16.1-94.

The information presented is primarily directed towards onshore building construction, and this is the basis for the load/resistance calibration. However, sections of the book also lend themselves to other applications (e.g., fatigue strength for the design of bridges, light and sign poles, and crane booms). Complete design examples are performed for statically loaded trusses fabricated from square HSS and from circular HSS, utilizing various connection design aids. Included in the book are design aids in both graphical and tabular formats, plus mention is made of appropriate computer-based design packages where appropriate.

Users of this book will be aware that the design of HSS *members* is well covered by CAN/CSA-S16.1-94 "Limit States Design of Steel Structures", Sections 13 and 18 (Concrete-filled HSS columns). Moreover, the Canadian Institute of Steel Construction "Handbook of Steel Construction" provides column resistance tables for both Class C and Class H HSS.

Since the first edition of "Design Guide for Hollow Structural Section Connections" in 1992, the content has been revised to conform with all current relevant specifications, and brought up-to-date with the latest research on the topic. The scope has been increased by over 30% and the title expanded to more accurately reflect the focus of the book.

The authors therefore hope that practising engineers will find this book on HSS *connections and trusses* a practical reference manual for the design of structures in HSS into the next millenium, thereby further increasing the already popular use of this elegant structural material.

A Chinese version (with J.J. Cao) of the first edition of this Design Guide was published, with permission, in Beijing, China in April of 1997.

Grateful acknowledgement is made for assistance received from many during the preparation and review of both editions of this book.

J.A. Packer
J.E. Henderson

June, 1997

Note: In June of 2003, the Canadian Commission on Building and Fire Codes amended the National Building Code of Canada to reference CSA-S16-01 "Limit States Design of Steel Structures." - Editor

FOREWORD

The Canadian Institute of Steel Construction (CISC) is a national industry organization representing the structural steel, open-web steel joist and steel plate fabricating industries in Canada. Formed in 1930 and granted a Federal charter in 1942, the CISC functions as a non-profit organization promoting the efficient and economic use of fabricated steel in construction.

CISC is pleased to publish this design guide — an important part of a continuing effort to provide current, practical information to assist designers, fabricators, educators, and others interested in the use of steel in construction.

The development of this guide has been generously supported by the Steel Structures Education Foundation (SSEF), the Canadian Welding Bureau (CWB) and the American Iron and Steel Institute (AISI).

The SSEF was incorporated in 1985 to advance the application and use of, and interest in, steel in structures, through education.

The CWB is a federally incorporated, not-for-profit, certification organization with responsibility for the administration of certification programs associated with the welded fabrication, welding consumable, and inspection industries. Certification programs are to various CSA standards such as series W47, W48 and W178. The Bureau is accredited by the Standards Council of Canada for W47.1 and W47.2, and has offices across Canada and in Europe with a head office in Mississauga, Ontario.

The AISI is a non-profit association whose membership includes a broad range of steel producing companies throughout the western hemisphere. Institute activities embrace research and technology, engineering, collection and dissemination of statistics, public distribution of information about the industry and its products, public affairs, and discussion of industrial relations including health and safety.

CISC works in close co-operation with the Steel Structures Education Foundation to develop educational courses and programmes related to the design and construction of steel structures, such as this design guide. The CISC supports and actively participates in the work of the Standards Council of Canada, the Canadian Standards Association, the Canadian Commission on Building and Fire Codes and numerous other organizations, in Canada and other countries, involved in research work and the preparation of codes and standards.

Preparation of engineering plans is not a function of the CISC. The Institute does provide technical information through its professional engineering staff, through the preparation and dissemination of publications, through the medium of seminars, courses, meetings, video tapes, and computer programs. Architects, engineers and others interested in steel construction are encouraged to make use of CISC information services.

Although no effort has been spared in an attempt to ensure that all data in this book is factual and that the numerical values are accurate to a degree consistent with current structural design practice, the Canadian Institute of Steel Construction does not assume responsibility for errors or oversights resulting from the use of the information contained herein. Anyone making use of the contents of this book assumes all liability arising from such use. All suggestions for improvement of this book will be forwarded to the authors for their consideration for future printings.

CISC is located at:

201 Consumers Road, Suite 300, Willowdale, Ontario, Canada, M2J 4G8

Telephone: +1-416-491-4552
Fax: +1-416-491-6461
Email: info@cisc-icca.ca
Web site: http://www.cisc-icca.ca

CONTENTS

Chapter 1 INTRODUCTION

1.1	Building with Hollow Structural Sections	1
1.2	Historical Perspective on HSS Connection Design	6
1.3	Contemporary International Guides and Specifications	10
1.4	Contemporary National Guides and Specifications	12
1.5	Canadian HSS Manufacturing Processes and Standards	14
1.6	Interpretation of Connection Resistance Expressions	17
1.7	Seismic Applications	18
1.8	Geometric Properties of HSS	19
1.9	Symbols and Abbreviations	20
References		28

Chapter 2 STANDARD TRUSS DESIGN

2.1	Practical Aspects and Costs	45
	2.1.1 Costs Relative to Alternative Structural Steel Sections	45
	2.1.2 Costs Among HSS Alternatives	46
2.2	Determination of Truss Forces for Design	48
	2.2.1 Truss Configurations	49
	2.2.2 Truss Analysis	50
2.3	Effective Lengths for Compression Members	53
	2.3.1 Simplified Rules	53
	2.3.2 Empirical Method for HSS Web Members	54
	2.3.3 Long Laterally Unsupported Compression Chords	55
2.4	Truss Deflections	60
2.5	Critical Considerations for Static Strength	60

ix

	2.5.1	General Considerations	60
	2.5.2	Specific Considerations	61
2.6	Truss Design Procedure	62
References	64

Chapter 3   STANDARD TRUSS WELDED CONNECTIONS

3.1	Terminology and Eccentricity	67
3.2	Trusses with HSS Chords and HSS Web Members	68
	3.2.1 K and N Connections	69
	3.2.1.1 Failure Modes for K and N Connections	70
	3.2.1.2 Design Formulae for K and N Connections	73
	3.2.2 T, Y and X Connections	76
	3.2.2.1 Design Formulae for T,Y and X Connections	76
	3.2.3 KT Connections	77
3.3	Trusses with Wide Flange Chords and HSS Web Members	79
3.4	Overlapped "Hidden Toe" Welding and Connection Sequence	79
3.5	Bending Moments Combined with Axial Forces in Web Members	92
3.6	Limits of Validity	92
	3.6.1 Circular HSS Chords and Web Members	93
	3.6.2 Square HSS Chords with Square or Circular HSS Web Members	93
	3.6.3 Rectangular HSS Chords with Rectangular, Square or Circular HSS Web Members	95
	3.6.4 Wide Flange Chords with Rectangular, Square or Circular HSS Web Members	97
3.7	Design Charts and Tables	98
	3.7.1 Design Charts	99
	Charts	100
	3.7.2 Design Tables	110
	Tables	115
3.8	Chord Splices	121
3.9	Design of Reinforced Connections	121
	3.9.1 With Stiffening Plates	121
	3.9.1.1 K and N Connections with Rectangular (or Square) HSS Chords	122
	3.9.1.2 K and KK Connections with Circular HSS Chords	124
	3.9.1.3 T, Y and X Connections with Rectangular (or Square) HSS Chords	125
	3.9.2 With Concrete Filling	127
	3.9.2.1 Compression-Loaded Rectangular (or Square) HSS X Connections	128
	3.9.2.2 Compression-Loaded Rectangular (or Square) HSS T and Y Connections	130

		3.9.2.3	Tension-Loaded Rectangular (or Square) HSS T, Y and X Connections	130
		3.9.2.4	Gap K Connections with Rectangular (or Square) HSS	130
		3.9.2.5	Design Example	132
3.10	Cranked-Chord Connections			135
References				138

Chapter 4 NON-STANDARD TRUSS DESIGN

4.1	Trusses with Double HSS Chords		141
	4.1.1	Types of Double Chord Trusses	141
	4.1.2	Experimental Studies on Double Chord Trusses	145
	4.1.3	Design Guidance for Double Chord Trusses	145
		4.1.3.1 Welded Separated Double Chord Trusses	146
		4.1.3.2 Bolted Separated Double Chord Trusses	146
		4.1.3.3 Back-to-Back Double Chord Trusses	147
	4.1.4	Connection Design for Separated Double Chord Trusses	147
4.2	Trusses with Cropped or Flattened Web Members		148
	4.2.1	Circular HSS Chords	149
	4.2.2	Square HSS Chords	151
4.3	Trusses with Web Members Framing onto Chord Corners		155
4.4	Joists Fabricated from HSS Sections		158
References			161

Chapter 5 MULTIPLANAR WELDED CONNECTIONS

5.1	Introduction		163
5.2	Connections between Circular HSS		163
	5.2.1	XX Connections	165
	5.2.2	TX (XT) Connections	167
	5.2.3	TT and Gap KK Connections under Symmetrical Loading	169
	5.2.4	Gap KK Connections under Non-Symmetrical Loading	171
	5.2.5	Fabrication Aspects	174
5.3	Connections between Rectangular HSS		174
	5.3.1	XX, TX, and TT Connections	174
	5.3.2	KK Connections	177
References			180

Chapter 6   HSS TO HSS MOMENT CONNECTIONS

6.1 Vierendeel Connections between Square or Rectangular HSS ... 183
    6.1.1 Introduction to Vierendeel Trusses 183
    6.1.2 Connection Behaviour and Strength 186
    6.1.3 Connection Flexibility 192
    6.1.4 Design Example 192
        6.1.4.1 Plastic Design of Members 193
        6.1.4.2 Elastic Design of Members 199
6.2 Knee Connections 201
6.3 In-plane and Out-of-plane Moments for T and X Connections ... 205
    6.3.1 In-Plane Bending Moments for Rectangular HSS 205
    6.3.2 In-Plane Bending Moments for Circular HSS 206
    6.3.3 Out-of-Plane Bending Moments for Rectangular HSS ... 207
    6.3.4 Out-of-Plane Bending Moments for Circular HSS 209
6.4 Reinforced Connections between Rectangular/Square HSS 211
References 212

Chapter 7   BOLTED HSS CONNECTIONS

7.1 Circular HSS Flange-Plate Connections 215
7.2 Rectangular HSS Flange-Plate Connections 220
    7.2.1 Connections Bolted along Two Sides of the HSS 220
    7.2.2 Connections Bolted along Four Sides of the HSS 224
7.3 Other Bolted Splice Connections 225
7.4 Blind Bolting into HSS Sections 227
    7.4.1 Huck Ultra-Twist Blind Bolts 227
    7.4.2 Flowdrilling 229
    7.4.3 Connection Failure Modes with Blind Bolts 231
7.5 Rectangular HSS to Gusset-Plate Connections 233
    7.5.1 Net Area, Effective Net Area and Reduced Effective
           Net Area 236
7.6 Other Bolted Connections 241
7.7 Design Examples 244
    7.7.1 Bolted Gusset-Plate Connection 244
    7.7.2 Connection with Plate Slotted into End of HSS 249
References 254

Chapter 8 FABRICATION, WELDING AND INSPECTION

8.1 Fabrication	257
8.1.1 Flattening the Ends of HSS Sections	258
8.1.1.1 Flattening Procedures	258
8.1.2 Simplified End Cuts on Circular HSS Connections	259
8.1.3 Bending HSS Members	260
8.1.3.1 Cold Bending HSS Members	260
8.1.3.2 Hot Bending HSS Members	266
8.1.4 Power Nailing HSS Connections	267
8.1.5 Drainage and Interior Corrosion	270
8.2 Welding and Inspection	271
8.2.1 Codes and Standards	271
8.2.2 Welding Processes	273
8.2.3 Types of Welded Joints	274
8.2.3.1 Fillet Welds	274
8.2.3.2 Partial Joint Penetration Groove Welds	276
8.2.3.3 Complete Joint Penetration Groove Welds	278
8.2.3.4 Prequalified Weld Joints	279
8.2.4 Weld Joint Applications to HSS Construction	279
8.2.4.1 Joint Types	280
8.2.4.2 Weld Effectiveness	288
8.2.4.2.1 Fillet Weld Size Considerations	288
8.2.4.2.2 Effective Weld Length Considerations	290
8.2.5 Inspection	294
8.2.5.1 Methods of Inspection	294
8.2.5.2 Application of Inspection Methods to HSS Construction	296
8.2.6 Avoiding Problem Areas	297
References	299

Chapter 9 BEAM TO HSS COLUMN CONNECTIONS

9.1 Introduction	301
9.2 Rectangular HSS Beams to Rectangular HSS Columns	302
9.2.1 Bolted Connections	302
9.2.2 Unreinforced Welded Connections	306
9.2.3 Reinforced Welded Connections	306
9.2.3.1 Connections Reinforced with Chord (Column) Plate	306
9.2.3.2 Connections Reinforced with Haunches	308
9.3 Wide Flange Beams to Square and Rectangular HSS Columns	309
9.3.1 Simple Shear Connections	309

9.3.2 Moment Connections	313
9.3.2.1 Continuous Beams	314
9.3.2.2 Continuity from Beam to Beam	316
9.3.2.3 Column Face Reinforcement	320
9.3.2.4 Stiffening by Tees on the Beam Flanges	326
9.3.2.5 Concrete Filling of the Column	328
9.4 Wide Flange Beams to Circular HSS Columns	328
References	332

Chapter 10   TRUSSES AND BASE PLATES TO HSS COLUMNS

10.1 Trusses to HSS Columns	335
10.2 Base Plates	335
10.2.1 Base Plates with Axial Loads only	338
10.2.1.1 Axial Compression	338
10.2.1.2 Axial Tension	340
10.2.2 Moment Resisting Base Plates	343
References	344

Chapter 11   PLATE TO HSS CONNECTIONS

11.1 Plate to Square and Rectangular HSS Connections	345
11.2 Plate to Circular HSS Connections	349
References	352

Chapter 12   HSS WELDED CONNECTIONS SUBJECTED TO FATIGUE LOADING

12.1 Introduction	353
12.2 Hot Spot Stress Approach for Fatigue Design	356
12.3 Fatigue Design Procedure for Planar Truss Connections	358
12.4 *SCFs* for Square HSS Connections	365
12.5 *SCFs* for Circular HSS Connections	366
12.6 Materials and Welding	370
12.7 Design Example	371
12.8 HSS to Base Plate Connections	376
References	378

Chapter 13 STANDARD TRUSS EXAMPLES

13.1 General 381
13.2 Warren Truss with Square HSS 381
    13.2.1 Considerations when Making Preliminary Member
            Selections 383
    13.2.2 Preliminary Member Selections 384
    13.2.3 Resistance of Gap K Connections 388
    13.2.4 Top Chord as a Beam-Column between Panel Points 2 and 6 393
    13.2.5 Resistance of an X Connection 396
    13.2.6 Resistance of KT Overlap Connections 397
    13.2.7 Design of Bolted Flange-Plate Splice Connection 401
    13.2.8 Design of Welded Joints 405
    13.2.9 Truss to Column Connection 410
    13.2.10 Truss Deflection 414
13.3 Warren Truss with Circular HSS 416
    13.3.1 Considerations when Making Preliminary Member
            Selections 416
    13.3.2 Preliminary Member Selections 417
    13.3.3 Resistance of Gap K Connections 420
    13.3.4 Resistance of an Overlap K Connection 427
    13.3.5 Top Chord as a Beam-Column between Panel Points 2 and 6 429
    13.3.6 Resistance of an X Connection 432
    13.3.7 Resistance of a KT Gap Connection 433
    13.3.8 Design of Bolted Flange-Plate Splice Connection 436
    13.3.9 Truss to Column Connection 438
References 441

INDEX 443

1
INTRODUCTION

1.1 Building with Hollow Structural Sections

The use of hollow structural sections (HSS) in Canada has expanded significantly over the years. Traditionally employed for aesthetic considerations in exposed structures, HSS are now increasingly selected for routine structural applications because of their other attributes. Such an endorsement of HSS is a result of distinct advantages available to designers who work with this versatile product.

Architects love the visual impact of HSS, and they regularly achieve their aesthetic goals by creating elegant structures which incorporate exposed steel. Atria, airport terminals, arenas and convention centres are examples which quickly come to mind. Hollow structural sections are also employed in all manner of other structures: schools, pavilions, telecommunication towers, electricity transmission line towers, sign supports, pedestrian bridges and canopies, hospitals, industrial buildings, conveyor galleries, highrise buildings and exhibition halls, to name just a few. Fig. 1.1 shows several examples.

The popularity of HSS construction results not just from aesthetic considerations but also from solid economic advantages, even though unit material costs are higher for HSS. The standard steel in Canada for HSS is CAN/CSA-G40.21-M92, Grade 350W, which has a specified yield strength of 350 MPa rather than the 300 MPa strength of Grade 300W steel, which is used for general structural purposes. This strength and a large radius of gyration about each axis provide superior column performance and dramatic weight savings. With this reduced weight come still other savings in transportation and erection.

(a) Roof structure of the SkyDome, Toronto, Ontario

(b) HSS bridge under construction

FIGURE 1.1
Examples of construction with HSS

(c) Atrium of Markham Suites Hotel, Ontario

(d) Roof of TGV railway station, Paris

FIGURE 1.1
Examples of construction with HSS

Columns made with HSS have a much lower material cost compared to those made from open sections, as is clearly evident in the example in Table 1.1. Therein, a five metre effective length column is designed to resist a factored compression load of 390 kN using a CAN/CSA-G40.21, Grade 300W wide flange section, along with alternative columns using circular and square Class C HSS from Grade 350W steel. This dramatically shows that HSS are the <u>less</u> expensive choice for compression members.

Column Type	Wide Flange	Circular HSS	Square HSS
Member	W200x36	HSS168x4.8	HSS152x152x4.8
Material	G40.21, 300W	G40.21, 350W, Class C	G40.21, 350W, Class C
Compressive resistance, C_r (kN)	437	396	463
Mass (kg/m)	35.6 (100%)	19.3 (54%)	21.7 (61%)
Surface area (m^2/m)	1.05 (100%)	0.529 (50%)	0.593 (56%)
Cost* ($/tonne)	$805 (100%)	$904 (112%)	$904 (112%)
Cost of 5 metre column	$143 (100%)	$87 (61%)	$98 (69%)

*Steel prices for budgeting were supplied by Russel Drummond, Mississauga, based on stock lengths.

TABLE 1.1

Material cost comparison—wide flange vs. HSS columns

Further, square and rectangular sections have about two thirds the surface area of similarly sized I-shaped members, and reduced surface area means less expensive painting and fire proofing. Over the years, structural steel designers have increasingly recognized the combined net saving from all these factors.

Closed sections offer torsionally rigid members, providing degrees of stiffness not readily available with open sections. Where cleanliness is a requirement, as in chemical or food processing facilities, HSS construction furnishes an environment relatively easy to clean, and free of protruding ledges, connection details and other dust collecting areas. The same smooth surface profiles cause lower aerodynamic loads on exposed structures fabricated from HSS than are present on similar structures made from conventional sections.

Hollow structural sections are particularly suitable for telescoping applications such as crane booms, adjustable seating stands, trailer hitches, jacks and scaffold legs. (In such applications, the removal of the inside flash of the weld must be specified when placing the order.) They have also been filled with concrete to enhance column strength and/or to

(a) Filling HSS columns with concrete on site, Hamilton, Ontario

(b) Rebars inside an HSS truss chord member, awaiting concrete filling

FIGURE 1.2
Concrete filling of HSS

provide fire protection, as shown in Fig.1.2. Recently, research has been completed on HSS members filled locally with concrete to enhance the connection performance.

In this age of resource conservation, the total recyclability of steel is demonstrated by the steel mills with their emphasis on gathering and remelting used steel for new production. Recycling is a powerful energy saver; it takes at least 60% less energy to produce steel from scrap than it does from iron ore (International Iron and Steel Institute).

The connecting of HSS members has been perceived by some as burdensome and expensive, and by others as both practical and economical. Much depends on the manner of application and the degree of understanding by practitioners. The large number of innovative HSS structures in Canada is testimony to the ingenuity and competence of Canadian architects and engineers. It is hoped that dissemination of the information in this handbook will promote their endeavours.

1.2 Historical Perspective on HSS Connection Design

After the advent of Hollow Structural Sections (HSS) in England, experimental and theoretical studies on welded connections with square and round members took place at Sheffield University, leading to the

FIGURE 1.3 (start ...)
Progression of design publications for HSS

landmark design recommendations of Eastwood and Wood (1970a, 1970b). These recommendations were quickly implemented in Canada and publicized by Stelco (1971) in the world's first HSS connections manual. Eastwood and Wood's connection strength formulae were also included in the Canadian Institute of Steel Construction's *Limit States Design Steel Manual* (CISC 1977), but have not appeared in later CISC publications.

A large amount of research and development work on HSS took place internationally during the 1970s, particularly with regard to connection behaviour and static strength. Much of this was co-ordinated by the Comité International pour le Développement et l'Étude de la Construction Tubulaire (CIDECT), which is a group of HSS producers with the aim of collectively developing the market for manufactured tubing. At present, the only Canadian member is IPSCO Inc. of Regina, Saskatchewan. The CIDECT Technical Secretariat is currently located in Paris, France.

Under preparation in 1980, CIDECT Monograph No. 6 (CIDECT 1986) was regarded as a new "state-of-the-art" approach to welded HSS connection design. However since its publication was continually being deferred, Stelco proceeded with its second connections manual (Stelco 1981). This was the first English-language HSS design guide to be expressed in a limit states format.

In early 1981, a new set of ultimate strength formulae for HSS welded truss connections was published in Canada by Packer and Haleem (1981),

FIGURE 1.3 (... continued ...)
Progression of design publications for HSS

applicable to trusses having square, rectangular or round web members but only square or rectangular chord members. These formulae were based on a wide range of European connection tests, but did not gain international acceptance; yet they did serve to point out some of the limitations (as well as areas of conservatism) of the Eastwood and Wood rules. As a result, Stelco issued supplementary provisions to its 1981 connections manual (Cran 1982), based on the Packer and Haleem (1981) recommendations.

The 1980s saw a period of consolidation of research knowledge and experience commencing with the treatise by Wardenier (1982). "The CIDECT book" (CIDECT 1984) on HSS design and construction, and CIDECT Monograph No. 6 (CIDECT 1986) on welded connection static design followed.

In 1983, a trans-Canada lecture tour, or "Canadian Symposium on HSS" was undertaken with the sponsorship of CIDECT, the National Research Council of Canada and the Canadian Steel Construction Council, to educate structural engineers and promote the use of HSS in Canada.

The International Institute of Welding (IIW) is a learned group comprised of national welding societies from around the world with headquarters also in Paris. It has played a major role in assessing and assimilating HSS connection design knowledge into specification format. This function is executed by volunteer members of IIW's Subcommission XV-E on Welded

FIGURE 1.3   (... continued ...)
Progression of design publications for HSS

Joints in Tubular Structures. Currently, IIW is approved as an official body for drafting ISO standards, so Subcommission XV-E will likely play a key role in influencing international standards relating to HSS connection design.

To date, the two principal connection design documents that this subcommission has issued relate to static (IIW 1981, 1989) and fatigue (IIW 1985) design of welded, truss-type connections. Predominantly in English, IIW documents can be obtained from Mr. M. Bramat, Secretary General, International Institute of Welding, c/o Institut de Soudure, B.P. 50362, F95942 Roissy CDG Cedex, France.

The IIW (1981) and Wardenier (1982) connection design recommendations were considerably more complicated than those originally used by North American engineers (Stelco 1981), so it was decided that their implementation in Canada would be facilitated by the production of design aids. A series of explanatory articles, design charts and design procedures was hence published by Packer (1983, 1985, 1986) and by Packer *et al.* (1983, 1986).

The intention in writing this book is to present the current state of learned opinion on as wide a range of HSS connection types as possible, in a consistent manner compatible with Canadian design practice. A pictorial progression of many of these HSS design publications is given in Fig. 1.3.

FIGURE 1.3 (... concluded)
Progression of design publications for HSS

1.3 Contemporary International Guides and Specifications

IIW

The current second edition design recommendations for statically-loaded, welded, planar, truss-type, HSS connections (IIW 1989) achieved a wide international consensus and have since been adopted worldwide by all national or regional specifications and guides for square and rectangular sections. For circular sections, the same is true except for the U.S. (AWS 1996). They are used in detailed CIDECT HSS design guides (Wardenier *et al.* 1991, Packer *et al.* 1992) and other prominent treatises published in German (Dutta and Würker 1988) and in Japanese (AIJ 1990).

The fatigue design recommendations (IIW 1985) are based on the modern approach of using the hot-spot stress method rather than the classification method, and are scheduled for updating in the near future (1997/98). Another recent valuable IIW publication dealing with fatigue definitions, analysis methods and recommendations (although not limited solely to HSS connections) is the IIW special report edited by Niemi (1995).

CIDECT

In recent years, CIDECT has adopted the policy of promoting and disseminating its wealth of accrued advice by publishing a series of design guides on various aspects of HSS construction. These guides supersede all previous CIDECT literature. To date the following have been published:

1. *Design guide for circular hollow section (CHS) joints under predominantly static loading* (Wardenier *et al.* 1991)

2. *Structural stability of hollow sections* (Rondal *et al.* 1992)

3. *Design guide for rectangular hollow section (RHS) joints under predominantly static loading* (Packer *et al.* 1992)

4. *Design guide for structural hollow section columns exposed to fire* (Twilt *et al.* 1994)

5. *Design guide for concrete-filled hollow section columns* (Bergmann *et al.* 1995). (This is based on Eurocode 4 for Composite Steel and Concrete Structures).

6. *Design guide for structural hollow sections in mechanical applications* (Wardenier *et al.* 1995).

These six guides (shown in Fig. 1.3) have been published in Germany in separate English, French and German editions and can be purchased either directly from the publisher (Verlag TÜV Rheinland GmbH, Köln, Germany) or from Canadian CIDECT member, IPSCO Inc. of Regina,

Saskatchewan. Spanish editions will also be forthcoming very soon. At least two further design guides are planned for the near future:

7. *Design guide for fabrication, assembly and erection of hollow section structures.*

8. *Design guide for circular and rectangular hollow section joints under fatigue loading.*

Another recent initiative has been to produce a computer program for performing checks on the LSD/LRFD resistance of planar, welded and bolted, truss-type, statically-loaded, connections made from circular, square or rectangular HSS. This program called CIDJOINT (Parik *et al.* 1994) follows the rules set out in the two relevant CIDECT design guides above (Wardenier *et al.* 1991, Packer *et al.* 1992). Version 1.1 is available for DOS and Windows 3.1, is in LSD/LRFD format, and has a choice of section databases for different countries, including Canada.

Eurocode

Soon to be adopted throughout Western Europe, Eurocode 3 for steel structures (European Committee for Standardization 1992a) will prove to be a very influential force in international standardization. Like the CIDECT design guides for statically-loaded, welded, connections (Wardenier *et al.* 1991, Packer *et al.* 1992), it conforms closely in Annex K to the recommendations set out by IIW Subcommission XV-E (IIW 1989). On the other hand, for fatigue design of HSS welded connections in the practical wall thickness range up to 12.5mm, the current version of EC3 permits the use of both the classification and the hot-spot stress methods. This generates some serious inconsistencies in the EC3 rules (van Wingerde *et al.* 1995), so this specification should be treated with caution for fatigue design.

Research

Although not in the coherent form of a guide or specification, advice and guidance resulting from new or innovative research in HSS construction can be best found in the Proceedings of the International Symposia on Tubular Structures. This series of symposia began in Boston, U.S.A. (1984) and have since been held in Tokyo, Japan (1986), Lappeenranta, Finland (1989), Delft, The Netherlands (1991), Nottingham, England (1993), Melbourne, Australia (1994) and Miskolc, Hungary (1996), under the organization of IIW Subcommission XV-E and the sponsorship of CIDECT.

The single-volume proceedings from each symposium acts as an excellent collation of the latest leading-edge research on HSS worldwide. The Proceedings of the 5th. Symposium (Nottingham) were published by E. & F.N. Spon, London, England (ISBN 0 419 18770 7), the 6th. Symposium

(Melbourne) by A.A. Balkema, Rotterdam, The Netherlands (ISBN 90 5410 520 8), and the 7th. Symposium (Miskolc), also by A.A. Balkema (ISBN 90 5410 828 2).

1.4 Contemporary National Guides and Specifications

U.S.A.

Considering the size of the market, there has been surprisingly little direction given to designing onshore structures with HSS in the U.S., and technical marketing and promotion have been very modest. At present, the American Welding Society D1.1 code (AWS 1996) covers the static design of welded truss-type connections—in both LRFD and ASD formats—between "box sections" (square and rectangular HSS) and tubular sections. As mentioned previously, the connection static design rules for square and rectangular HSS generally conform to IIW/CIDECT/EC3, but those for circular HSS do not. Fatigue design is also covered, by both the hot-spot stress and punching shear methods. But a recent comparison between these two design methods in AWS D1.1 shows that connections can have very different allowable force (or stress) ranges depending on which method is used (van Wingerde *et al.* 1996).

The design of members, including ties, columns and beam-columns (both unfilled and concrete-filled) is covered by the *LRFD Specification for Structural Steel Buildings* by the American Institute of Steel Construction (AISC 1993). AISC is now in the process of producing a separate *LRFD Specification for the Design of Steel Hollow Structural Sections*, which is being drafted by AISC Subcommittee 118. This will cover both member and connection design, and will be available in 1997.

Some HSS promotional material, mainly consisting of safe-load tables and case studies, has been published by the Pittsburgh-based American Institute for Hollow Structural Sections (AIHSS). This Institute represented several American tube maufacturers but has now been closed. Its role has been largely assumed by the Cleveland-based Steel Tube Institute of North America (STI), which has the support of many tube manufacturers across the U.S. and Canada. Structural design aids from STI have not yet been generated but a connection design guide conforming to the pending AISC *LRFD Specification for the Design of Steel Hollow Structural Sections* will be available in 1997.

One should also be aware of ASTM Standard A500 (ASTM 1993), the American specification regulating the geometric and mechanical properties of cold-formed HSS used in the U.S. The range of HSS grades produced to ASTM A500 has yield stresses from 228 to 317 MPa for round HSS and from 250 to 345 MPa for square and rectangular sections. ASTM A500 permits a hollow section wall thickness to be as much as 10% below the nominal wall thickness t without specifying any mass (or weight or cross-sectional area) tolerance. Consequently, most HSS manufacturers tend to produce undersized sections, but still within these excessively-generous ASTM tolerances.

This can have a major negative effect on assumed (nominal) structural properties (Packer 1993). Also impacted are connection resistances, which are frequently a function of wall thickness squared.

In recognition of the serious degradation to safety indices (when HSS wall thicknesses approach $0.90t$), reduced geometric properties based on wall thicknesses of $0.93t$ have been recommended by AISC (U.S.) and have been generated by the Steel Tube Institute for use with ASTM A500 sections (STI 1996a). Alternatively, closer conformity to nominal member dimensions can be ensured by adding supplementary requirements to contract documents, such as a minimum mass of 96.5% of nominal mass (as per Canadian specifications for HSS). Then the existing geometric properties would be appropriate.

Japan

The design of tubular structures in Japan is regulated by the Architectural Institute of Japan (AIJ 1990). It is notable that Japanese standards for cold-formed HSS permit a wall thickness tolerance of –10%, for the common range of thicknesses between 3mm and 12mm, with no mass/weight/area tolerance (JIS 1988a, 1988b, JSSC 1988, AIJ 1991).

China

A Chinese translation of the first edition of this book, supplemented by product information and engineering properties for European HSS, has recently been published in Beijing (Packer et al. 1997) to address the Asian market.

Germany

A prominent reference source has been the handbook by Dutta and Würker (1988), although the recent CIDECT Guides (Wardenier et al. 1991, Rondal et al. 1992, Packer et al. 1992, Twilt et al. 1994, Bergmann et al. 1995, Wardenier et al. 1995) have been very popular in Germany. There has been a German standard for steel structures made from hollow sections but, as in most other Western European countries, this is destined for replacement by parts of Eurocode 3 (European Committee for Standardization 1992a).

Draft European standards are already in place for the manufacturing requirements of hot-formed and cold-formed hollow sections (European Committee for Standardization 1992b, 1992c), and these allow for local thickness tolerances of up to –10% (depending on size) but are accompanied by a mass tolerance of –6%. Considering the broad influence that these EuroNorms will have, this mass tolerance is still far too liberal, especially in view of today's modern manufacturing capabilities.

Australia

The national limit states steel structures specification (SAA 1990) prescribes the design of HSS members in addition to other structural components. As an aid to HSS connection design, the Australian Institute of Steel Construction (AISC) is currently in the process of producing a "pre-engineered" connections manual. This will be published in two volumes, the first dealing with "Design Models", which is imminent (AISC 1997), and the second dealing with "Design Tables".

Cold-formed HSS are produced in Australia to minimum specified yield strengths of 250, 350 and 450 MPa, with a permitted local thickness tolerance of −10% but accompanied by a mass tolerance of −4% (SAA 1991). The 450 MPa yield strength is only available at present for square and rectangular HSS with perimeters up to 400 mm. This grade (C450/C450L0) is manufactured by Palmer Tube Mills and BHP Structural & Pipeline Products, with the latter producing it by in-line galvanizing to a mechanically (shot-blasted) and chemically-cleaned, bright metal (Tubemakers 1994). Innovative products such as this, combining high strength steels with surface pre-treatment should quickly increase the popularity, market share and export potential for Australian HSS.

South Africa

The South African structural steelwork specification (SABS 1993) closely follows the Canadian standard for structural steelwork. Hence it prescribes class categories for HSS based on wall slenderness and gives member resistances for HSS columns and concrete-filled HSS columns. However, HSS produced in South Africa are generally smaller and of lower strength than those produced in Canada. Advice on HSS connection design is provided by the South African Institute of Steel Construction (SAISC 1989).

1.5 Canadian HSS, Manufacturing Processes and Standards

The principal sources of HSS in Canada are Atlas Tube, Ipsco Inc., Sonco Steel Tube, Standard Tube Canada Inc., and Welded Tube of Canada Ltd. Tubes are produced to CAN/CSA-G40.20-M92 Class C (cold-formed) or Class H (either hot formed to final shape, or cold formed to final shape and stress relieved). Residual stresses in the Class H product are relatively small, which gives superior structural performance in compression. The lower residual stresses in the corners of thick-walled Class H sections are also reported to result in a product with enhanced ductility for local corner strains.

Some producers in Canada also manufacture tubing in accordance with ASTM A500 (1993), and users of this product in Canada should heed the earlier warning about strength and dimensional tolerances (see Section 1.4 dealing with U.S.A.).

The full range of official (CAN/CSA-G312.3-M92) HSS sizes (rectangular, 51x25 mm to 305x204 mm; square, 25 mm to 305 mm; circular, 27 mm to 406 mm) are presently produced in Canada. Rectangular HSS to 356x254 mm (Sonco) and circular HSS to 610 mm (Ipsco) are also available in Canada. Custom made sections are manufactured in the U.S., up to 32 inches on a side.

Specification	Grade	Minimum Yield Strength (MPa)	Tensile Strength (MPa)	Minimum Elongation Percent (50 mm)
METRIC				
CAN/CSA-G40.21-M92	300W	300	410-590	23
	350W	350	450-620	22
	380W	380	480-650	21
	350WT	350	480-650	22
	380WT	380	480-650	21
	350A	350	480-650	21
	350AT	350	480-650	21

Specification	Grade	Minimum Yield Strength (ksi)	Tensile Strength (ksi)	Minimum Elongation Percent[1] (2 inches)
IMPERIAL				
CAN/CSA-G40.21-92	44W	44	60-85	23
	50W	50	65-90	22
	55W	55	70-95	21
	50WT	50	70-95	22
	55WT	55	70-95	21
	50A	50	70-95	21
	50AT	50	70-95	21
ASTM A500-93	A	39	45 min	25
	B	46	58 min	23
Square and rectangular HSS	C	50	62 min	21
	D[2]	36	58 min	23
	A	33	45 min	25
Circular HSS	B	42	58 min	23
	C	46	62 min	21
	D[2]	36	58 min	23

(1) Varies with thickness for ASTM A500-93
(2) Must be stress relieved

TABLE 1.2

Metric and imperial properties of HSS

Hollow structural sections can be specified in a wide variety of types and grades in conformance with CAN/CSA-G40.20/G40.21-M92, CAN/CSA-G40.20/G40.21-92, or ASTM A500, as shown in Table 1.2.

In Canada, Type W steels "suitable for general welded construction where notch toughness at low temperature is not a design requirement" are the standard for HSS. These are normally produced as Grade 350W which has a specified minimum yield strength of 350 MPa (50 ksi), and all design examples given later use this standard grade.

Hollow structural sections can also be supplied with certified impact properties for structures exposed to dynamic loadings at low temperature. Type WT steels "suitable for welded construction where notch toughness at low temperature is a design requirement" provide four categories of notch toughness which meet specified Charpy V-Notch impact requirements at temperatures ranging from 0°C to −45°C, plus a fifth category to be negotiated between customer and supplier.

When structures are to be left unpainted, Type A steels that "display an atmospheric corrosion resistance approximately four times that of plain carbon steels" are available. They meet the chemical requirements of specifications such as ASTM A588 Grade C (ASTM 1991), and are used for applications similar to those for Type W steels.

Type AT steels are "suitable for welded construction where notch toughness at low temperature and improved corrosion resistance are required". They correspond to Type WT steels, but for structures that are to be left unpainted.

As the metric sizes produced in Canada are merely "soft conversions" of imperial sizes, HSS may be ordered using either metric or imperial dimensions. For example:

For Rectangular HSS

    Metric: HSS 203x152x9.5 Class H (or Class C)

                  CAN/CSA-G40.21-M92 Grade 350W

    Imperial: HSS 8x6x0.375 Class H (or Class C)

                  CAN/CSA-G40.21-92 Grade 50W.

The outside radius of rectangular HSS corners is generally taken as two times the wall thickness for design considerations and for the calculation of physical properties. Production corner radii may vary from this within the CAN/CSA-G40.20-M92 tolerances shown in Table 1.3. The actual outside corner radius can become a practical consideration when details of joint preparation for welding procedures are being developed for connections, as discussed in Section 8.2.4.

Wall Thickness (mm)	Maximum Outside Corner Radii (mm)	
	Perimeter to 700 mm	Perimeter over 700 mm
To 3 inclusive	6	—
Over 3 to 4 inclusive	8	—
Over 4 to 5 inclusive	15	—
Over 5 to 6 inclusive	18	18
Over 6 to 8 inclusive	21	24
Over 8 to 10 inclusive	27	30
Over 10 to 13 inclusive	36	39
Over 13	—	3 times wall thickness

TABLE 1.3

Maximum outside corner radii for rectangular sections
(CAN/CSA-G40.20-M92, Table 34)

Permissible variations for HSS mass, wall thickness, dimensions, squareness, straightness, and length are also listed in CAN/CSA-G40.20-M92. The tolerance for mass is −3.5% to +10% (Table 31), while the tolerance for wall thickness is −5% to +10% (Table 32). These are not contradictory, since the local wall thickness is allowed to be a little less than the overall restriction for mass.

In recent years, designs incorporating large size rounds (greater than 508 mm O.D.) have become increasingly popular. Strictly speaking, such rounds are not HSS but they may be purchased in the form of pipe. Pipe is produced to several ASTM and CSA Standards, and is available in a wide range of sizes, wall thicknesses and grades.

Canadian manufacturers produce HSS by cold forming a continuous coil of sheet or plate into a closed shape and then fusing the open seam by electric resistance welding (ERW). If the seam weld is likely to be highly stressed in a transverse direction, the presence of the seam should be accounted for in design, as the mechanical properties across the seam may not be equivalent to those of the parent material. In the U.S., particularly for some very large HSS sizes, square and rectangular HSS may be manufactured from flat plate in a press brake to channel-shaped half sections, then welded by submerged arc (SAW) to produce the final sections (STI 1996b).

1.6 Interpretation of Connection Resistance Expressions

Chapter 3 and succeeding chapters in this Design Guide contain a considerable number of connection resistance formulae expressed in a Limit States Design (LSD) or a Load and Resistance Factor Design (LRFD) format. It is important to note that all expressions are <u>resistances</u>, and the appropriate resistance factor (ϕ) is included in the resistance formulae either explicitly or indirectly. As such, <u>no addditional resistance factors should be added</u>.

Within reasonable engineering accuracy, the connection resistance formulae in this Design Guide can also be used directly with the American Institute of Steel Construction LRFD Specification for Structural Steel Buildings (AISC 1993) to obtain design strengths. If Allowable Stress Design (ASD) is used, the connection resistance (design strength) expressions should be divided by an appropriate load factor to obtain allowable connection resistances. In this case a load factor of 1.5 is recommended.

The nominal (or specified) values are to be used for mechanical or geometric properties listed in the Symbols and Abbreviation (Section 1.9) in limit states *design* equations, or in conjunction with *design* charts.

1.7 Seismic Applications

The design recommendations in this book are primarily directed at structures that are predominantly statically loaded. An exception to this is Chapter 12 where provisions for the design of fatigue-critical welded HSS connections are given.

Under cyclic axial loading during earthquakes, HSS braces have been known to fracture catastrophically. To avoid this result, the formation of local buckles must be prevented and tests have shown that tight limits must be placed on the tube wall slenderness (Lui and Goel 1987, Sherman 1996). Recent AISC Seismic Provisions (AISC 1992) stipulate, for members in axial compression in seismic zones, that the flat width-to-thickness ratio for square or rectangular HSS be limited to

$$(b_i - 4t_i)/t_i \text{ and } (h_i - 4t_i)/t_i \leq 0.646\sqrt{E/F_y} = 289/\sqrt{F_y}, \text{ with } F_y \text{ in MPa}$$

This is considerably less than the CAN/CSA-S16.1-94 Class 1 limit of

$$(b_i - 4t_i)/t_i \text{ and } (h_i - 4t_i)/t_i = 420/\sqrt{F_y}$$

Filling rectangular HSS members with concrete stiffens the tube walls and also improves their performance under cyclic loading.

Fracture at a low number of cycles is also possible with cold-formed circular HSS braces that form local buckles. This form of premature failure is controlled (AISC 1992) by limiting the diameter-to-thickness ratio for circular HSS, in axial compression or flexure in seismic zones, to

$$d_i/t_i \leq 0.0448 E/F_y = 8960/F_y \quad \text{with } F_y \text{ in MPa}$$

This is considerably less than the CAN/CSA-S16.1-94 Class 1 limit of

$$d_i/t_i = 13\,000/F_y$$

1.8 Geometric Properties of HSS

The following pages list the designations, sizes and geometric properties of rectangular, square and circular HSS that are common to present Canadian usage, manufactured in accordance with CAN/CSA-G40.20-M92. <u>Note that these engineering properties are not appropriate for HSS manufactured to ASTM A500 (1993)</u>, as discussed in Section 1.4.

Data are first presented in metric units, then in imperial (inch-pound) units. The metric designations and sizes are from CAN/CSA-G312.3-M92. Properties are from the Handbook of Steel Construction (CISC 1995).

RECTANGULAR Hollow Structural Sections — METRIC Dimensions and Properties

Designation mm x mm x mm	Mass kg/m	Dead Load kN/m	Area mm²	I_x 10⁶ mm⁴	S_x 10³ mm³	r_x mm	Z_x 10³ mm³	I_y 10⁶ mm⁴	S_y 10³ mm³	r_y mm	Z_y 10³ mm³	Torsion J 10³ mm⁴	Shear C_{rt} mm²
HSS 305x203x13	93.0	0.912	11 800	147	964	111	1 190	78.2	769	81.2	897	167 000	6 450
x 11	82.4	0.808	10 500	132	867	112	1 060	70.5	694	82.0	802	149 000	5 790
x 9.53	71.3	0.699	9 090	116	762	113	926	62.1	611	82.7	701	130 000	5 080
x 7.95	60.1	0.590	7 660	99.4	652	114	787	53.3	525	83.4	596	111 000	4 340
x 6.35	48.6	0.476	6 190	81.5	535	115	640	43.8	431	84.1	486	89 800	3 550
HSS254x152x13	72.7	0.713	9 260	75.2	592	90.1	747	33.6	442	60.3	522	78 200	5 160
x 11	64.6	0.634	8 230	68.2	537	91.0	671	30.6	402	61.0	470	70 200	4 660
x 9.53	56.1	0.550	7 150	60.4	475	91.9	589	27.2	357	61.7	413	61 600	4 110
x 7.95	47.5	0.465	6 050	52.0	410	92.7	503	23.6	309	62.4	354	52 600	3 530
x 6.35	38.4	0.377	4 900	42.9	338	93.6	411	19.5	256	63.1	290	43 000	2 900
HSS 203x152x13	62.6	0.614	7 970	43.0	423	73.4	528	27.3	359	58.6	432	56 400	3 870
x 11	55.7	0.547	7 100	39.2	385	74.3	476	25.0	327	59.3	390	50 800	3 530
x 9.53	48.5	0.476	6 180	34.8	343	75.1	420	22.3	292	60.0	344	44 600	3 150
x 7.95	41.1	0.403	5 240	30.2	297	75.9	360	19.3	254	60.8	295	38 200	2 730
x 6.35	33.4	0.327	4 250	25.0	246	76.7	295	16.1	211	61.5	243	31 200	2 260
x 4.78	25.5	0.250	3 250	19.5	192	77.5	228	12.6	165	62.2	188	24 100	1 760
HSS 203x102x13	52.4	0.514	6 680	31.3	308	68.4	405	10.2	201	39.1	246	27 000	3 870
x 11	46.9	0.460	5 970	28.7	283	69.4	368	9.48	187	39.8	224	24 600	3 530
x 9.53	40.9	0.401	5 210	25.8	254	70.3	326	8.57	169	40.5	199	21 900	3 150
x 7.95	34.8	0.341	4 430	22.5	221	71.2	281	7.54	148	41.3	172	18 900	2 730
x 6.35	28.3	0.278	3 610	18.8	185	72.2	232	6.35	125	42.0	143	15 600	2 260
x 4.78	21.7	0.213	2 760	14.7	145	73.1	180	5.03	99.0	42.7	111	12 200	1 760
HSS 178x127x13	52.4	0.514	6 680	26.4	297	62.9	378	15.5	244	48.1	298	33 600	3 230
x 11	46.9	0.460	5 970	24.2	273	63.7	343	14.3	225	48.9	271	30 400	2 970
x 9.53	40.9	0.401	5 210	21.7	244	64.6	303	12.8	202	49.6	240	26 900	2 660
x 7.95	34.8	0.341	4 430	19.0	213	65.4	261	11.2	177	50.3	207	23 100	2 320
x 6.35	28.3	0.278	3 610	15.8	178	66.2	216	9.40	148	51.1	171	19 000	1 940
x 4.78	21.7	0.213	2 760	12.4	140	67.1	168	7.41	117	51.8	133	14 700	1 520
HSS 152x102x11.1	38.0	0.373	4 840	13.6	179	53.1	230	7.15	141	38.4	173	16 300	2 400
x 9.53	33.3	0.327	4 240	12.4	162	54.0	206	6.51	128	39.2	155	14 500	2 180
x 7.95	28.4	0.279	3 620	10.9	143	54.8	179	5.76	113	39.9	135	12 600	1 920
x 6.35	23.2	0.228	2 960	9.19	121	55.7	148	4.88	96.2	40.6	112	10 500	1 610
x 4.78	17.9	0.175	2 280	7.28	95.6	56.5	116	3.89	76.6	41.3	87.8	8 160	1 270

RECTANGULAR Hollow Structural Sections — METRIC Dimensions and Properties

Designation	Size	Mass	Dead Load	Area	I_x	S_x	r_x	Z_x	I_y	S_y	r_y	Z_y	Torsion J	Shear C_{rt}
mm × mm × mm	mm × mm × mm	kg/m	kN/m	mm²	10^6 mm⁴	10^3 mm³	mm	10^3 mm³	10^6 mm⁴	10^3 mm³	mm	10^3 mm³	10^3 mm⁴	mm²
HSS 127×76.2×9.53		25.7	0.252	3 280	6.13	96.5	43.3	126	2.70	70.8	28.7	87.8	6 600	1 690
	× 7.95	22.1	0.217	2 820	5.49	86.5	44.2	111	2.44	63.9	29.4	77.4	5 810	1 510
	× 6.35	18.2	0.178	2 320	4.70	74.1	45.1	93.4	2.10	55.2	30.1	65.3	4 890	1 290
	× 4.78	14.1	0.138	1 790	3.78	59.6	45.9	73.8	1.71	44.8	30.8	51.8	3 860	1 030
HSS 127×64×9.5	×9.53	23.8	0.234	3 030	5.29	83.4	41.8	112	1.71	53.9	23.8	67.7	4 640	1 690
	× 7.95	20.5	0.201	2 610	4.77	75.2	42.7	99.1	1.56	49.2	24.5	60.2	4 130	1 510
	× 6.35	16.9	0.166	2 150	4.12	64.8	43.7	83.7	1.36	43.0	25.2	51.1	3 510	1 290
	× 4.78	13.1	0.129	1 670	3.33	52.4	44.6	66.4	1.12	35.2	25.9	40.8	2 800	1 030
HSS 127×51×9.5	×9.53	21.9	0.215	2 790	4.46	70.2	40.0	98.0	0.971	38.2	18.7	49.2	2 930	1 690
	× 7.95	18.9	0.186	2 410	4.06	63.9	41.0	87.0	0.902	35.5	19.3	44.2	2 650	1 510
	× 6.35	15.6	0.153	1 990	3.53	55.6	42.1	73.9	0.800	31.5	20.0	37.9	2 290	1 290
	× 4.78	12.2	0.119	1 550	2.87	45.3	43.1	59.0	0.666	26.2	20.7	30.6	1 850	1 030
HSS 102×76×9.5	×9.53	21.9	0.215	2 790	3.42	67.4	35.0	87.9	2.16	56.6	27.8	71.6	4 710	1 210
	× 7.95	18.9	0.186	2 410	3.10	61.1	35.9	77.9	1.96	51.5	28.5	63.6	4 170	1 110
	× 6.35	15.6	0.153	1 990	2.69	52.9	36.7	66.0	1.71	44.8	29.3	54.0	3 530	968
	× 4.78	12.2	0.119	1 550	2.18	43.0	37.5	52.6	1.39	36.6	30.0	43.1	2 800	789
HSS 102×51×8.0	×7.95	15.8	0.155	2 010	2.21	43.6	33.2	59.0	0.714	28.1	18.9	35.6	1 950	1 110
	× 6.4	13.1	0.129	1 670	1.95	38.5	34.2	50.7	0.640	25.2	19.6	30.8	1 690	968
	× 4.8	10.3	0.101	1 310	1.61	31.8	35.1	40.8	0.537	21.1	20.3	25.0	1 370	789
	× 3.8	8.37	0.082	1 070	1.36	26.8	35.7	33.9	0.457	18.0	20.7	20.8	1 140	658
	× 3.2	7.09	0.069	903	1.17	23.1	36.1	29.0	0.397	15.6	21.0	17.9	979	565
HSS 89×64×8.0	×7.95	15.8	0.155	2 010	1.88	42.2	30.6	55.1	1.09	34.4	23.3	43.3	2 450	908
	× 6.35	13.1	0.129	1 670	1.65	37.1	31.4	47.2	0.968	30.5	24.1	37.3	2 100	806
	× 4.78	10.3	0.101	1 310	1.36	30.6	32.3	38.0	0.803	25.3	24.8	30.1	1 690	667
	× 3.81	8.37	0.082	1 070	1.15	25.8	32.8	31.5	0.679	21.4	25.2	25.0	1 400	561
	× 3.18	7.09	0.069	903	0.990	22.3	33.1	27.0	0.588	18.5	25.5	21.4	1 190	485
HSS 76×51×8.0	×7.95	12.6	0.123	1 600	1.02	26.7	25.2	36.1	0.527	20.7	18.1	26.9	1 270	706
	× 6.35	10.6	0.104	1 350	0.919	24.1	26.1	31.5	0.479	18.9	18.9	23.6	1 110	645
	× 4.78	8.35	0.082	1 060	0.775	20.3	27.0	25.8	0.408	16.1	19.6	19.4	911	546
	× 3.81	6.85	0.067	872	0.660	17.3	27.5	21.6	0.350	13.8	20.0	16.3	762	465
HSS 51×25×3.2	×3.18	3.28	0.032	418	0.122	4.81	17.1	6.34	0.0400	3.15	9.78	3.85	106	242
	× 2.54	2.71	0.026	345	0.105	4.15	17.5	5.35	0.0349	2.75	10.1	3.27	89.8	206

21

SQUARE Hollow Structural Sections — METRIC Dimensions and Properties

Designation mm x mm x mm	Size mm x mm x mm	Mass kg/m	Dead Load kN/m	Area mm²	I 10⁶ mm⁴	S 10³ mm³	r mm	Z 10³ mm³	Torsion J 10³ mm⁴	Surface Area m²/m	Shear C_{rt} mm²
HSS 305x305x13	HSS 305x305x12.7	113	1.110	14 400	202	1 330	118	1 560	324 000	1.18	6 450
x 11	x 11.1	100	0.982	12 800	181	1 190	119	1 390	288 000	1.18	5 790
x 9.5	x 9.53	86.5	0.848	11 000	158	1 040	120	1 210	250 000	1.19	5 080
x 8.0	x 7.95	72.8	0.714	9 280	135	886	121	1 030	211 000	1.19	4 340
x 6.4	x 6.35	58.7	0.576	7 480	110	723	121	833	171 000	1.20	3 550
HSS 254x254x13	HSS 254x254x12.7	93.0	0.912	11 800	113	889	97.6	1 060	183 000	0.972	5 160
x 11	x 11.1	82.4	0.808	10 500	102	800	98.4	946	163 000	0.978	4 660
x 9.5	x 9.53	71.3	0.699	9 090	89.3	703	99.1	825	142 000	0.983	4 110
x 8.0	x 7.95	60.1	0.590	7 660	76.5	602	99.9	702	121 000	0.989	3 530
x 6.4	x 6.35	48.6	0.476	6 190	62.7	494	101	571	97 900	0.994	2 900
HSS 203x203x13	HSS 203x203x12.7	72.7	0.713	9 260	54.7	538	76.9	651	90 700	0.769	3 870
x 11	x 11.1	64.6	0.634	8 230	49.6	488	77.6	585	81 200	0.775	3 530
x 9.5	x 9.53	56.1	0.550	7 150	43.9	432	78.4	513	71 000	0.780	3 150
x 8.0	x 7.95	47.5	0.465	6 050	37.9	373	79.2	439	60 500	0.786	2 730
x 6.4	x 6.35	38.4	0.377	4 900	31.3	308	79.9	359	49 300	0.791	2 260
HSS 178x178x13	HSS 178x178x12.7	62.6	0.614	7 970	35.2	396	66.5	484	59 200	0.668	3 230
x 11	x 11.1	55.7	0.547	7 100	32.1	361	67.2	437	53 200	0.673	2 970
x 9.5	x 9.53	48.5	0.476	6 180	28.6	322	68.0	385	46 700	0.678	2 660
x 8.0	x 7.95	41.1	0.403	5 240	24.8	279	68.8	330	39 900	0.684	2 320
x 6.4	x 6.35	33.4	0.327	4 250	20.6	231	69.6	271	32 700	0.689	1 940
x 4.8	x 4.78	25.5	0.250	3 250	16.1	181	70.3	210	25 200	0.695	1 520
HSS 152x152x13	HSS 152x152x12.7	52.4	0.514	6 680	21.0	276	56.1	342	36 000	0.566	2 580
x 11	x 11.1	46.9	0.460	5 970	19.3	253	56.9	310	32 500	0.571	2 400
x 9.5	x 9.53	40.9	0.401	5 210	17.3	227	57.6	275	28 700	0.577	2 180
x 8.0	x 7.95	34.8	0.341	4 430	15.1	198	58.4	237	24 600	0.582	1 920
x 6.4	x 6.35	28.3	0.278	3 610	12.6	166	59.2	196	20 300	0.588	1 610
x 4.8	x 4.78	21.7	0.213	2 760	9.93	130	59.9	152	15 700	0.593	1 270
HSS 127x127x11	HSS 127x127x11.1	38.0	0.373	4 840	10.5	165	46.5	205	18 000	0.470	1 840
x 9.5	x 9.53	33.3	0.327	4 240	9.48	149	47.3	183	16 000	0.475	1 690
x 8.0	x 7.95	28.4	0.279	3 620	8.36	132	48.0	159	13 900	0.481	1 510
x 6.4	x 6.35	23.2	0.228	2 960	7.05	111	48.8	132	11 500	0.486	1 290
x 4.8	x 4.78	17.9	0.175	2 280	5.60	88.1	49.6	103	8 920	0.492	1 030

SQUARE Hollow Structural Sections — METRIC Dimensions and Properties

Designation mm x mm x mm	Size mm x mm x mm	Mass kg/m	Dead Load kN/m	Area mm²	I 10⁶ mm⁴	S 10³ mm³	r mm	Z 10³ mm³	Torsion J 10³ mm⁴	Surface Area m²/m	Shear C_{rt} mm²
HSS 102x102x9.5	HSS 102x102x9.53	25.7	0.252	3 280	4.45	87.6	36.9	110	7 740	0.374	1210
x 8.0	x 7.95	22.1	0.217	2 820	3.99	78.4	37.6	96.8	6 780	0.379	1110
x 6.4	x 6.35	18.2	0.178	2 320	3.42	67.3	38.4	81.4	5 670	0.385	968
x 4.8	x 4.78	14.1	0.138	1 790	2.75	54.2	39.2	64.3	4 450	0.390	789
HSS 89x89x9.5	HSS 88.9x88.9x9.53	21.9	0.215	2 790	2.80	63.0	31.7	80.5	4 970	0.323	968
x 8.0	x 7.95	18.9	0.186	2 410	2.54	57.1	32.4	71.4	4 390	0.328	908
x 6.4	x 6.35	15.6	0.153	1 990	2.20	49.5	33.2	60.5	3 700	0.334	806
x 4.8	x 4.78	12.2	0.119	1 550	1.79	40.3	34.0	48.2	2 930	0.339	667
HSS 76x76x8.0	HSS 76.2x76.2x7.95	15.8	0.155	2 010	1.49	39.1	27.2	49.8	2 630	0.278	706
x 6.4	x 6.35	13.1	0.129	1 670	1.31	34.5	28.0	42.8	2 250	0.283	645
x 4.8	x 4.78	10.3	0.101	1 310	1.08	28.5	28.8	34.4	1 800	0.288	546
HSS 64x64x6.4	HSS 63.5x63.5x6.35	10.6	0.104	1 350	0.703	22.2	22.8	28.1	1 230	0.232	484
x 4.8	x 4.78	8.35	0.082	1 060	0.594	18.7	23.6	23.0	1 000	0.238	424
x 3.8	x 3.81	6.85	0.067	872	0.506	16.0	24.1	19.2	836	0.241	368
x 3.2	x 3.18	5.82	0.057	741	0.441	13.9	24.4	16.6	717	0.243	323
HSS 51x51x6.4	HSS 50.8x50.8x6.35	8.05	0.079	1 030	0.319	12.6	17.6	16.4	580	0.181	323
x 4.8	x 4.78	6.45	0.063	821	0.279	11.0	18.4	13.8	485	0.187	303
x 3.8	x 3.81	5.33	0.052	679	0.243	9.55	18.9	11.7	410	0.190	271
x 3.2	x 3.18	4.55	0.045	580	0.214	8.42	19.2	10.2	355	0.192	242
x 2.8	x 2.79	4.05	0.040	516	0.194	7.64	19.4	9.16	318	0.194	221
HSS 38x38x4.8	HSS 38.1x38.1x4.78	4.54	0.044	578	0.101	5.30	13.2	6.95	184	0.136	181
x 3.8	x 3.81	3.81	0.037	485	0.0912	4.79	13.7	6.06	160	0.139	174
x 3.2	x 3.18	3.28	0.032	418	0.0822	4.31	14.0	5.35	141	0.141	161
x 2.5	x 2.54	2.71	0.026	345	0.0708	3.72	14.3	4.52	118	0.144	142
HSS 32x32x3.8	HSS 31.8x31.8x3.81	3.06	0.030	389	0.0482	3.03	11.1	3.94	87.0	0.114	126
x 3.2	x 3.18	2.65	0.026	338	0.0442	2.78	11.4	3.52	77.6	0.116	121
x 2.5	x 2.54	2.20	0.022	281	0.0388	2.44	11.8	3.01	66.1	0.118	110
HSS 25x25x3.2	HSS 25.4x25.4x3.18	2.01	0.020	257	0.0199	1.57	8.82	2.06	36.3	0.091	80.6
x 2.5	x 2.54	1.69	0.017	216	0.0180	1.42	9.14	1.80	31.6	0.093	77.4

CIRCULAR Hollow Structural Sections — METRIC Dimensions and Properties

Designation	Size	Mass	Dead Load	Area	I	S	r	Z	Torsion J	Surface Area	Shear C_{rt}
mm x mm	mm x mm	kg/m	kN/m	mm²	10^6 mm⁴	10^3 mm³	mm	10^3 mm³	10^3 mm⁴	m²/m	mm²
HSS 406x13	HSS 406x12.7	123	1.21	15 700	305	1 500	139	1 970	609 000	1.28	7 860
x 11	x 11.1	108	1.06	13 800	270	1 330	140	1 740	540 000	1.28	6 910
x 9.5	x 9.53	93.3	0.915	11 900	234	1 150	140	1 500	468 000	1.28	5 940
x 8.0	x 7.95	78.1	0.766	9 950	198	972	141	1 260	395 000	1.28	4 970
x 6.4	x 6.35	62.6	0.614	7 980	160	786	141	1 020	319 000	1.28	3 990
HSS 356x13	HSS 356x12.7	107	1.05	13 700	201	1 130	121	1 490	403 000	1.12	6 850
x 11	x 11.1	94.6	0.927	12 000	179	1 010	122	1 320	358 000	1.12	6 030
x 9.5	x 9.53	81.3	0.798	10 400	155	873	122	1 140	310 000	1.12	5 180
x 8.0	x 7.95	68.2	0.668	8 680	131	738	123	961	262 000	1.12	4 340
x 6.4	x 6.35	54.7	0.536	6 970	106	598	123	775	213 000	1.12	3 480
HSS 324x13	HSS 324x12.7	97.5	0.956	12 400	151	930	110	1 230	301 000	1.02	6 210
x 11	x 11.1	85.9	0.842	10 900	134	827	111	1 090	268 000	1.02	5 470
x 9.5	x 9.53	73.9	0.725	9 410	116	719	111	942	233 000	1.02	4 710
x 8.0	x 7.95	61.9	0.607	7 890	98.5	608	112	794	197 000	1.02	3 950
x 6.4	x 6.35	49.7	0.488	6 330	79.9	493	112	640	160 000	1.02	3 170
HSS 273x13	HSS 273x12.7	81.6	0.800	10 400	88.3	646	92.2	862	177 000	0.858	5 200
x 11	x 11.1	71.9	0.705	9 160	78.7	577	92.7	764	157 000	0.858	4 590
x 9.5	x 9.53	61.9	0.607	7 890	68.6	502	93.2	662	137 000	0.858	3 950
x 8.0	x 7.95	52.0	0.510	6 620	58.2	427	93.8	559	116 000	0.858	3 310
x 6.4	x 6.35	41.8	0.410	5 320	47.4	347	94.3	452	94 700	0.858	2 660
HSS 219x13	HSS 219x12.7	64.6	0.634	8 230	44.0	402	73.1	542	88 000	0.688	4 130
x 11	x 11.1	57.1	0.560	7 270	39.4	360	73.6	482	78 900	0.688	3 640
x 9.5	x 9.53	49.3	0.483	6 270	34.5	315	74.2	419	69 000	0.688	3 140
x 8.0	x 7.95	41.4	0.406	5 270	29.4	269	74.7	355	58 900	0.688	2 640
x 6.4	x 6.35	33.3	0.327	4 240	24.0	219	75.3	288	48 100	0.688	2 120
x 4.8	x 4.78	25.3	0.248	3 220	18.5	169	75.8	220	37 000	0.688	1 610
HSS 168x9.5	HSS 168x9.53	37.3	0.366	4 750	15.0	179	56.2	241	30 100	0.529	2 380
x 8.0	x 7.95	31.4	0.308	4 000	12.9	153	56.8	205	25 800	0.529	2 010
x 6.4	x 6.35	25.4	0.249	3 230	10.6	126	57.3	167	21 200	0.529	1 620
x 4.8	x 4.78	19.3	0.189	2 460	8.21	97.6	57.8	128	16 400	0.529	1 230
HSS 141x9.5	HSS 141x9.53	31.0	0.304	3 950	8.61	122	46.7	166	17 200	0.444	1 980
x 8.0	x 7.95	26.1	0.256	3 330	7.43	105	47.2	142	14 900	0.444	1 670
x 6.4	x 6.35	21.1	0.207	2 690	6.14	86.9	47.8	116	12 300	0.444	1 350
x 4.8	x 4.78	16.1	0.158	2 050	4.78	67.7	48.3	89.1	9 560	0.444	1 030

CIRCULAR Hollow Structural Sections — METRIC Dimensions and Properties

Designation	Size	Mass	Dead Load	Area	I	S	r	Z	Torsion J	Surface Area	Shear C_{rt}
mm x mm	mm x mm	kg/m	kN/m	mm²	10^6 mm⁴	10^3 mm³	mm	10^3 mm³	10^3 mm⁴	m²/m	mm²
HSS 114x8.0	HSS 114x7.95	20.9	0.204	2 660	3.78	66.1	37.7	90.1	7 550	0.359	1 330
x 6.4	x 6.35	16.9	0.166	2 150	3.15	55.1	38.2	74.1	6 300	0.359	1 080
x 4.8	x 4.78	12.9	0.127	1 640	2.47	43.2	38.8	57.4	4 940	0.359	823
HSS 102x8.0	HSS 102x7.95	18.4	0.180	2 340	2.58	50.8	33.2	69.9	5 170	0.319	1 180
x 6.4	x 6.35	14.9	0.146	1 900	2.16	42.6	33.8	57.7	4 330	0.319	954
x 4.8	x 4.78	11.4	0.112	1 450	1.71	33.6	34.3	44.8	3 420	0.319	728
x 3.8	x 3.81	9.19	0.090	1 170	1.40	27.6	34.6	36.5	2 800	0.319	586
HSS 89x8.0	HSS 88.9x7.95	15.9	0.156	2 020	1.67	37.6	28.8	52.3	3 340	0.279	1 020
x 6.4	x 6.35	12.9	0.127	1 650	1.41	31.7	29.3	43.4	2 820	0.279	828
x 4.8	x 4.78	9.92	0.097	1 260	1.12	25.2	29.8	33.9	2 240	0.279	633
x 3.8	x 3.81	8.00	0.078	1 020	0.924	20.8	30.1	27.6	1 850	0.279	510
HSS 73x6.4	HSS 73.0x6.35	10.4	0.102	1 330	0.745	20.4	23.7	28.3	1 490	0.229	670
x 4.8	x 4.78	8.04	0.079	1 020	0.599	16.4	24.2	22.3	1 200	0.229	514
x 3.8	x 3.81	6.50	0.064	828	0.497	13.6	24.5	18.3	994	0.229	415
x 3.2	x 3.18	5.48	0.054	698	0.426	11.7	24.7	15.5	852	0.229	349
HSS 60x6.4	HSS 60.3x6.35	8.45	0.083	1080	0.397	13.2	19.2	18.6	794	0.189	545
x 4.8	x 4.78	6.54	0.064	834	0.324	10.7	19.7	14.8	647	0.189	420
x 3.8	x 3.81	5.31	0.052	676	0.271	8.99	20.0	12.2	542	0.189	339
x 3.2	x 3.18	4.48	0.044	571	0.233	7.74	20.2	10.4	467	0.189	286
HSS 48x4.8	HSS 48.3x4.78	5.13	0.050	654	0.157	6.48	15.5	9.09	313	0.152	331
x 3.8	x 3.81	4.18	0.041	533	0.133	5.50	15.8	7.56	265	0.152	268
x 3.2	x 3.18	3.54	0.035	451	0.115	4.77	16.0	6.48	231	0.152	226
x 2.8	x 2.79	3.13	0.031	399	0.104	4.29	16.1	5.79	207	0.152	200
HSS 42x3.2	HSS 42.2x3.18	3.06	0.030	390	0.075	3.54	13.8	4.85	149	0.133	196
x 2.5	x 2.54	2.48	0.024	316	0.062	2.96	14.1	4.00	125	0.133	159
HSS 33x3.2	HSS 33.4x3.18	2.37	0.023	302	0.035	2.09	10.7	2.91	69.7	0.105	153
x 2.5	x 2.54	1.93	0.019	246	0.029	1.77	10.9	2.42	59.0	0.105	124
HSS 27x3.2	HSS 26.7x3.18	1.84	0.018	235	0.016	1.24	8.39	1.77	33.1	0.084	120
x 2.5	x 2.54	1.51	0.015	193	0.014	1.07	8.59	1.49	28.4	0.084	97.4

RECTANGULAR Hollow Structural Sections — IMPERIAL Dimensions and Properties

Outside Dimensions in. x in.	Wall Thickness in.	Mass lb./ft	Area in.²	I_x in.⁴	S_x in.³	r_x in.	Z_x in.³	I_y in.⁴	S_y in.³	r_y in.	Z_y in.³	Torsion J in.⁴	Shear C_{rt} in.²
12 x 8	.500	62.5	18.4	353	58.8	4.39	72.4	188	46.9	3.20	54.7	402	10.0
	.438	55.3	16.3	318	52.9	4.42	64.7	169	42.3	3.23	48.9	359	8.98
	.375	47.9	14.1	279	46.5	4.45	56.5	149	37.3	3.25	42.7	313	7.88
	.313	40.4	11.9	239	39.8	4.49	48.0	128	32.0	3.28	36.4	266	6.73
	.250	32.6	9.59	196	32.6	4.52	39.1	105	26.3	3.31	29.6	216	5.50
10 x 6	.500	48.9	14.4	181	36.1	3.55	45.6	80.7	26.9	2.37	31.8	188	8.00
	.438	43.4	12.8	164	32.7	3.58	40.9	73.5	24.5	2.40	28.7	169	7.23
	.375	37.7	11.1	145	29.0	3.62	35.9	65.4	21.8	2.43	25.2	148	6.38
	.313	31.9	9.37	125	25.0	3.65	30.7	56.6	18.9	2.46	21.6	126	5.48
	.250	25.8	7.59	103	20.6	3.69	25.1	46.9	15.6	2.49	17.7	103	4.50
8 x 6	.500	42.2	12.4	103	25.8	2.89	32.2	65.6	21.9	2.30	26.3	136	6.00
	.438	37.5	11.0	94.0	23.5	2.92	29.0	59.9	20.0	2.33	23.8	122	5.47
	.375	32.6	9.58	83.7	20.9	2.96	25.6	53.5	17.8	2.36	21.0	107	4.88
	.313	27.6	8.12	72.5	18.1	2.99	21.9	46.4	15.5	2.39	18.0	91.7	4.22
	.250	22.4	6.59	60.1	15.0	3.02	18.0	38.6	12.9	2.42	14.8	75.1	3.50
	.188	17.1	5.04	46.9	11.7	3.05	13.9	30.2	10.1	2.45	11.5	57.9	2.73
8 x 4	.500	35.2	10.4	75.0	18.8	2.69	24.7	24.5	12.3	1.54	15.0	65.0	6.00
	.438	31.5	9.25	68.9	17.2	2.73	22.4	22.7	11.4	1.57	13.6	59.2	5.47
	.375	27.5	8.08	61.9	15.5	2.77	19.9	20.6	10.3	1.60	12.1	52.6	4.88
	.313	23.4	6.87	54.0	13.5	2.80	17.1	18.1	9.05	1.62	10.5	45.4	4.22
	.250	19.0	5.59	45.1	11.3	2.84	14.1	15.3	7.63	1.65	8.71	37.6	3.50
	.188	14.6	4.28	35.4	8.86	2.88	11.0	12.1	6.04	1.68	6.79	29.2	2.73
7 x 5	.500	35.2	10.4	63.4	18.1	2.47	23.0	37.1	14.8	1.89	18.2	80.8	5.00
	.438	31.5	9.25	58.2	16.6	2.51	20.9	34.2	13.7	1.92	16.5	73.1	4.60
	.375	27.5	8.08	52.2	14.9	2.54	18.5	30.8	12.3	1.95	14.6	64.6	4.13
	.313	23.4	6.87	45.5	13.0	2.57	15.9	27.0	10.8	1.98	12.6	55.6	3.60
	.250	19.0	5.59	38.0	10.9	2.61	13.2	22.6	9.03	2.01	10.4	45.8	3.00
	.188	14.6	4.28	29.9	8.53	2.64	10.2	17.8	7.12	2.04	8.12	35.4	2.35
6 x 4	.438	25.5	7.50	32.7	10.9	2.09	14.0	17.1	8.57	1.51	10.5	39.2	3.72
	.375	22.4	6.58	29.7	9.90	2.12	12.5	15.6	7.81	1.54	9.42	34.9	3.38
	.313	19.1	5.62	26.2	8.72	2.16	10.9	13.8	6.92	1.57	8.21	30.3	2.97
	.250	15.6	4.59	22.1	7.36	2.19	9.05	11.7	5.87	1.60	6.84	25.1	2.50
	.188	12.0	3.53	17.5	5.83	2.23	7.08	9.35	4.67	1.63	5.36	19.6	1.97

26

RECTANGULAR Hollow Structural Sections
IMPERIAL Dimensions and Properties

Outside Dimensions in. x in.	Wall Thickness in.	Mass lb./ft	Area in.²	I_x in.⁴	S_x in.³	r_x in.	Z_x in.³	I_y in.⁴	S_y in.³	r_y in.	Z_y in.³	Torsion J in.⁴	Shear C_{rt} in.²
5 x 3	.375	17.3	5.08	14.7	5.88	1.70	7.70	6.46	4.30	1.13	5.34	15.9	2.63
	.313	14.8	4.36	13.2	5.27	1.74	6.77	5.84	3.89	1.16	4.71	14.0	2.35
	.250	12.2	3.59	11.3	4.52	1.77	5.69	5.05	3.36	1.19	3.98	11.7	2.00
	.188	9.45	2.78	9.08	3.63	1.81	4.50	4.09	2.73	1.21	3.16	9.27	1.60
5 x 2½	.375	16.0	4.70	12.7	5.08	1.64	6.83	4.09	3.27	.932	4.12	11.2	2.63
	.313	13.8	4.05	11.5	4.58	1.68	6.04	3.74	2.99	.961	3.66	9.92	2.35
	.250	11.4	3.34	9.88	3.95	1.72	5.10	3.27	2.62	.990	3.11	8.43	2.00
	.188	8.81	2.59	7.99	3.20	1.76	4.05	2.68	2.15	1.02	2.49	6.72	1.60
5 x 2	.375	14.7	4.33	10.7	4.27	1.57	5.96	2.31	2.31	.730	2.99	7.03	2.63
	.313	12.7	3.74	9.74	3.90	1.61	5.30	2.15	2.15	.750	2.69	6.37	2.35
	.250	10.5	3.09	8.47	3.39	1.66	4.51	1.92	1.92	.788	2.31	5.50	2.00
	.188	8.17	2.40	6.90	2.76	1.70	3.60	1.60	1.60	.816	1.86	4.44	1.60
4 x 3	.375	14.7	4.33	8.20	4.10	1.38	5.34	5.16	3.44	1.09	4.35	11.3	1.88
	.313	12.7	3.74	7.44	3.72	1.41	4.74	4.70	3.14	1.12	3.87	10.0	1.72
	.250	10.5	3.09	6.45	3.22	1.44	4.02	4.10	2.73	1.15	3.29	8.47	1.50
	.188	8.17	2.40	5.24	2.62	1.48	3.21	3.35	2.23	1.18	2.63	6.72	1.22
4 x 2	.313	10.6	3.11	5.31	2.65	1.31	3.59	1.70	1.70	.740	2.16	4.68	1.72
	.250	8.81	2.59	4.69	2.34	1.35	3.09	1.53	1.53	.769	1.87	4.06	1.50
	.188	6.89	2.03	3.88	1.94	1.38	2.49	1.29	1.29	.797	1.52	3.29	1.22
	.150	5.62	1.65	3.27	1.63	1.41	2.07	1.10	1.10	.815	1.27	2.74	1.02
	.125	4.76	1.40	2.82	1.41	1.42	1.77	.954	.954	.826	1.09	2.35	.876
3½ x 2½	.313	10.6	3.11	4.49	2.57	1.20	3.35	2.61	2.09	.916	2.63	5.89	1.41
	.250	8.81	2.59	3.96	2.26	1.24	2.88	2.32	1.86	.947	2.27	5.05	1.25
	.188	6.89	2.03	3.27	1.87	1.27	2.32	1.93	1.54	.976	1.83	4.05	1.03
	.150	5.62	1.65	2.75	1.57	1.29	1.92	1.63	1.30	.994	1.52	3.35	.870
	.125	4.76	1.40	2.38	1.36	1.30	1.65	1.41	1.13	1.01	1.31	2.86	.751
3 x 2	.313	8.46	2.49	2.43	1.62	.990	2.19	1.25	1.25	.710	1.63	3.06	1.09
	.250	7.11	2.09	2.20	1.47	1.03	1.92	1.15	1.15	.741	1.44	2.68	1.00
	.188	5.61	1.65	1.86	1.24	1.06	1.57	.978	.978	.770	1.18	2.19	.846
	.150	4.60	1.35	1.59	1.06	1.08	1.32	.839	.839	.788	.992	1.83	.720
2 x 1	.125	2.21	.648	.293	.293	.673	.386	.096	.192	.385	.234	.254	.375
	.100	1.82	.534	.253	.253	.688	.326	.084	.168	.396	.199	.216	.320

27

SQUARE Hollow Structural Sections   IMPERIAL Dimensions and Properties

Outside Dimensions	Wall Thickness	Mass	Area	I_x	S_x	r_x	Z_x	Torsion J	Surface Area	Shear C_{rt}
in. x in.	in.	lb./ft	in.^{2}	in.^{4}	in.^{3}	in.	in.^{3}	in.^{4}	ft.^{2}/ft.	in.^{2}
12 x 12	.500	76.1	22.4	485	80.9	4.66	95.4	778	3.86	10.0
	.438	67.3	19.8	435	72.5	4.69	85.0	692	3.87	8.98
	.375	58.1	17.1	381	63.4	4.72	73.9	600	3.89	7.88
	.313	48.9	14.4	324	54.1	4.75	62.6	507	3.91	6.73
	.250	39.4	11.6	265	44.1	4.78	50.8	411	3.93	5.50
10 x 10	.500	62.5	18.3	271	54.2	3.84	64.6	441	3.19	8.00
	.438	55.3	16.3	244	48.8	3.87	57.7	393	3.21	7.23
	.375	47.9	14.1	215	42.9	3.90	50.4	342	3.23	6.38
	.313	40.4	11.9	184	36.7	3.93	42.8	290	3.24	5.48
	.250	32.6	9.59	151	30.1	3.96	34.9	235	3.26	4.50
8 x 8	.500	48.9	14.4	131	32.8	3.02	39.7	218	2.52	6.00
	.438	43.4	12.8	119	29.8	3.06	35.7	195	2.54	5.47
	.375	37.7	11.1	106	26.4	3.09	31.3	171	2.56	4.88
	.313	31.9	9.37	91.0	22.8	3.12	26.8	145	2.58	4.22
	.250	25.8	7.59	75.1	18.8	3.15	21.9	119	2.60	3.50
7 x 7	.500	42.0	12.4	84.5	24.2	2.62	29.5	142	2.19	5.00
	.438	37.5	11.0	77.1	22.0	2.65	26.6	128	2.21	4.60
	.375	32.6	9.58	68.7	19.6	2.68	23.5	112	2.23	4.13
	.313	27.6	8.12	59.5	17.0	2.71	20.1	95.9	2.24	3.60
	.250	22.4	6.59	49.4	14.1	2.74	16.5	78.5	2.26	3.00
	.188	17.1	5.04	38.6	11.0	2.77	12.8	60.4	2.28	2.35
6 x 6	.500	35.2	10.4	50.4	16.8	2.21	20.8	86.4	1.86	4.00
	.438	31.5	9.25	46.3	15.4	2.24	18.9	78.1	1.87	3.72
	.375	27.5	8.08	41.6	13.9	2.27	16.8	69.0	1.89	3.38
	.313	23.4	6.87	36.3	12.1	2.30	14.4	59.2	1.91	2.97
	.250	19.0	5.59	30.3	10.1	2.33	11.9	48.7	1.93	2.50
	.188	14.6	4.28	23.9	7.95	2.36	9.27	37.6	1.95	1.97
5 x 5	.438	25.5	7.50	25.1	10.0	1.83	12.5	43.3	1.54	2.85
	.375	22.4	6.58	22.7	9.10	1.86	11.2	38.5	1.56	2.63
	.313	19.1	5.62	20.1	8.03	1.89	9.70	33.3	1.58	2.35
	.250	15.6	4.59	16.9	6.78	1.92	8.07	27.5	1.60	2.00
	.188	12.0	3.53	13.4	5.38	1.95	6.31	21.4	1.61	1.60

SQUARE Hollow Structural Sections — IMPERIAL Dimensions and Properties

Outside Dimensions in. x in.	Wall Thickness in.	Mass lb./ft	Area in.²	I_x in.⁴	S_x in.³	r_x in.	Z_x in.³	Torsion J in.⁴	Surface Area ft.²/ft.	Shear C_{rt} in.²
4 x 4	.375	17.3	5.08	10.7	5.34	1.45	6.70	18.6	1.23	1.88
	.313	14.8	4.36	9.57	4.79	1.48	5.90	16.3	1.24	1.72
	.250	12.2	3.59	8.21	4.11	1.51	4.96	13.6	1.26	1.50
	.188	9.45	2.78	6.61	3.31	1.54	3.92	10.7	1.28	1.22
3½ x 3½	.375	14.7	4.33	6.70	3.83	1.24	4.90	11.9	1.06	1.50
	.313	12.7	3.74	6.09	3.48	1.28	4.35	10.5	1.08	1.41
	.250	10.5	3.09	5.28	3.02	1.31	3.69	8.89	1.10	1.25
	.188	8.17	2.40	4.30	2.46	1.34	2.94	7.04	1.11	1.03
3 x 3	.313	10.6	3.11	3.57	2.38	1.07	3.03	6.33	.910	1.09
	.250	8.81	2.59	3.15	2.10	1.10	2.60	5.40	.928	1.00
	.188	6.89	2.03	2.60	1.74	1.13	2.10	4.32	.946	.846
2½ x 2½	.250	7.11	2.09	1.68	1.35	.898	1.71	2.97	.762	.750
	.188	5.61	1.65	1.42	1.14	.929	1.40	2.41	.779	.658
	.150	4.60	1.35	1.22	.973	.948	1.17	2.01	.790	.570
	.125	3.91	1.15	1.06	.848	.961	1.01	1.72	.798	.501
2 x 2	.250	5.41	1.59	.761	.761	.692	.998	1.39	.595	.500
	.188	4.33	1.27	.668	.666	.725	.840	1.17	.613	.469
	.150	3.58	1.05	.582	.582	.744	.714	.985	.624	.420
	.125	3.06	.899	.514	.514	.756	.621	.852	.631	.375
	.110	2.72	.799	.466	.466	.764	.558	.764	.635	.343
1½ x 1½	.188	3.05	.896	.241	.322	.519	.422	.442	.446	.281
	.150	2.56	.752	.218	.291	.530	.369	.384	.457	.270
	.125	2.21	.648	.197	.263	.551	.326	.338	.464	.250
	.100	1.82	.534	.170	.227	.564	.275	.284	.471	.220
1¼ x 1¼	.150	2.05	.603	.115	.184	.437	.239	.209	.374	.195
	.125	1.78	.524	.106	.169	.450	.214	.187	.381	.188
	.100	1.48	.435	.093	.149	.463	.184	.159	.389	.170
1 x 1	.125	1.35	.398	.048	.095	.346	.125	.087	.298	.125
	.100	1.14	.334	.043	.086	.359	.109	.076	.305	.120

CIRCULAR Hollow Structural Sections

IMPERIAL Dimensions and Properties

Outside Dimensions in.	Wall Thickness in.	Mass lb./ft	Area in.²	I_x in.⁴	S_x in.³	r_x in.	Z_x in.³	Torsion J in.⁴	Surface Area ft.²/ft.	Shear C_{rt} in.²
16.00	.500	82.8	24.3	732	91.5	5.48	120	1460	4.19	12.2
	.438	72.9	21.4	649	81.1	5.50	106	1300	4.19	10.7
	.375	62.7	18.4	562	70.3	5.53	91.6	1120	4.19	9.21
	.313	52.5	15.4	475	59.3	5.55	77.0	949	4.19	7.71
	.250	42.1	12.4	384	48.0	5.57	62.0	767	4.19	6.18
14.00	.500	72.2	21.2	484	69.1	4.78	91.2	968	3.67	10.6
	.438	63.5	18.7	430	61.4	4.80	80.6	859	3.67	9.34
	.375	54.6	16.1	373	53.3	4.82	69.7	746	3.67	8.03
	.313	45.8	13.5	315	45.0	4.84	58.6	631	3.67	6.73
	.250	36.7	10.8	255	36.5	4.86	47.3	511	3.67	5.40
12.75	.500	65.5	19.2	362	56.7	4.34	75.1	723	3.34	9.63
	.438	57.7	17.0	322	50.5	4.36	66.5	643	3.34	8.48
	.375	49.6	14.6	280	43.9	4.38	57.5	589	3.34	7.30
	.313	41.6	12.2	237	37.1	4.40	48.4	473	3.34	6.12
	.250	33.4	9.82	192	30.1	4.42	39.1	384	3.34	4.91
10.75	.500	54.8	16.1	212	39.4	3.63	52.6	424	2.81	8.07
	.438	48.3	14.2	189	35.2	3.65	46.6	378	2.81	7.11
	.375	41.6	12.2	165	30.7	3.67	40.4	330	2.81	6.12
	.313	34.9	10.3	140	26.0	3.69	34.1	280	2.81	5.13
	.250	28.1	8.25	114	21.2	3.71	27.6	228	2.81	4.12
8.625	.500	43.4	12.8	106	24.5	2.88	33.1	212	2.26	6.40
	.438	38.4	11.3	94.7	22.0	2.90	29.4	189	2.26	5.65
	.375	33.1	9.73	82.9	19.2	2.92	25.6	166	2.26	4.87
	.313	27.8	8.17	70.7	16.4	2.94	21.6	141	2.26	4.09
	.250	22.4	6.58	57.7	13.4	2.96	17.5	115	2.26	3.29
	.188	17.0	4.99	44.4	10.3	2.98	13.4	88.8	2.26	2.49
6.625	.375	25.1	7.37	36.1	10.9	2.21	14.7	72.2	1.73	3.70
	.313	21.1	6.21	31.0	9.36	2.23	12.5	62.0	1.73	3.11
	.250	17.0	5.01	25.5	7.69	2.26	10.2	51.0	1.73	2.51
	.188	13.0	3.81	19.7	5.96	2.28	7.80	39.5	1.73	1.90
5.562	.375	20.8	6.11	20.7	7.43	1.84	10.1	41.4	1.46	3.07
	.313	17.6	5.16	17.8	6.42	1.86	8.64	35.7	1.46	2.59
	.250	14.2	4.17	14.8	5.30	1.88	7.06	29.5	1.46	2.09
	.188	10.8	3.18	11.5	4.13	1.90	5.44	23.0	1.46	1.59

CIRCULAR Hollow Structural Sections

IMPERIAL Dimensions and Properties

Outside Dimensions in.	Wall Thickness in.	Mass lb./ft	Area in.^{2}	I_x in.^{4}	S_x in.^{3}	r_x in.	Z_x in.^{3}	Torsion J in.^{4}	Surface Area ft.^{2}/ft.	Shear C_{rt} in.^{2}
4.50	.313	14.0	4.12	9.07	4.03	1.48	5.50	18.1	1.18	2.07
	.250	11.4	3.34	7.56	3.36	1.51	4.52	15.1	1.18	1.67
	.188	8.67	2.56	5.94	2.64	1.53	3.50	11.9	1.18	1.28
4.00	.313	12.3	3.63	6.20	3.10	1.31	4.27	12.4	1.05	1.82
	.250	10.0	2.95	5.20	2.60	1.33	3.52	10.4	1.05	1.48
	.188	7.67	2.25	4.10	2.05	1.35	2.74	8.21	1.05	1.13
	.150	6.17	1.81	3.37	1.68	1.36	2.22	6.73	1.05	.908
3.50	.313	10.7	3.13	4.02	2.30	1.13	3.19	8.03	.916	1.58
	.250	8.69	2.56	3.39	1.94	1.15	2.65	6.78	.916	1.28
	.188	6.66	1.96	2.69	1.54	1.17	2.07	5.39	.916	.982
	.150	5.37	1.58	2.22	1.27	1.19	1.68	4.44	.916	.790
2.875	.250	7.01	2.06	1.79	1.25	.933	1.73	3.58	.752	1.04
	.188	5.40	1.59	1.44	1.00	.952	1.36	2.88	.752	.797
	.150	4.37	1.28	1.19	.831	.965	1.11	2.39	.752	.643
	.125	3.68	1.08	1.02	.712	.973	.947	2.05	.752	.541
2.375	.250	5.68	1.67	.954	.804	.756	1.13	1.91	.622	.845
	.188	4.40	1.29	.778	.655	.776	.901	1.56	.622	.651
	.150	3.57	1.05	.651	.548	.788	.743	1.30	.622	.526
	.125	3.01	.885	.561	.473	.796	.634	1.12	.622	.443
1.90	.188	3.45	1.01	.376	.396	.609	.555	.752	.498	.512
	.150	2.81	.825	.319	.335	.622	.461	.638	.498	.415
	.125	2.38	.699	.277	.291	.630	.396	.554	.498	.351
	.110	2.10	.618	.249	.262	.635	.353	.498	.498	.310
1.66	.125	2.06	.604	.179	.216	.545	.296	.359	.435	.304
	.100	1.67	.491	.150	.181	.553	.244	.300	.435	.246
1.315	.125	1.59	.468	.084	.127	.423	0.178	.167	.344	.236
	.100	1.30	.382	.071	.108	.431	0.148	.142	.344	.192
1.05	.125	1.24	.364	.040	.076	.330	0.108	.079	.275	.185
	.100	1.02	.299	.034	.065	.338	0.091	.068	.275	.151

1.9 Symbols and Abbreviations

A	Area
A_g	Gross area
A_i	Cross sectional area of member i
A_m	Area of fusion face
A_n	Net area; Weld fusion area of base metal normal to a tensile load
A_{ne}	Effective net area
A'_{ne}	Effective net area reduced for shear lag
A_V	Effective shear area of the chord
A_w	Effective throat area of a weld
B	Width; width of a doubler plate
B_r	Factored bearing resistance of a member or component
C	Coefficient used with τ to determine SCF_{ax-ch} and SCF_{ax-w} for square HSS
C_e	Euler buckling strength
C_f	Compressive force in a member or component under factored loads
C_{fi}	Compressive force in member i under factored loads
C_{ipb}, C_{opb}	Uncorrected connection moment efficiencies (expressed as the connection moment resistance divided by the plastic moment capacity M_{pi}) of a web member
C_K, C_T, C_X	Uncorrected connection efficiencies (expressed as the connection resistance divided by the yield load $A_i F_{yi}$) of a web member
C_r	Factored compressive resistance
C_{ri}	Factored compressive resistance of member i
C_{rt}	Shear constant for hollow section member
C_y	Axial compressive load at yield stress
E	Elastic modulus of steel (200 000 MPa assumed)
F_k	Web buckling stress
F_u	Specified minimum tensile strength
F_{ui}	Specified minimum tensile strength of member i
F_y	Specified minimum yield stress
F_{yc}	Specified minimum yield stress of a column
F_{yi}	Specified minimum yield stress of member i

F_{0p}	Additional stress in a truss chord at a panel point, other than that required to maintain equilibrium with web member forces (i.e., the chord "prestress"); Axial stress in a column above a beam connection
G	Shear modulus of steel (77 000 MPa assumed)
I	Moment of Inertia
J	St. Venant torsion constant
K	Effective length factor
K_a	Relative length factor used by AWS to compute the contact perimeter of an inclined HSS web member
L	Length
L_{cr}	Maximum unbraced length adjacent to a plastic hinge
L_n	Net length
M_f	Bending moment under factored loads
M_{fi}	Bending moment under factored loads applied to member i
M_{opi}	Bending moment, out of the plane of the structure, under factored loads applied to member i
M^*_{opri}	Factored moment <u>connection</u> resistance for member i, for bending moments out of the plane of the structure
M_{pi}	Plastic moment capacity of member i
M_{ri}	Factored moment resistance of <u>member i</u>
M^*_{ri}	Factored moment <u>connection</u> resistance for member i, for bending moments in the plane of the structure
M_{yi}	Yield moment of member i
N	Axial force (tension or compression) under factored loads in a member; Fatigue life (number of load cycles)
N_i	Axial force under factored loads applied to member i
N^*_i	Connection resistance, as an axial force in member i
N_{ri}	Axial resistance of member i
$N^*_{0(ingap)}$	Reduced axial load resistance due to shear in the cross section of the chord at the gap
N_{0p}	Additional axial force in a truss chord at a panel point, other than that required to maintain equilibrium with web member forces (i.e., the "prestress" force); Axial force in a column above a beam connection
O_v	Overlap, $(O_v = (q/p) \times 100\%)$
P	Load

P_f	External factored tensile load applied to a bolt
R	Stress ratio
S	Size of a fillet weld
S_i	Elastic section modulus of member i (also S_x, S_y)
S_r	Stress range
$S_{r\,ax-nom}$	Nominal axial stress range
$S'_{r\,ax}$	Nominal axial stress range factored to include effects of secondary bending stresses
$S'_{r\,ax-w}$	Nominal axial stress range in a web member factored to include effects of secondary bending stresses
$S_{r\,b-nom-ch}$	Nominal bending stress range in the *chord* member produced by primary bending moments
$S_{r\,b-nom-w}$	Nominal bending stress range in the *web* (or branch) member produced by primary bending moments
$S_{r-h.s.}$	Hot spot stress range
S_{r-nom}	Nominal stress range
SCF	Stress Concentration Factor
SCF_{ax-ch}	Stress concentration factor for an axial load in the *web* member causing cracking in the *chord* member
SCF_{ax-w}	Stress concentration factor for an axial load in the *web* member causing cracking in the *web* member
SCF_{b-ch}	Stress concentration factor for a primary in-plane bending moment in the *web* (or branch) member causing cracking in the *chord* member
SCF_{b-w}	Stress concentration factor for a primary in-plane bending moment in the *web* (or branch) member causing cracking in the *web* (or branch) member
SCF^*	Stress concentration factor for $\gamma = 12.5$ and $\tau = 0.5$
$SNCF$	Strain Concentration Factor
T_f	Tensile force in a bolt under factored load
T_{fi}	Tensile force in member i under factored loads
T_r	Factored tensile resistance
T_u	Ultimate tensile strength of a bolt
U_1	Factor to account for moment gradient and for second-order effects of axial force acting on the deformed member
V_f	Shear force under factored loads
V_{fi}	Shear force in member i under factored loads
V_p, V_{pi}	Shear yield strength of a section, or of member i

V_r	Factored shear resistance
X_u	Ultimate strength of weld metal
Z_i	Plastic section modulus of member i
a	Edge distance of plate from centre of bolt hole
b	Width of a compression element; Distance from bolt line to the HSS face (Fig.7.1); Subscript to denote a *beam*
b'	$b - d/2 + t_i$
b_e	Effective width of a member
$b_{e(ov)}$	Effective width for *overlapping* member connected to *overlapped* member
b_{ep}	Effective punching shear width
b_i	External width of rectangular HSS member i (90° to plane of the truss).
b_m	Effective width of the web of an I-shaped chord
d	Nominal diameter of bolt
d'	Bolt hole diameter; Hole diameter allowance used for net area when holes are punched
d'_1	Transverse distance between outsides of adjacent braces in a multi-planar connection
d_i	External diameter of circular HSS member i
d_l	Combined width of overlapping web members from 2 planes of a multiplanar structure
e	Noding eccentricity at a truss panel point (positive towards the outside of the truss)
f	Width of the weld fusion face
f'_c	Specified 28-day crushing strength of concrete
$f(n), f(n')$	Functions in connection resistance formulae that incorporate the detrimental influence of compressive stresses in HSS chord or column members (applicable when n or n' is negative)

$$f(n) = 1.3 + \frac{0.4}{\beta} n$$

(applied to axially-loaded T, Y, X, K and N connections between square and rectangular HSS)

$$f_2(n) = 1.2 + \frac{0.5}{\beta} n \quad \text{where} \quad n = \frac{N_0}{A_0 F_{y0}}$$

(applied to moment-loaded T and X connections (e.g., Vierendeel trusses) between square and rectangular HSS)

$$f_3(n) = 1.0 + 0.06 \left(\frac{b_0}{t_0}\right)^{0.52} n - 0.02 \left(\frac{b_0}{t_0}\right)^{0.69} n^2 \qquad \text{where } n = \frac{N_0}{A_0 F_{y0}}$$
(applied to transverse plate to square and rectangular HSS connections)

$$f(n') = 1.0 + 0.3\, n' - 0.3\, n'^2$$
(applied to:
axially-loaded T, Y, X, K and N connections between circular HSS;
moment-loaded T, Y and X connections between circular HSS;
longitudinal plate to square and rectangular HSS connections; and
"bird mouth" T and K connections between square HSS)

$$f_2(n') = 1.0 + 0.2\, n'$$
(applied to certain truss connections using circular HSS cropped or flattened web members)

$$f_3(n') = f(n'), \text{ or}$$
$$= 1.0 - n'^2 (\beta - \beta^2)(2\gamma)^{0.303} \qquad \text{where } n' = \frac{N_{0p}}{A_0 F_{y0}}$$
(applied to wide flange beam to circular HSS column moment connections)

g	Gap between web members at a truss panel point (ignoring welds) on the chord face; Transverse spacing between fastener gauge lines
g_t	Transverse gap between adjacent braces in a multiplanar connection
g'	Gap divided by chord wall thickness
h	Clear height of web between flanges
h_b	Height of a beam between flange centres; Distance between flange plate centres
h_i	External height of member i (in plane of the truss)
h_w	Depth of the web of an I-shaped chord, $(h_w = h_0 - 2(t_0 + r_0))$
i	Subscript to denote member of the connection: $i = 0$ designates chord; $i = 1$ refers to a web member, either tension or compression, for T, Y and X connections, or it refers to the compression member for K, N and KT connections; $i = 2$ refers to the tension member for K, N and KT connections; $i = 3$ refers to the vertical for KT connections: $i = i$, refers to the overlapping member for K and N connections.
j	Subscript to denote the *overlapped* member for K and N connections
k	Distance from outer face of member flange to web-toe of fillet ($k = t_0 + r_0$)
n	$\dfrac{N_0}{A_0 F_{y0}} + \dfrac{M_{f0}}{S_0 F_{y0}};$

	Number of bolts; Subscript used to denote a *net*, rather than a *gross*, amount
n'	$\dfrac{N_{0p}}{A_0 F_{y0}} + \dfrac{M_{f0}}{S_0 F_{y0}}$
n_i	Number of applied load cycles at stress range i
p	Length of projected contact area between *overlapping* web member and chord without presence of the *overlapped* web member; Length of *flange plate* tributary to each bolt, or bolt pitch; Subscript to denote a *plate*
q	Length of overlap of web members on the chord face
r	Radius of gyration (also r_x, r_y)
r_0	Flange/web radius of an I-shaped chord member
s	Centre-to-centre longitudinal spacing of any two successive fastener holes (Fig. 7.17)
t	Thickness
t_c	Thickness of column flange
t_i	Thickness of wall of HSS member i; Thickness of flange of I-shaped member i
t_w	Throat thickness of a weld
w	Width; Thickness of web of an I-shaped member
w_n	Net width
w_c	Thickness of a column web
$x1, x2$	Exponents used to determine SCF_{ax-ch} for circular HSS
α	Ratio of the equilibrating moment per unit plate width at the bolt line, to the flange moment at the inner plastic hinge; Coefficient used to determine effective shear area of a chord member, either rectangular HSS or I-shaped; Stress reduction factor used for mitred knee connections; Stiffness of supports to a long laterally unsupported compression chord
β	Width or diameter ratio between web member(s) and chord

$$\beta = \frac{d_1}{d_0}, \frac{d_1'}{b_0}, \frac{b_1}{b_0} \quad (\text{T, Y, X})$$

$$\beta = \frac{d_1+d_2}{2d_0}, \frac{d_1+d_2}{2b_0}, \frac{b_1+b_2+h_1+h_2}{4b_0} \quad (\text{K, N})$$

$$\beta = \frac{d_1+d_2+d_3}{3d_0}, \frac{d_1+d_2+d_3}{3b_0}, \frac{b_1+b_2+b_3+h_1+h_2+h_3}{6b_0} \quad \text{(KT)}$$

γ Half width to thickness ratio of the chord,

$$\gamma = \frac{d_0}{2t_0} \text{ or } \frac{b_0}{2t_0}$$

η Web or branch member depth to chord width ratio,

$$\eta = \frac{h_i}{b_0} \text{ or } \frac{h_i}{d_0}$$

θ Included angle

θ'_1 Angle between the chord axis and the plane in which two adjacent braces lie

θ_i Included angle between web member i ($i = 1,2,3$) and the chord

κ Ratio of the smaller factored moment to the larger, at opposite ends of the unbraced length, positive for double curvature, negative for single

λ Non dimensional slenderness ratio from CAN/CSA-S16.1-94, Clause 13.3

τ ratio of web wall thickness to chord wall thickness

ϕ Resistance factor

ϕ_c Resistance factor for concrete in bearing, taken as 0.60.

ϕ_w Resistance factor for weld metal, taken as 0.67

φ Angle between web planes in multiplanar strutures

ω_1 Coefficient used to determine equivalent uniform bending effect in beam-columns

AWS   American Welding Society

CIDECT   Comité International pour le Développement et l'Etude de la Construction Tubulaire

HSS   Hollow Structural Section

IIW   International Institute of Welding

ISO   International Standards Organization

K or N   Connection involving two branch members on the same side of a main member, with the force components from each branch normal to the main member substantially balancing each other

KT   Connection involving three branch members on the same side of a main member

MPa   Megapascal (one newton per square millimetre)

T or Y	Connection involving one branch member and a main member
X	Connection involving two or more branch members on opposite sides of a main member, with force being transferred through the main member
kN	Kilonewton
m	Metre
mm	Millimetre

REFERENCES

AIJ. 1990. Recommendations for the design and fabrication of tubular structures in steel, 3rd. ed. Architectural Institute of Japan, Tokyo, Japan.

AIJ. 1991. Japanese architectural standard specification, JASS 6 steelwork. Architectural Institute of Japan, Tokyo, Japan.

AISC. 1992. Seismic provisions for structural steel buildings. American Institute of Steel Construction, Chicago, Illinois, U.S.A.

AISC. 1993. Load and resistance factor design specification for structural steel buildings. American Institute of Steel Construction, Chicago, Illinois, U.S.A.

AISC. 1997. Pre-engineered connections for structural steel hollow sections. Volume 1: design models, 1st. ed. Australian Institute of Steel Construction, North Sydney, New South Wales, Australia.

ASTM. 1991. Standard specification for high-strength low-alloy structural steel with 50 ksi [345 MPa] minimum yield point to 4 in. [100 mm] thick, ASTM A588-91a. American Society for Testing and Materials, Philadelphia, Pennsylvania, U.S.A.

ASTM. 1993. Standard specification for cold-formed welded and seamless carbon steel structural tubing in rounds and shapes, ASTM A500-93. American Society for Testing and Materials, Philadelphia, Pennsylvania, U.S.A.

AWS. 1996. Structural welding code—steel. ANSI/AWS D1.1-96, American Welding Society, Miami, Florida, U.S.A.

BERGMANN, R., DUTTA, D., MATSUI, C., MEINSMA, C., and TSUDA, T. 1995. Design guide for concrete-filled hollow section columns. CIDECT (ed.) and Verlag TÜV Rheinland GmbH, Köln, Federal Republic of Germany.

CIDECT. 1984. Construction with hollow steel sections. British Steel plc., Corby, Northants., England.

CISC. 1977. Limit states design steel manual. Canadian Institute of Steel Construction, Willowdale, Ontario.

CISC. 1995. Handbook of steel construction, 6th. ed. Canadian Institute of Steel Construction, Willowdale, Ontario.

CRAN, J.A. 1982. Hollow structural sections—Warren and Pratt truss connections—weld gap and overlap joints using rectangular chord members. Technical Bulletin No. 22, Stelco Inc., Hamilton, Ontario.

CSA. 1992. General requirements for rolled or welded structural quality steel, CAN/CSA-G40.20-M92. Canadian Standards Association, Rexdale, Ontario.

CSA. 1992. Structural quality steels, CAN/CSA-G40.21-M92. Canadian Standards Association, Rexdale, Ontario.

CSA. 1992. Metric dimensions of structural steel shapes and hollow structural sections, CAN/CSA-G312.3-M92. Canadian Standards Association, Rexdale, Ontario.

CSA. 1994. Limit states design of steel structures, CAN/CSA-S16.1-94. Canadian Standards Association, Rexdale, Ontario.

DUTTA, D., and WÜRKER, K. 1988. Handbuch hohlprofile in stahlkonstruktionen. Verlag TÜV Rheinland GmbH, Köln, Federal Republic of Germany.

EASTWOOD, W., and WOOD, A.A. 1970a. Welded joints in tubular structures involving rectangular sections. Proceedings, Conference on Joints in Structures, Session A Paper 2, Sheffield, England.

EASTWOOD, W., and WOOD, A.A. 1970b. Recent research on joints in tubular structures. Proceedings, Canadian Structural Engineering Conference, Toronto, Ontario.

EUROPEAN COMMITTEE for STANDARDIZATION. 1992a. Eurocode 3: design of steel structures. Part 1.1—general rules and rules for buildings. ENV 1993-1-1:1992E, British Standards Institution, London, England.

EUROPEAN COMMITTEE for STANDARDIZATION. 1992b. Hot finished structural hollow sections of non-alloy and fine grain structural steels. Part 2—tolerances, dimensions and sectional properties. prEN10210-2 (Draft Doc. No. 92/46922), British Standards Institution, London, England.

EUROPEAN COMMITTEE for STANDARDIZATION. 1992c. Cold formed structural hollow sections of non-alloy and fine grain structural steels. Part 2—tolerances, dimensions and sectional properties. prEN10219-2 (Draft Doc. No. 92/46923), British Standards Institution, London, England.

GIDDINGS, T.W., and WARDENIER, J. (eds.). 1986. The strength and behaviour of statically loaded welded connections in structural hollow sections. CIDECT Monograph No. 6, British Steel plc., Corby, Northants., England.

IIW. 1981. Design recommendations for hollow section joints—predominantly statically loaded, 1st. ed. International Institute of Welding, Subcommission XV-E, IIW Doc. XV-491-81 (revised), IIW Annual Assembly, Oporto, Portugal.

IIW. 1985. Recommended fatigue design procedure for hollow section joints. Part I—hot spot stress method for nodal joints. International Institute of Welding, Subcommission XV-E, IIW Doc. XV-582-85, IIW Annual Assembly, Strasbourg, France.

IIW. 1989. Design recommendations for hollow section joints—predominantly statically loaded, 2nd. ed. International Institute of Welding, Subcommission XV-E, IIW Doc. XV-701-89, IIW Annual Assembly, Helsinki, Finland.

JIS. 1988a. Carbon steel tubes for general structural purposes, JIS G3444-1988,. Japanese Industrial Standards, Tokyo, Japan.

JIS. 1988b. Carbon steel square pipes for general structural purposes, JIS G3466-1988. Japanese Industrial Standards, Tokyo, Japan.

JSSC. 1988. Cold-formed carbon steel square and rectangular hollow sections (box section columns). Japanese Society of Steel Construction, JJSS II-10-1988, Tokyo, Japan.

LUI, Z. and GOEL, S.C. 1987. Investigation of concrete-filled steel tubes under cyclic bending and buckling. UMCE Report No. 87-3, University of Michigan, Ann Arbor, Michigan, U.S.A.

NIEMI, E. (ed.). 1995. Stress determination for fatigue analysis of welded components. Abington Publishing, Abington, Cambridge, England.

PACKER, J.A. 1983. Developments in the design of welded HSS truss joints with RHS chords. Canadian Journal of Civil Engineering, **10**(1): 92–103.

PACKER, J.A. 1985. Welded connections with rectangular tubes. Proceedings, Symposium on Hollow Structural Sections in Building Construction, ASCE Structural Engineering Congress, Chicago, Illinois, U.S.A., pp. 6–1 to 6–26.

PACKER, J.A. 1986. Design examples for HSS trusses. Canadian Journal of Civil Engineering, **13**(4): 460–473.

PACKER, J.A. 1993. Overview of current international design guidance on hollow structural section connections. Proceedings, 3rd. International Offshore and Polar Engineering Conference, Singapore, Vol. 4, pp. 1–7.

PACKER, J.A., and HALEEM, A.S. 1981. Ultimate strength formulae for statically-loaded welded HSS joints in lattice girders with RHS chords. Proceedings, Canadian Society for Civil Engineering Annual Conference, Fredericton, New Brunswick, Vol.1, pp. 331–343.

PACKER, J.A., BIRKEMOE, P.C., and TUCKER, W.J. 1983. Design of gap and overlap joints in single chord HSS trusses. Canadian Symposium on Hollow Structural Sections, lecture tour to 10 Canadian cities, 32 pp.

PACKER, J.A., BIRKEMOE, P.C., and TUCKER, W.J. 1986. Design aids and design procedures for HSS trusses. Journal of Structural Engineering, Proceedings of the American Society of Civil Engineers, **112**(7): 1526–1543.

PACKER, J.A., HENDERSON, J.E., and CAO, J.J. 1997. Design guide for hollow structural section connections—Chinese edition. Science Press, Beijing, P.R. China.

PACKER, J.A., WARDENIER, J., KUROBANE, Y., DUTTA, D., and YEOMANS, N. 1992. Design guide for rectangular hollow section (RHS) joints under predominantly static loading. CIDECT (ed.) and Verlag TÜV Rheinland GmbH, Köln, Federal Republic of Germany.

PARIK, J., DUTTA, D., and YEOMANS, N. 1994. User guide for PC-program CIDJOINT for hollow section joints under predominantly static loading. CIDECT (ed.) and Ing.-Software Glubal GmbH, Tiefenbach, Federal Republic of Germany.

RONDAL, J., WÜRKER, K.-G., DUTTA, D., WARDENIER, J., and YEOMANS, N. 1992. Structural stability of hollow sections. CIDECT (ed.) and Verlag TÜV Rheinland GmbH, Köln, Federal Republic of Germany.

SAA. 1990. Steel structures, AS4100-1990. Standards Association of Australia, North Sydney, New South Wales, Australia.

SAA. 1991. Structural steel hollow sections, AS1163-1991. Standards Association of Australia, North Sydney, New South Wales, Australia.

SABS. 1993. The structural use of steel. Part 1—limit-states design of hot-rolled steelwork. SABS 0162-1: 1993, South African Bureau of Standards, Pretoria, Republic of South Africa.

SAISC. 1989. South African structural hollow sections handbook, 1st. ed. South African Institute of Steel Construction, Johannesburg, Republic of South Africa.

SHERMAN, D.R. 1996. Designing with structural tubing. Engineering Journal, American Institute of Steel Construction, **33**, 3rd. quarter: 101–109.

STELCO. 1971. Hollow structural sections—design manual for connections, 1st. ed. Stelco Inc., Hamilton, Ontario.

STELCO. 1981. Hollow structural sections—design manual for connections, 2nd. ed. Stelco Inc., Hamilton, Ontario.

STI. 1996a. Hollow structural sections—dimensions and section properties. Steel Tube Institute of North America, Mentor, Ohio, U.S.A.

STI. 1996b. Hollow structural sections—principal producers and capabilities. Steel Tube Institute of North America, Mentor, Ohio, U.S.A.

TUBEMAKERS. 1994. Design capacity tables for Duragal steel hollow sections. Tubemakers Structural and Engineering Products, Tubemakers of Australia Ltd., Newcastle, New South Wales, Australia.

TWILT, L., HASS, R., KLINGSCH, W., EDWARDS, M., and DUTTA, D. 1994. Design guide for structural hollow section columns exposed to fire. CIDECT (ed.) and Verlag TÜV Rheinland GmbH, Köln, Federal Republic of Germany.

WARDENIER, J. 1982. Hollow section joints. Delft University Press, Delft, The Netherlands.

WARDENIER, J., KUROBANE, Y., PACKER, J.A., DUTTA, D., and YEOMANS, N. 1991. Design guide for circular hollow section (CHS) joints under predominantly static loading. CIDECT (ed.) and Verlag TÜV Rheinland GmbH, Köln, Federal Republic of Germany.

WARDENIER, J., DUTTA, D., YEOMANS, N., PACKER, J.A., and BUCAK, Ö. 1995. Design guide for structural hollow sections in mechanical applications. CIDECT (ed.) and Verlag TÜV Rheinland GmbH, Köln, Federal Republic of Germany.

van WINGERDE, A.M., PACKER, J.A., and WARDENIER, J. 1995. Criteria for the fatigue assessment of hollow structural section connections. Journal of Constructional Steel Research, **35**: 71–115.

van WINGERDE, A.M., PACKER, J.A., and WARDENIER, J. 1996. New guidelines for fatigue design of HSS connections. Journal of Structural Engineering, American Society of Civil Engineers, **122**(2): 125–132.

2
STANDARD TRUSS DESIGN

2.1 Practical Aspects and Costs

When framing trusses from hollow structural sections, the designer needs to consider certain practical aspects which bear directly on the cost of the structure. Some of these features relate to a comparison with alternative structural steel sections, and others relate to options within the usage of HSS.

2.1.1 Costs Relative to Alternative Structural Steel Sections

Some of the factors which contribute to the total cost of a truss follow.

The material cost for hollow structural sections may be up to 25% higher (Class H) than that for open rolled sections or plate, as is to be expected for a secondary product manufactured from the primary plate. However, this cost premium is frequently offset by the more efficient use of material to provide a shape superior in compression and torsion. In addition, the 350 MPa standard yield strength of HSS provides a 17% advantage over the 300 MPa steel generally used for non-tubular sections. This is well demonstrated by the column design example in Table 1.1 of Chapter 1.

Fabrication costs are primarily a function of the labour hours required to produce a given truss. These hours need not be greater with HSS designs than with alternative designs, and can even be less, depending on connection configurations.

It is essential that the designer realize that the selection of HSS truss components (chords and webs) determines the complexity of the joints at the panel points. One should not normally expect that members selected for minimum mass can be connected for minimum labour time. This will seldom be the case because the efficiency of HSS connections is a subtle function of parameters whose values are determined by relative dimensions of the connecting members. Some combinations of member sizes can be connected directly, while other combinations carrying the same loads require expensive reinforcing detail material.

Since the cost of a fabrication hour is roughly equivalent to 50 kilograms of HSS material, an apparent saving by virtue of minimum material used may turn out to be a considerable cost penalty. It has been reported that tonnages for different designs performing the same function usually vary by only 20% to 30%, whereas fabrication costs can vary by a factor of three or four (Firkins and Hemphill 1990).

Handling and erecting costs can be less for HSS trusses than for alternative trusses. Their greater stiffness and lateral strength mean that they are easier to pick up and more stable to erect. Furthermore, trusses comprised of HSS are likely to be lighter than their counterparts fabricated from non-tubular sections, as truss members are primarily axially loaded and HSS represent the most efficient use of a steel cross-section in compression.

Surface protection costs are appreciably lower for HSS trusses than for other steel trusses. A square hollow section generally has less than two thirds the surface area of the same capacity wide flange shape in compression. An example is a W200x36 with 1.05 m^2 of surface area per metre of length compared with an HSS 152x152x4.8 with 0.593 m^2 per metre. The absence of re-entrant corners makes the application of paint or fire protection easier. Rectangular HSS have only four surfaces to be painted, whereas wide flange sections have eight flat surfaces. Thus HSS require both less material and less labour in applying surface coatings than do trusses from other shapes.

Regardless of the type of shape used to design a truss, it is generally false economy to attempt to minimize mass by selecting a multitude of sizes for web members. The increased cost to source and to separately handle the various shapes more than offsets the apparent savings in material. It is therefore better to use the same section size for a group of web members.

2.1.2 Costs Among HSS Alternatives

The designer will also wish to bear in mind certain practical aspects relating to hollow structural sections, as well as some detail considerations which influence the costs of truss structures.

Circular HSS are more expensive to fabricate than rectangular (or square) HSS. Connections of circular HSS require that the tube ends be profile cut when the tubes are to be fitted directly together, unless the web tubes are much smaller than the chords. More than that, the bevel of the end cut must generally be varied for welding access as one progresses around the tube. Automated equipment for this purpose is not available in most shops, and semi automatic or manual profile cutting is much more expensive than straight bevel cuts on rectangular HSS.

Since circular HSS are used less frequently than rectangular HSS, both producers and fabricators carry less inventory, with the result that procurement may take longer. In structures where deck or panelling is laid directly on the top chord of trusses, rectangular HSS offer superior surfaces for attaching and supporting the deck. Fitting backing bars to rectangular HSS for welding can be less onerous than with circular HSS. Still another aspect to consider when choosing between circular and rectangular HSS is the relative ease of handling and of stacking the latter, an important item because material handling is said to be the highest cost in the shop (Jensen and Busch 1989).

Joint configurations are increasingly expensive progressing from gap to complete overlap to partial overlap as illustrated in Fig. 2.1. Gap joints have the advantage of a single bevel cut, if the chord is a rectangular HSS, and complete ease of fitting. Partial overlap joints have double cuts with minimum flexibility in fitting (especially if *both* ends are partial overlaps). Watson *et al.* (1996) note that the fabrication time (and hence cost) of a

(a) Gap connection noding
$e = 0$

(b) Gap connection with positive eccentricity
$e > 0$

(c) Partial overlap connection with negative eccentricity
$e < 0$

(d) Total overlap connection with negative eccentricity
$e < 0$

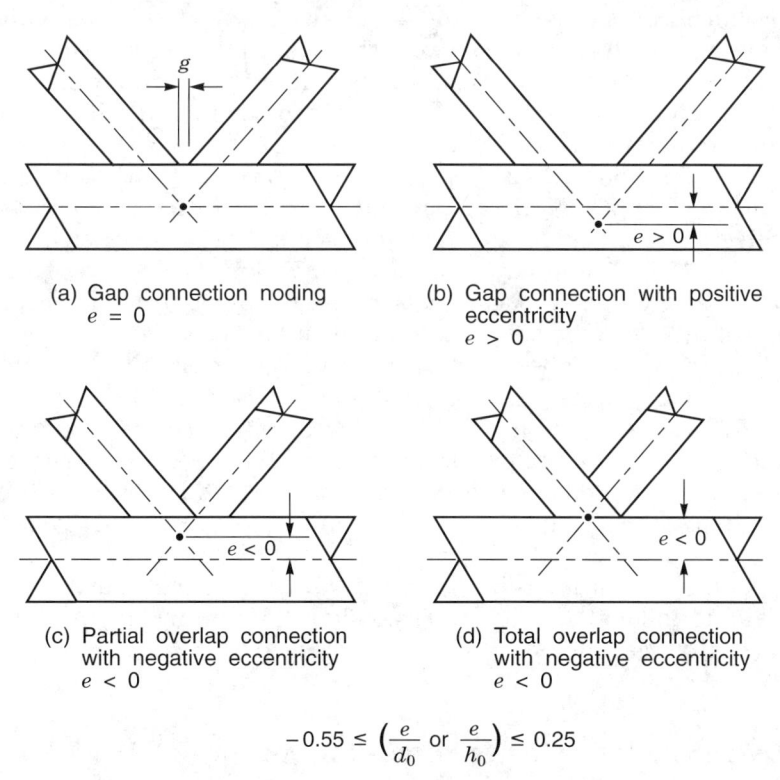

$$-0.55 \leq \left(\frac{e}{d_0} \text{ or } \frac{e}{h_0}\right) \leq 0.25$$

FIGURE 2.1

Noding eccentricity, with permitted limits between which the resulting moment on the connection can be ignored, for connection design

circular HSS connection is approximately three times that for a rectangular (or square) HSS connection. The former involves profiling the ends of the web members, whereas the latter only involves straight cutting of the web members.

These relative costs detract from the initial appeal of overlap connections, which the designer will come to recognize as usually having superior static and fatigue connection strength compared to gap connections. Also, when the concealed portion of an overlapped web member needs to be welded, it must be done before the overlapping web member is fitted. This condition prevents fitting and tack welding all members prior to final structural welding, an economical sequence which is preferred by many fabricators.

Welding costs are sensitive to joint geometry, weld type and weld size. Fillet welds usually do not require the preparation of bevel surfaces that is inherent in almost all partial or full penetration groove welds. A 90° cross-section, 12 mm fillet weld has twice the resistance of a 6 mm fillet weld; however, it has four times the volume. Therefore cost per unit resistance is clearly lower with smaller size welds.

A rectangular HSS web member whose width is the same as a rectangular HSS chord member presents a condition where the side walls of the web line up with the round corners of the chord. Depending on the corner radius and the wall thicknesses, at best there is a flare bevel joint (more awkward than a fillet), or more likely a flare bevel joint with a gap which requires custom fitted backing. Furthermore, flare bevel joints require special qualification procedures. Thus, web members with widths slightly less than the flat width of rectangular chords are the economical choice. Weld selection and design are considered in detail in Chapter 8.

Connection pieces such as gusset plates obviously add material and labour costs. Welding is essentially doubled because loads are transferred twice, first from a member to the connecting piece, then from that piece to another member. Watson *et al.* (1996) claim that the fabrication cost of a circular HSS truss K connection is greater when the web members are slotted onto a gusset plate projecting from the chord than when the ends of the tubular web members are profiled and welded directly to the chord. Hence, direct connection of one HSS to another is preferred, even then.

Stiffeners and other reinforcement (which similarly increase costs of material and labour) should always be kept to a minimum, and used only when actually needed.

2.2 Determination of Truss Forces for Design

The remainder of this chapter deals with the design philosophy applicable to triangulated (e.g., Warren or Pratt) planar HSS trusses with web members directly welded to single-section chord members.

2.2.1 Truss Configurations

Some of the common truss types are shown in Fig. 2.2. Warren trusses will generally provide the most economical solution, since their long compression web members can take advantage of the fact that HSS are very efficient in compression. They have about one half the number of web members and one half the number of connections compared to Pratt trusses, resulting in considerable labour and cost savings. The panel points of a Warren truss can be located at the load application points on the chord, if necessary with an irregular truss geometry. However, even if the chord is loaded in bending, that disadvantage is usually less significant with HSS chords than with alternatives. See example trusses in Chapter 13.

If support is required at all load points to a chord (for example, to reduce the unbraced length), a modified Warren truss could be used rather than a Pratt truss by adding vertical members as shown in Fig. 2.2(a).

Warren trusses provide greater opportunities to use gap joints, the preferred arrangement at panel points. Also, when possible, a regular

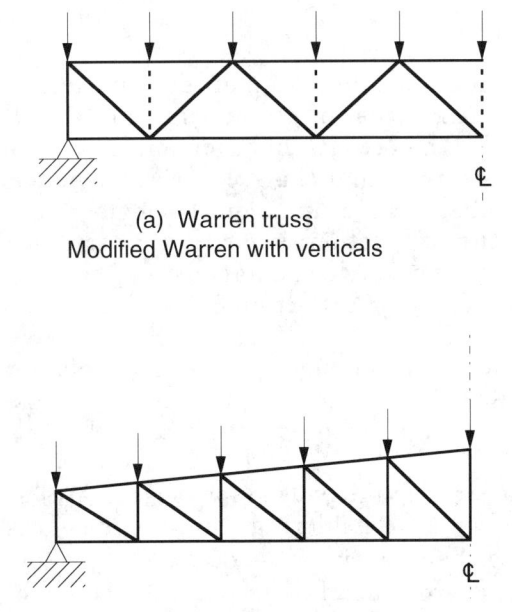

(a) Warren truss
Modified Warren with verticals

(c) Fink truss

(b) Pratt truss
Shown with a sloped roof,
but may have parallel chords

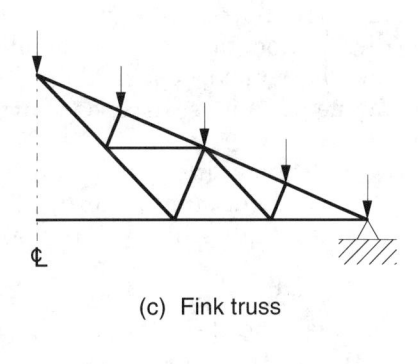

FIGURE 2.2
Common planar triangulated HSS trusses

Warren truss achieves a more "open" truss suitable for practical placement of mechanical, electrical, and other services.

Truss depth is determined in relation to the span, loads, maximum deflection, etc., with increased truss depth reducing the loads in the chords and increasing the lengths of the web members. The ideal span to depth ratio is usually found to be between 10 and 15 (CIDECT 1984).

2.2.2 Truss Analysis

Elastic analysis of HSS trusses is frequently performed by assuming that all members are pin connected. Nodal eccentricities between the centrelines of intersecting members at panel points should preferably be kept within the limits shown in Fig. 2.1.

These eccentricities produce primary bending moments which, for a pinned joint analysis, need only be taken into account in *member* design when proportioning the *compression chord*, by treating it as a beam-column. This is done by distributing the panel point moment, (sum of the horizontal components of the web member forces multiplied by the nodal eccentricity), to the chord on the basis of relative chord stiffness either side of the connection (i.e., in proportion to the values of moment of inertia divided by chord length to the next panel point, on either side of the connection). The eccentricity moments can be ignored for the design of the tension chord and web members.

Eccentricity moments can be ignored for the design of the *connections* provided that the eccentricities are within the limits shown in Fig. 2.1. If these eccentricity limits are exceeded, the eccentricity moment may have a detrimental effect on connection strength and the eccentricity moment must be distributed among the members at a connection. If moments are distributed to the web members, the connection capacity for each web member must be checked for the interaction between axial load and bending moment. This is discussed further in Chapter 3, Section 3.2.4.

A rigid joint frame analysis is not recommended for most planar, triangulated, single-chord, directly-welded trusses, as the axial force distribution will still be similar to that for a pin-jointed analysis.

Computer plane frame programs are regularly used for truss analyses. In this case the truss can be modelled by considering continuous chords with web members pin connected to them at distances of $+e$ or $-e$ from them (e being the distance from the chord centreline to the intersection of the web member centrelines). The links to the pins are treated as being extremely stiff as indicated in Fig. 2.3. The advantage of this model is that a sensible distribution of bending moments is automatically generated throughout the truss, for cases in which bending moments need to be taken into account in the design of the chords.

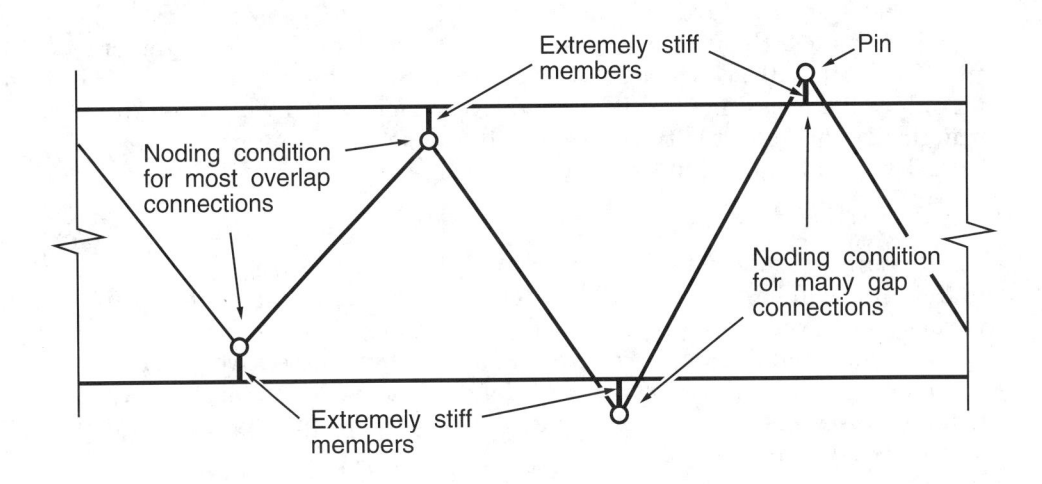

FIGURE 2.3

Plane frame connection modelling assumptions
to obtain realistic forces for member design

Transverse loads applied to either chord away from the panel points produce primary moments which must always be taken into account when designing the chords, and also the connections when the terms f(n) or f(n') apply (see Chapter 3, Tables 3.1, 3.2 and 3.3).

Secondary moments (resulting from end fixity of the web members to a flexible chord wall) can generally be ignored for *both* members and connections, provided that there is deformation and rotation capacity adequate to redistribute stresses after some local yielding at the connections. This is the case when the prescribed geometric limits of validity for design formulae given in the next chapter are followed.

Secondary moments due to connection flexibility can also only be neglected providing the ratio of member length (between nodal intersection points) to member depth (in the plane of the truss) is at least six (i.e., $L/h_i \geq 6$ for web members and for chord members). This limit was chosen by the International Institute of Welding Subcommission XV-E (Doc. XV-E-93-199) after Eurocode 3 decided to introduce a limit in Annex K (dealing with HSS connections) to prevent the application of this waiver to very stocky trusses. This relatively low limit of six was justified on the basis of extensive experimental evidence behind the IIW design recommendations, and was subsequently adopted by EC3. For the rare case where $L/h_i < 6$, it is recommended that the secondary moments be accomodated by using plastic design (Class 1) sections and designing the welds to develop the full capacity of the connected web member wall.

Welds in particular need to have potential for adequate stress redistribution without premature failure. To this end, Eurocode 3 (European Committee for Standardization 1992) requires that fillet welds around web members have a throat thickness at least 1.10 times the web member wall thickness for 350 MPa material. Canadian Standard CAN/CSA-S16.1-94 introduced expressions for fillet weld resistances (CSA 1994) that are now practically consistent with this generally conservative approach, if the full capacity of the connected web member wall is to be developed. (See Section 8.2.4.2 in this Design Guide for details.)

The aim of this rule is to avoid any need to confirm the strength of individual weld joints where weld effectiveness may change due to stiffness variations. If welds are proportioned on the basis of particular web member loads, the designer must recognize that less than the entire length of the weld may be effective, and the model for the weld resistance must be justified in terms of strength and deformation capacity. Further guidance on this is given in Chapter 8.

Table 2.1 summarizes when moments need to be considered for designing an HSS truss.

Plastic design could be used to proportion the chords of a truss by considering them as continuous beams with pin supports from the web members. In such a design the plastically designed members must be Class 1 sections and the welds must be sized to develop the capacity of the connected web members.

DESIGN of:	TYPE AND ORIGIN OF MOMENT		
	PRIMARY	PRIMARY	SECONDARY
	Nodal Eccentricity	Transverse Member Loading	Secondary Effects such as Local Deformations
Compression chord	Yes	Yes	No, provided $L/h_0 \geq 6$
Other members	No	Yes	No, provided $L/h_i \geq 6$
Connections	No, provided eccentricity limits are not exceeded	Yes, influences $f(n)$ and $f(n')$	No, provided parametric limits of validity are met and $L/h_i \geq 6$

TABLE 2.1

Moment considerations for design of HSS trusses

2.3 Effective Lengths for Compression Members

To determine the effective length KL for a compression member in a truss, the effective length factor K can always be conservatively taken as 1.0. However, considerable end restraint is generally present for compression members in an HSS truss, and it has been shown that K is generally appreciably less than 1.0 (Mouty 1981).

This restraint offered by members framing into a connection could disappear, or be greatly reduced, if all members were designed optimally for minimum mass, thereby achieving ultimate capacity simultaneously under static loading. This could be critical in a statically determinate structure such as a triangulated truss (Galambos 1988). In practice, design for optimal or minimum mass will rarely coincide with minimum cost; the web members are usually standardized to a few selected dimensions (perhaps even two), to minimize the number of section sizes for the truss.

CIDECT has sponsored and co-ordinated extensive research work to specifically address the determination of effective lengths in HSS trusses. Results are in reports from CIDECT Programs 3E-3G, Monograph No. 4, and a recent CIDECT Design Guide (CIDECT 1980; Mouty 1981; Rondal *et al.* 1992). A re-evaluation of all test results has been undertaken to produce recommendations for Eurocode 3 (Sedlacek *et al.* 1989; Rondal 1988, 1989), which has a member safety calibration level comparable to the Canadian steel structures standard (CAN/CSA-S16.1-94). This has resulted in the following effective length recommendations, which are implemented in the design examples later in this book (Chapter 13).

2.3.1 Simplified Rules

For HSS chord members:

In the plane of the truss,

$KL = 0.9\,L$ [2.1]

where L is the distance between chord panel points.

In the plane perpendicular to the truss,

$KL = 0.9\,L$ [2.2]

where L is the distance between points of lateral support for the chord.

For HSS web members:

In either plane,

$KL = 0.75\,L$ [2.3]

where L is the panel point to panel point length of the member.

These values of K are only valid for HSS members which are connected around the full perimeter of the member, without cropping or flattening of the members. The chord members should also be parallel, or approximately parallel. In addition, the smaller dimension of the web member must be at least one quarter of the chord width, and the wall thickness of the web member t_1 must not exceed that of the chord t_0. Compliance with the connection design requirements of Chapter 3 will likely place even more restrictive control on the member dimensions.

For compresssion web members with 100% overlap end connections, it is recommended that a value of KL equal to L still be used for such web members, in the absence of experimental evidence (Rondal et al. 1992).

2.3.2 Empirical Method for HSS Web Members

For trusses with identical width top and bottom chords, compression web member width to chord width ratio less than 0.6, and with a (Rondal et al. 1992):

circular web member welded to circular chords
$$K = 2.2\,(d_1^2/Ld_0)^{0.25} \quad \text{but} \leq 0.75 \qquad [2.4]$$

circular web member welded to rectangular chords
$$K = 2.35\,(d_1^2/Lb_0)^{0.25} \quad \text{but} \leq 0.75 \qquad [2.5]$$

rectangular web member welded to rectangular chords
$$K = 2.3\,(b_1^2/Lb_0)^{0.25} \quad \text{but} \leq 0.75 \qquad [2.6]$$

where L is again the panel point to panel point length of the web member and "rectangular" includes "square".

CIDECT Monograph No. 4 (Mouty 1981) presented a method for determining a web member effective length in trusses which had different width or section shape members for the top and bottom chords. This has not been addressed in the latest provisions for Eurocode 3 (Sedlacek et al. 1989), so it is recommended that the effective length factor K be calculated for the connection condition at each end of the web member, and the higher value be used. Another conservative rider from CIDECT Monograph No. 4, which impacts upon [2.5] and [2.6] above, should also be added to the above recommendations:

For rectangular chord members, b_0 is replaced by h_0 when $h_0 < b_0$.

For rectangular web members, b_1 is replaced by h_1 when $h_1 > b_1$.

2.3.3 Long Laterally Unsupported Compression Chords

Long, laterally unsupported compression chords can exist in pedestrian bridges, and in roof trusses subjected to large wind uplift (the bottom chord), as illustrated in Fig. 2.4. A pedestrian pony truss bridge is shown in Fig. 2.5.

The effective length of such laterally unsupported truss chords can be considerably less than the unsupported length. For example, the actual effective length of a bottom chord, loaded in compression by uplift (Fig. 2.4(b)), depends on the loading in the chord, the stiffness of the web members, the torsional rigidity of the tension chord, the purlin to truss

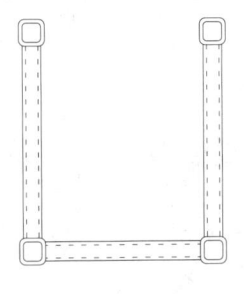

(a) Pony truss
Cross-section

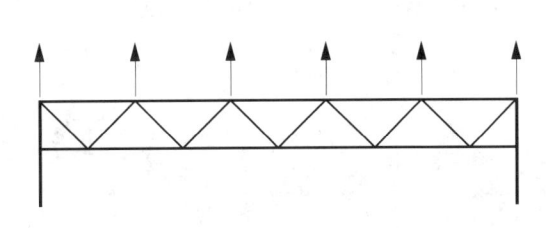

(b) Roof truss
Wind uplift at roof purlin locations

FIGURE 2.4
Cases of long, laterally unsupported truss compression chords

FIGURE 2.5
Pony truss for pedestrians, a frequent use for HSS
(Niagara, Ontario)

connections and the bending stiffness of the purlins. The web members act as local elastic supports at each panel point. When the stiffness of these elastic supports is known, the effective length of the compression chord can be calculated. Although other methods are in use (Herth 1994), a detailed method for effective length factor calculation has been given by CIDECT (Mouty 1981), and a summary of this procedure for a pedestrian bridge follows (see Fig. 2.6).

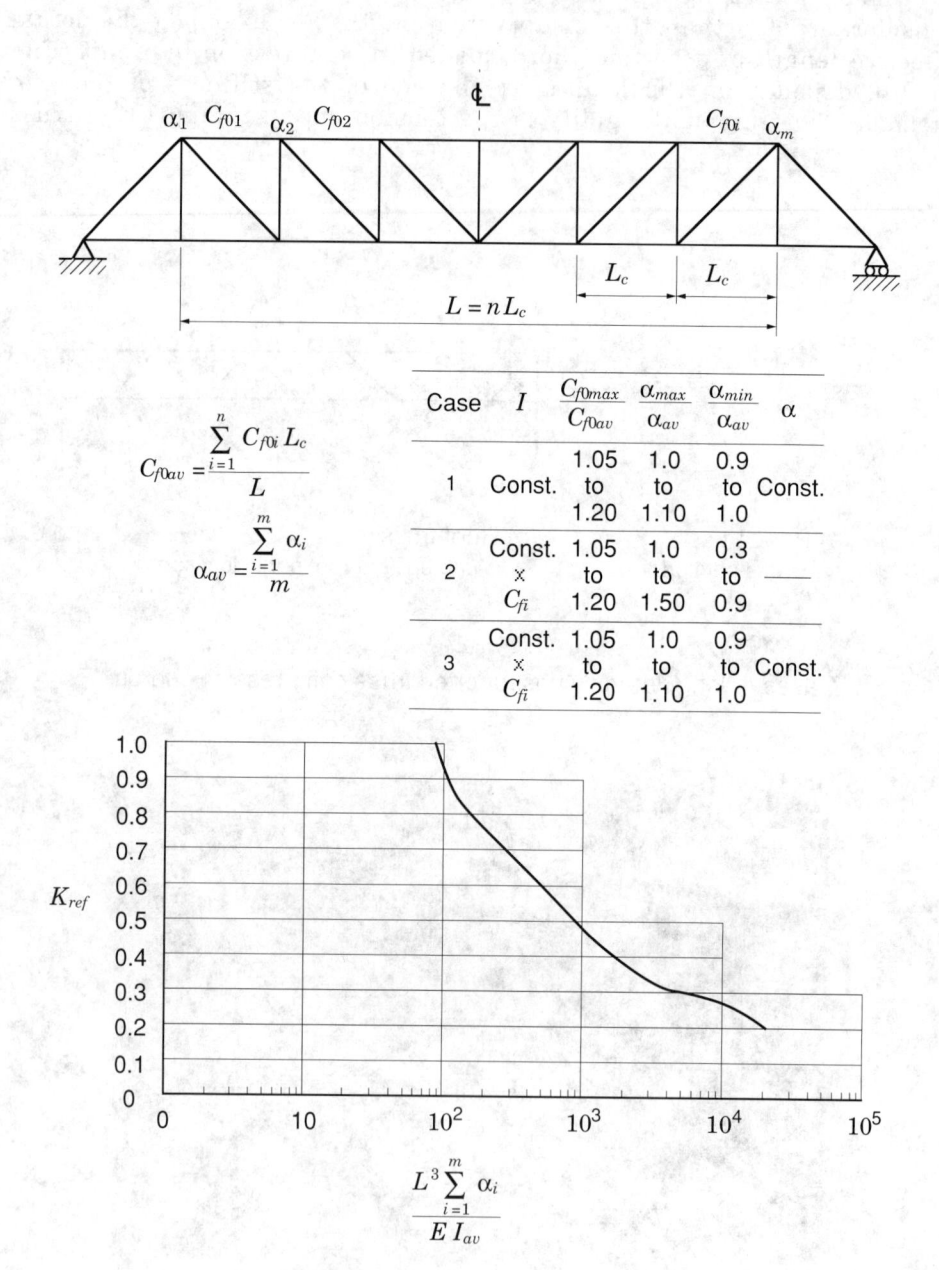

$$C_{f0av} = \frac{\sum_{i=1}^{n} C_{f0i} L_c}{L}$$

$$\alpha_{av} = \frac{\sum_{i=1}^{m} \alpha_i}{m}$$

Case	I	$\frac{C_{f0max}}{C_{f0av}}$	$\frac{\alpha_{max}}{\alpha_{av}}$	$\frac{\alpha_{min}}{\alpha_{av}}$	α
1	Const.	1.05 to 1.20	1.0 to 1.10	0.9 to 1.0	Const.
2	Const. × C_{fi}	1.05 to 1.20	1.0 to 1.50	0.3 to 0.9	—
3	Const. × C_{fi}	1.05 to 1.20	1.0 to 1.10	0.9 to 1.0	Const.

K_{ref} vs $\dfrac{L^3 \sum_{i=1}^{m} \alpha_i}{E\, I_{av}}$

FIGURE 2.6
Pedestrian bridge (pony truss) example for chord effective length calculation

The formulae used are based on simplifying assumptions:

Stiffnesses of the elastic supports are independent of each other.

Torsional rigidity of the web members and the influence of axial forces on their bending stiffness may be neglected.

Web members are assumed to be pin connected to the compression chord, and completely fixed to the tension chord.

The stiffness of each individual elastic support is determined by using the following elements:

bending stiffness of the floor beam, α_b

torsional stiffness of the tension chord, α_{tor}

stiffness of the web members at the connection, α_w.

At the tension chord, the first two are added to give the stiffness of the tension chord, α_t.

That is, $\quad \alpha_t = \alpha_b + \alpha_{tor}$ [2.7]

At the elastic support, the flexibility of the tension chord $1/\alpha_t$ and the flexibility of the web members at the connection $1/\alpha_w$ are added.

That is, $\quad \dfrac{1}{\alpha} = \dfrac{1}{\alpha_t} + \dfrac{1}{\alpha_w}$

or $\quad \dfrac{1}{\alpha} = \dfrac{1}{\alpha_b + \alpha_{tor}} + \dfrac{1}{\alpha_w}$

Hence, the general formula of the lateral support stiffness α is

$$\alpha = \dfrac{1}{\dfrac{1}{\alpha_b + \alpha_{tor}} + \dfrac{1}{\alpha_w}}$$ [2.8]

The value of each element of the general formula for α can be calculated by using [2.9], [2.10] and [2.11].

$$\alpha_b = \dfrac{2 E I_b}{u h_t^2}$$ [2.9]

provided that floor beams are located only at each panel point between the two trusses

57

where

u = distance between trusses
h_t = height of the truss at the connection considered
I_b = moment of inertia of the floor beam

$$\alpha_{tor} = \frac{2\,G\,J_t}{L_t\,h_t^2}\left(1 - \cos\frac{\pi L_t}{L}\right) \qquad [2.10]$$

where

G = shear modulus of steel (77 000 MPa)
J_t = torsional moment of inertia of the tension chord
L_t = distance between connections on the tension chord
L = length of compression chord.

If the chords are not parallel, α_{tor} is taken to be = 0.

$$\alpha_w = \sum_{i=1}^{j} \frac{3\,E\,I_{wi}}{L_{wi}^3} \qquad [2.11]$$

where

j = number of web members ending at the panel point being considered
I_{wi} = moment of inertia of the ith web member about the weak axis of the truss
L_{wi} = length of the ith web member.

Equation 2.8 gives the total stiffness at each panel point. These stiffnesses (α_i) are then added in the following manner to find the dimensionless parameter:

$$\frac{L^3 \sum_{i=1}^{m} \alpha_i}{E\,I_{av}} \qquad [2.12]$$

where

m = number of elastic supports (panel points on the compression chord)

$$I_{av} = \frac{\sum_{i=1}^{n} I_i\,L_{ci}}{L}$$

where

I_{av} = average moment of inertia of the compression chord about the weak axis of the truss

n = number of panel lengths of compression chord

I_i = moment of enertia of the ith length of the chord

L_{ci} = length, between panel points, of the ith length of chord.

The dimensionless parameter from [2.12] is used for the graph in Fig. 2.5 to determine K_{ref}, a reference effective length factor.

Once K_{ref} is known, the compression chord can be checked. Since the force in the chord varies, and the chord section itself may vary, it is necessary to consider the condition which exists in each individual length of chord L_c between panel points. Effective lengths $(KL)_i$ for the compression chord are calculated with the following equation by using the force and the moment of inertia of each length of chord in turn.

$$(KL)_i = K_{ref} L \sqrt{\frac{C_{f0av} I_i}{C_{f0i} I_{av}}} \qquad [2.13]$$

where

C_{f0av} = weighted average compressive axial force (factored) in the chord

$$= \frac{\sum_{i=1}^{n} C_{f0i} L_{ci}}{L}$$

C_{f0i} = compressive axial force (factored) in the individual chord length being considered.

The KL thus produced for each length of chord is used for the design of that length of the compression chord. Effective lengths of 0.3 times the total chord length are quite feasible.

In the case of a KT connection, the vertical member should be ignored.

The graph curve in Fig. 2.5 gives conservative solutions for the three listed cases which were developed for the web member layout illustrated. Solutions will be precise only if the truss to be designed corresponds to the model in all respects; however, an adequate approximation will be obtained for any layout of web members for which

$$I, \frac{C_{f0max}}{C_{f0av}}, \frac{\alpha_{max}}{\alpha_{av}}, \frac{\alpha_{min}}{\alpha_{av}} \text{ and } \alpha$$

fall within the limits of one of the cases shown in the figure.

2.4 Truss Deflections

For the purpose of checking the serviceability condition of overall truss deflection under specified (unfactored) loads, an analysis with all members being pin-jointed will provide a conservative (over) estimate of truss deflections when all the connections are overlapped (Coutie *et al.* 1987; Philiastides 1988). A better assumption for *overlap* conditions is to assume continuous chord members and pin-jointed web members.

However, for *gap-connected* trusses, a pin-jointed analysis still generally *under* estimates overall truss deflections, because of the flexibility of the connections (Czechowski *et al.* 1984; Coutie *et al.* 1987; Philiastides 1988; Frater and Packer 1992). At the service load level, gap-connected rectangular HSS truss deflections have been under estimated by around 12 to 15% (Philiastides 1988; Frater and Packer 1992). Thus, a conservative approach for gap-connected HSS trusses is to estimate the maximum truss deflection as 1.15 times that calculated from a pin-jointed analysis.

2.5 Critical Considerations for Static Strength

As discussed at the beginning of this chapter, it is essential that the designer have an appreciation of factors which make it possible for HSS members to be connected together at truss panel points without extensive (and expensive) reinforcement. Apparent economies from minimum-mass member selection will quickly vanish at the connections if a designer does not have a knowledge of the critical considerations which influence connection efficiency.

2.5.1 General Considerations

1. Chords should generally have thick walls rather than thin walls. The stiffer walls resist loads from the web members more effectively, and the connection resistance thereby increases as width (or diameter) to thickness ratio of the chord decreases. For the *compression* chord, however, a large thin section is more efficient in providing buckling resistance, so for this member the final HSS wall slenderness will be a compromise between connection strength and buckling strength, and relatively stocky sections will usually be chosen.

2. Web members should have thin walls rather than thick walls, as connection resistance increases as the ratio of chord wall thickness to web wall thickness increases. In addition, thin web walls will require smaller fillet welds for a full strength joint.

3. Web compression members should have diameter to thickness ratios, or "width of flat" to thickness ratios for rectangular HSS, which satisfy CAN/CSA-S16.1-94 requirements for Class 1 or Class 2 members for gap connections, but Class 1 requirements for overlap connections.

4. Ideally, rectangular (or square) or circular HSS web members should not be the same width as rectangular HSS chord members, as this presents an awkward flare bevel weld situation (possibly with backing bars) for the joint at the corner of the chord section. A preferred arrangement is web members just sufficiently narrower than the chord to permit the web member and some of the fillet weld to sit on the "flat" of the rectangular HSS chord member. The outside corner radius of a North American cold-formed rectangular HSS member is generally taken as $2t$, although Standard CAN/CSA-G40.20-M92 (Table 34) allows $3t$ for some of the HSS sizes listed in the tables in Section 1.7 of this Design Guide. See Table 1.2 for details.

5. Gap connections (for K and N situations) are preferred to overlap connections because the members are easier to prepare, fit and weld.

6. When overlap connections are used, at least a quarter of the height (dimension h_i in the plane of the truss) of the overlapping member needs to be engaged in the overlap.

7. An angle of less than 30° between a web member and a chord creates significant welding difficulties, and is not covered by the scope of these recommendations.

2.5.2 Specific Considerations

The next chapter outlines the behaviour of standard HSS truss connections. Internationally accepted design formulae are presented for connection factored resistances along with value ranges of dimensional parameters for which the formulae are valid.

Simplified graphical connection design charts and connection resistance tables are also given for easy sizing of truss members for economical connections.

2.6 Truss Design Procedure

In summary, the design of an HSS truss should be approached in the following way to obtain an efficient and economical structure:

1. Determine the truss layout, span, depth, panel lengths, truss spacing and bracing by the usual methods, but keep the number of connections to a minimum.

2. Determine loads at connections and on members; simplify these to equivalent loads at the panel points.

3. Determine axial forces in all members by assuming that joints are pinned and that all member centrelines are concentric.

4. Determine chord member sizes by considering axial loading, corrosion protection and tube wall slenderness. (Usual diameter to thickness ratios are 15 to 30, and width to thickness ratios are 15 to 25.) An effective length factor of $K = 0.9$ can be used for the design of the compression chord. (See Section 2.3.1.)

5. Determine web member sizes based on axial loading, preferably with thicknesses smaller than the chord thickness. The effective length factor for the web members can initially be assumed to be 0.75. (See Section 2.3.1.)

6. Standardize the web members to a few selected dimensions (perhaps even two) to minimize the number of section sizes for the structure. Consider availability of all sections when making member selections. For aesthetic reasons, a constant outside member width (or diameter) may be preferred for all web members, with wall thicknesses varying; but this may require special quality control procedures in the fabrication shop to ensure correct sizes at intended locations.

7. Layout the connections, trying gap connections first. Check that the connection geometry and member dimensions satisfy the validity ranges for the dimensional parameters given in Chapter 3, with particular attention to the eccentricity limits. Consider the fabrication procedure when deciding on a connection layout.

8. Check the connection efficiencies with the charts given in Chapter 3. In some instances (e.g., when using rectangular rather than square HSS members, or when greater precision is needed), direct use of the connection factored resistance equations given in Chapter 3 may be required.

9. If the connection resistances (efficiencies) are not adequate, modify the connection layout (for example, overlap rather than gap), or modify the web or chord members as appropriate, and recheck the

connection capacities. Generally, only a few connections will need checking.

10. Check the effect of primary moments on the design of the chords. For example, use the proper load positions rather than equivalent panel point loading. Determine the bending moments in the chords by assuming either: *a)* pinned joints everywhere or *b)* continuous chords with pin-ended web members. For the compression chord, also determine the bending moments produced by any noding eccentricities, by using either of the above analysis assumptions. Then check that the factored resistance of all chord members is still adequate, under the influence of both axial loads and primary bending moments.

11. Check truss deflections (see Section 2.4) at the specified (unfactored) load level, using the proper load positions.

12. Design welded connections (see Chapter 8).

REFERENCES

CIDECT. 1980. Buckling lengths of HSS web members welded to HSS chords. CIDECT Programs 3E-3G, Supplementary Report—Revised Version, CIDECT Doc. 80/3-E.

CIDECT. 1984. Construction with hollow steel sections. British Steel plc., Corby, Northants., England.

COUTIE, M.G., DAVIES, G., PHILIASTIDES, A., and YEOMANS, N. 1987. Testing of full-scale lattice girders fabricated with RHS members. Proceedings, Conference on Structural Assessment Based on Full and Large-Scale Testing, Building Research Station, Watford, England.

CSA. 1994. Limit states design of steel structures, CAN/CSA-S16.1-94. Canadian Standards Association, Rexdale, Ontario.

CSA. 1992. General requirements for rolled or welded structural quality steel, CAN/CSA-G40.20-M92. Canadian Standards Association, Rexdale, Ontario.

CZECHOWSKI, A., GASPARSKI, T., ZYCINSKI, J., and BRODKA, J. 1984. Investigation into the static behaviour and strength of lattice girders made of RHS. International Institute of Welding Doc. XV-562-84, Poland.

EUROPEAN COMMITTEE for STANDARDIZATION. 1992. Eurocode 3: design of steel structures. Part 1.1—general rules and rules for buildings. ENV 1993-1-1:1992E, British Standards Institution, London, England.

FIRKINS, A., and HEMPHILL, D. 1990. Fabrication cost of structural steelwork. Steel Construction, Australian Institute of Steel Construction, **24**(2): 2–14.

FRATER, G.S., and PACKER, J.A. 1992. Modelling of hollow structural section trusses. Canadian Journal of Civil Engineering, **19**(6): 947–959.

GALAMBOS, T.V. (ed.). 1988. Guide to stability design criteria for metal structures, 4th. edition. Structural Stability Research Council, John Wiley & Sons, New York, N.Y., U.S.A.

HERTH, S.J. 1994. Design of half-through or "pony" truss bridges using square or rectangular hollow structural sections. Proceedings, IASS–ASCE International Symposium on Spatial, Lattice and Tension Structures, Atlanta, Georgia, U.S.A., pp. 210–220.

JENSEN, C.B., and BUSCH, J.H. 1989. The economics of structural shop automation. Proceedings, Pacific Structural Steel Conference, Gold Coast, Australia.

MOUTY, J. (ed.). 1981. Effective lengths of lattice girder members. CIDECT Monograph No. 4, Boulogne, France.

PHILIASTIDES, A. 1988. Fully overlapped rolled hollow section welded joints in trusses. Ph. D. thesis, University of Nottingham, England.

RONDAL, J. 1988. Effective lengths of tubular lattice girder members—statistical tests. CIDECT Report 3K-88/9.

RONDAL, J. 1989. Addendum to Report 3K-88/9. CIDECT Report 3K-89/9.

RONDAL, J., WÜRKER, K.-G., DUTTA, D., WARDENIER, J., and YEOMANS, N. 1992. Structural stability of hollow sections. CIDECT (ed.) and Verlag TÜV Rheinland GmbH, Köln, Federal Republic of Germany.

SEDLACEK, G., WARDENIER, J., DUTTA, D., and GROTMANN, D. 1989. Evaluation of test results on hollow section lattice girder connections in order to obtain strength functions and suitable model factors. Background report to Eurocode 3: Common Unified Rules for Steel Structures. Document 5.07, Eurocode 3 Editorial Group.

WATSON, K.B., DALLAS, S. van der KREEK, N., and MAIN, T. 1996. Costing of steelwork from feasibility through to completion. Steel Construction, Australian Institute of Steel Construction, **30**(2): 2–47.

3

STANDARD TRUSS WELDED CONNECTIONS

3.1 Terminology and Eccentricity

This book uses terminology adopted by CIDECT (Comité International pour le Développement et l'Etude de la Construction Tubulaire) and IIW (International Institute of Welding) which has been widely accepted by the international community, slightly modified for conformance with traditional Canadian symbols. Fig. 3.1 illustrates its application to typical gap and overlap K connections in a Warren truss.

Eccentricity is positive when measured towards the outside of a chord, and negative towards the inside. The gap or overlap, g or q, may be represented by x in the following equations, which are useful for calculating eccentricities, gaps and overlaps.

$$x = \frac{e+D}{C} - (A+B) \qquad [3.1]$$

$$e = C(A+B+x) - D \qquad [3.2]$$

where

$x = g$ when there is a gap, and $x = -q$ when there is an overlap

$$A = \frac{h_1 \text{ or } d_1}{2\sin\theta_1} \qquad B = \frac{h_2 \text{ or } d_2}{2\sin\theta_2}$$

$$C = \frac{\sin\theta_1 \sin\theta_2}{\sin(\theta_1+\theta_2)} \qquad D = \frac{h_0 \text{ or } d_0}{2}$$

These equations also apply for panel points which have a stiffening plate on the surface of the chord. Then,

$$D = \frac{h_0 \text{ or } d_0}{2} + t_p \qquad \text{where } t_p \text{ is the stiffening plate thickness.}$$

3.2 Trusses with HSS Chords and HSS Web Members

Three classifications of HSS connections are discussed in the following sections, namely: K and N, T, Y and X, and TK.

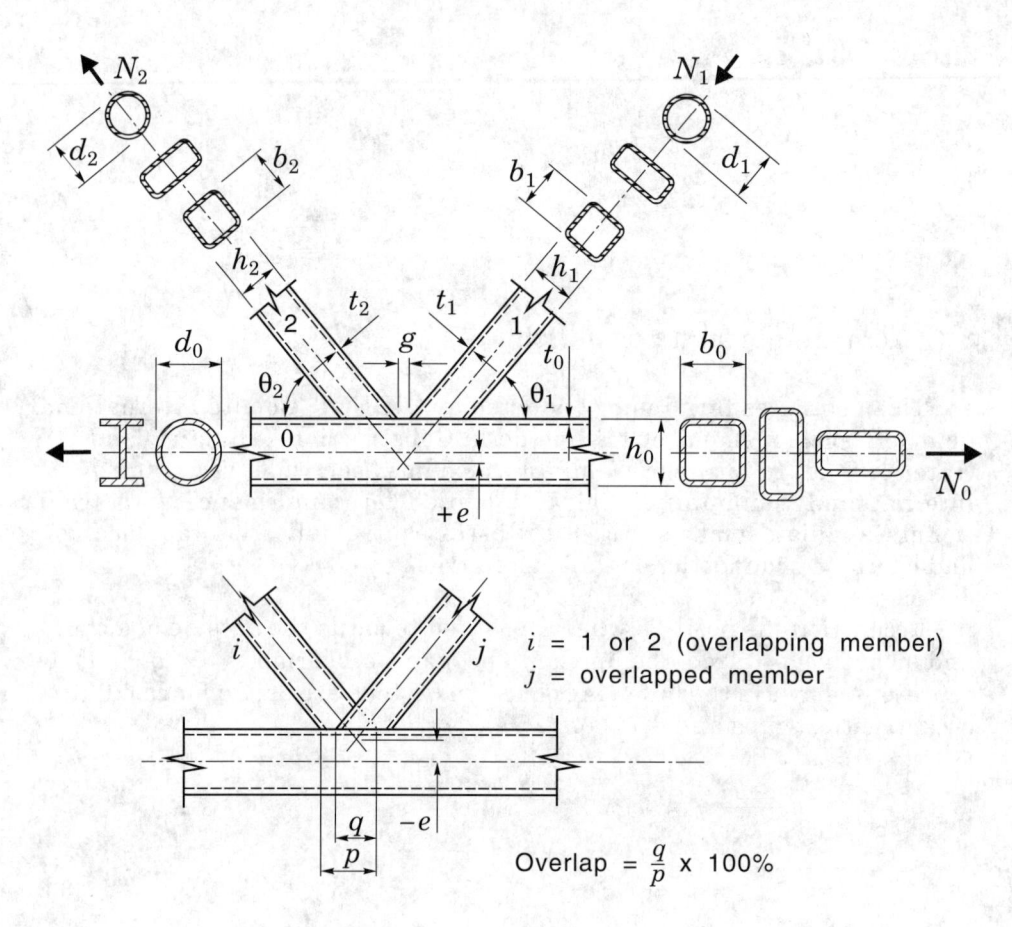

Limits of eccentricity for design formulae are $-0.55 \leq \dfrac{e}{h_0}$ and $\dfrac{e}{d_0} \leq 0.25$

FIGURE 3.1
Standard terminology for gap and overlap HSS K connections

3.2.1 K and N Connections

The majority of HSS truss connections have one compression web member and one tension web member welded to the chord as shown in Fig. 3.1. The Warren arrangement is commonly referred to as a K connection and the Pratt as an N connection. The latter is basically a particular case of the former; both can be either gap type or overlap type connections as can be seen in Fig. 3.2.

Classification as a K (or N) connection is contingent upon the normal component from one web member force being primarily balanced by the similar component from the other web member force. Thus for example, if a connection had a K configuration (with two web members on one side of a chord coming to the same node), but with the forces acting in the same sense balanced by one or more members on the other side of the chord, it would be classified as an X connection because the mechanism of force transfer would be <u>through</u> the chord member.

FIGURE 3.2
Typical HSS K (gap) and N (overlap) connections

3.2.1.1 Failure Modes for K and N Connections

Experimental research (for example by Wardenier and Stark 1978, Kurobane *et al.* 1980, and Kurobane 1981) on HSS welded truss connections has shown that different failure modes can exist depending on the type of joint, loading conditions, and various geometric parameters. Failure modes

Section view

Mode A: Plastic failure of the chord face

Mode C: Tension failure of the web member

Mode E: Overall shear failure of the chord

Section view

Mode B: Punching shear failure of the chord face

Mode D: Local buckling of the web member

Mode F: Local buckling of the chord walls

Mode G: Local buckling of the chord face

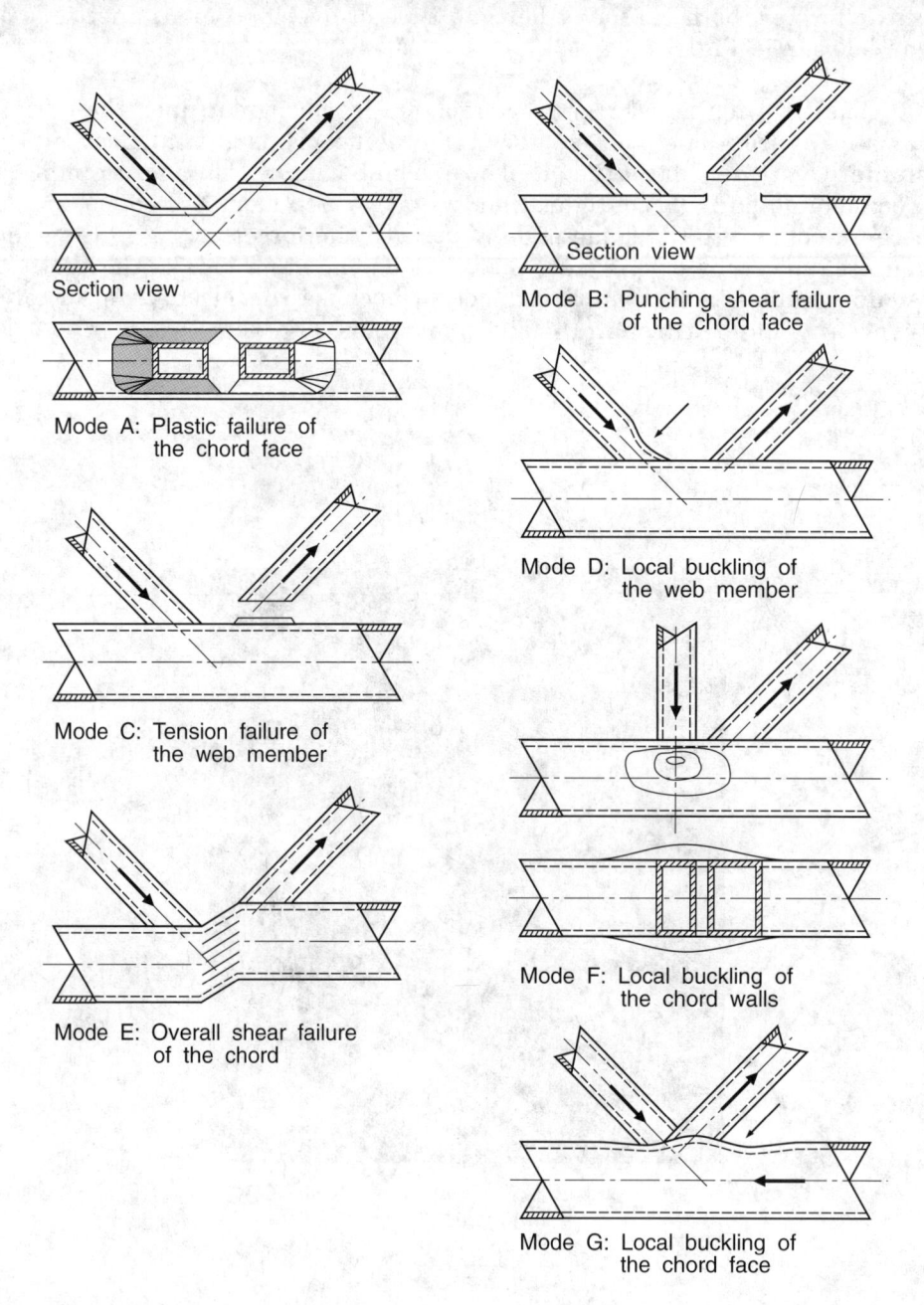

FIGURE 3.3
Failure modes for K and N rectangular HSS truss connections

FIGURE 3.4
Section view of a plastic failure of the chord face in a gap connection (Mode A)

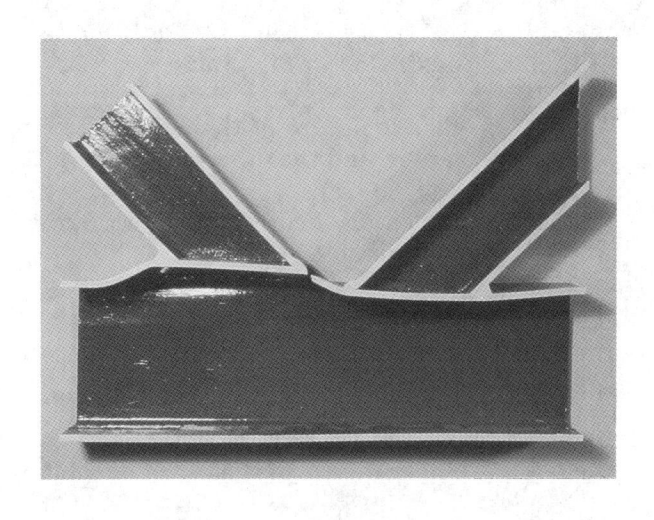

FIGURE 3.5
Local buckling of compression web member in an overlap connection (Mode D)

FIGURE 3.6
Local buckling of the chord face in an overlap connection (Mode G)

have been described (Wardenier 1982) for square and rectangular hollow sections as illustrated in Fig. 3.3:

Mode A   Plastic failure of the chord face (one web member pushing the face in and the other pulling it out)

Mode B   Punching shear of the chord face around a web member (either compression or tension)

Mode C   Rupture of the tension member or its weld

Mode D   Local buckling of the compression web member

Mode E   Shear failure of the chord member in the gap

Mode F   Chord wall bearing or local buckling under the compression web member

Mode G   Local buckling of the chord face behind the heel of the tension web member.

Failure in test specimens has also been observed to be a combination of more than one failure mode.

It should be noted here that Modes C and D are generally lumped together under the term "effective width failures" (or "uneven load distributions") and are treated identically, since the connection resistance in both cases is determined by the effective cross section of the critical web member, with some web member walls possibly being only partially effective.

Plastic failure of the chord face (Mode A) is the most common failure mode for gap connections with small to medium ratios of web member widths to chord width. This is illustrated in Fig. 3.4 which shows a longitudinal cross section through a gap connection. For medium width ratios ($\beta \approx 0.6$ to 0.8), this mode generally occurs together with tearing in the chord (Mode B) or the tension web member (Mode C), although the latter only occurs in connections with relatively thin walled web members. Mode D, involving local buckling of the compression web member, is the most common failure mode for overlap connections. This is shown in Fig. 3.5 for a Pratt truss overlap connection. Shear failure of the entire chord section (Mode E) is observed in gap connections where the width (or diameter) of the web members is close to that of the chord ($\beta \approx 1.0$). Local buckling failure (Modes F and G) occurs occasionally in square and rectangular HSS connections with high chord width (or depth) to thickness ratios (b_0/t_0 or h_0/t_0).

It has been found that in some cases one or two governing modes can be used to predict connection resistance (Wardenier 1982). For example, Mode G failure, shown in Fig. 3.6, is excluded from the design expressions in Section 3.2.1.2 by restrictions on the range permitted for geometric parameters.

For circular HSS K and N connections, the predominant failure mode is chord plastification (Kurobane *et al.* 1980, Kurobane 1981), similar to Fig. 3.3 Mode A, although premature local buckling of the compression web member is also a possibility (see Fig. 3.7).

3.2.1.2 Design Formulae for K and N Connections

Various formulae exist for the connection failure modes described in Section 3.2.1.1. Some have been derived theoretically, while others are primarily empirical. The general criterion for design is ultimate resistance, but the recommendations presented herein (and their limits of validity) have been set so that a limit state for deformation is not exceeded at specified (service) loads.

In accordance with IIW (1989), Wardenier *et al.* (1991) and Packer *et al.* (1992), limit states design recommendations are summarized in Tables 3.1 and 3.1(a) (for circular chords), Tables 3.2 and 3.2(a) (for square chords), and Tables 3.3 and 3.3(a) (for rectangular chords). A number of observations for K and N connections can be made from an examination of these tables:

FIGURE 3.7
Local buckling of compression web member in a circular HSS connection
(Photograph courtesy of Professor Y. Kurobane, Kumamoto Institute of Technology, Japan)

A common design criterion for all K and N gap connections is Mode A, plastic failure of the chord face. The constants in the resistance equations are derived from extensive experimental data. Other terms reflect ultimate strength parameters such as plastic moment capacity of the chord face per unit length ($F_{y0} t_0^2/4$), web to chord width ratio β, chord wall slenderness γ, (which considerably affects the amount of post-yield membrane action that can be generated in the chord), and the term $f(n')$ or $f(n)$, which accounts for the influence of compression chord longitudinal stresses.

The terms $f(n')$ for circular HSS, and $f(n)$ for square and rectangular HSS, are similar, except that $f(n')$ does not include the longitudinal forces introduced into the chord by the web members at the connection under consideration. In other words, only the "preload" in the chord is used with $f(n')$ for circular HSS.

Tables 3.2 and 3.3 show that the resistance of a gap K or N connection with a square or rectangular HSS chord is largely independent of the gap size (no gap size parameter) whereas Table 3.1 shows that a continuous function covers the range of both gap and overlap circular HSS K and N connections.

Table 3.2 which is restricted to square HSS chords was derived from the more general Table 3.3, and uses more confined geometric parameters. The result is that gap K and N connections with square HSS need only be examined for failure Mode A, whereas those with rectangular HSS must be considered for failure Modes B, C or D, and E as well. This approach has allowed the creation of helpful graphical design charts which are presented later for connections between square HSS.

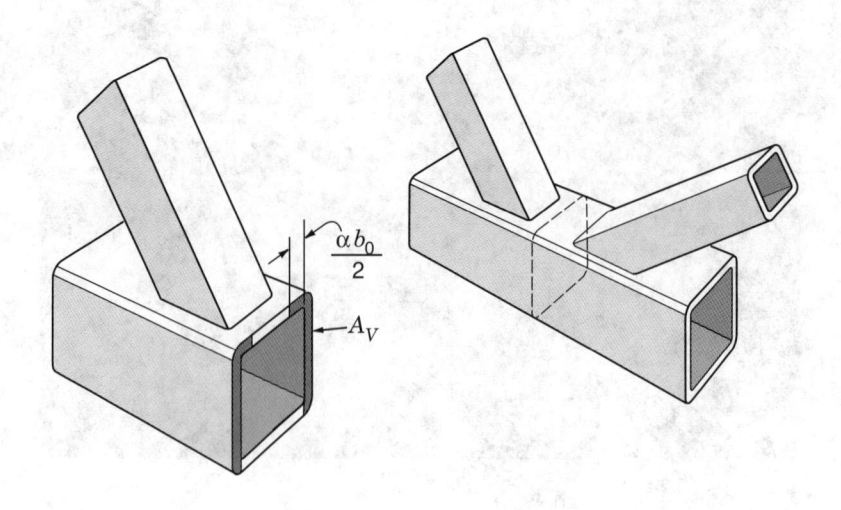

FIGURE 3.8

Shear area (A_V) of the chord in the gap region of a rectangular HSS K or N connection

In Table 3.3, the Mode E check for chord shear in the gap of K and N connections involves dividing the chord cross section into two portions. The first is a shear area A_V comprising the side walls plus part of the top flange, shown in Fig. 3.8, which can carry both shear and axial loads interactively. The second is the remaining area, $A_0 - A_V$, which is effective in carrying axial forces but not shear.

Overlap K and N connections are presented differently in Table 3.1 for circular HSS than in Tables 3.2 and 3.3 for square and rectangular HSS. The work of Kurobane and associates in Japan forms the basis of Table 3.1 where a continuous expression gives increasing resistance for the full range of gap to overlap configurations. Tables 3.2 and 3.3 present a range of resistances based on the concept of effective width for square and rectangular HSS overlap connections, starting with 25% overlap, which is the minimum to ensure overlap behaviour. The resistance increases linearly with overlap from 25% to 50%, is constant from 50% to 80%, then is constant above 80% at a higher level. Fig. 3.9 illustrates the physical interpretation of the expressions for effective width given in the tables.

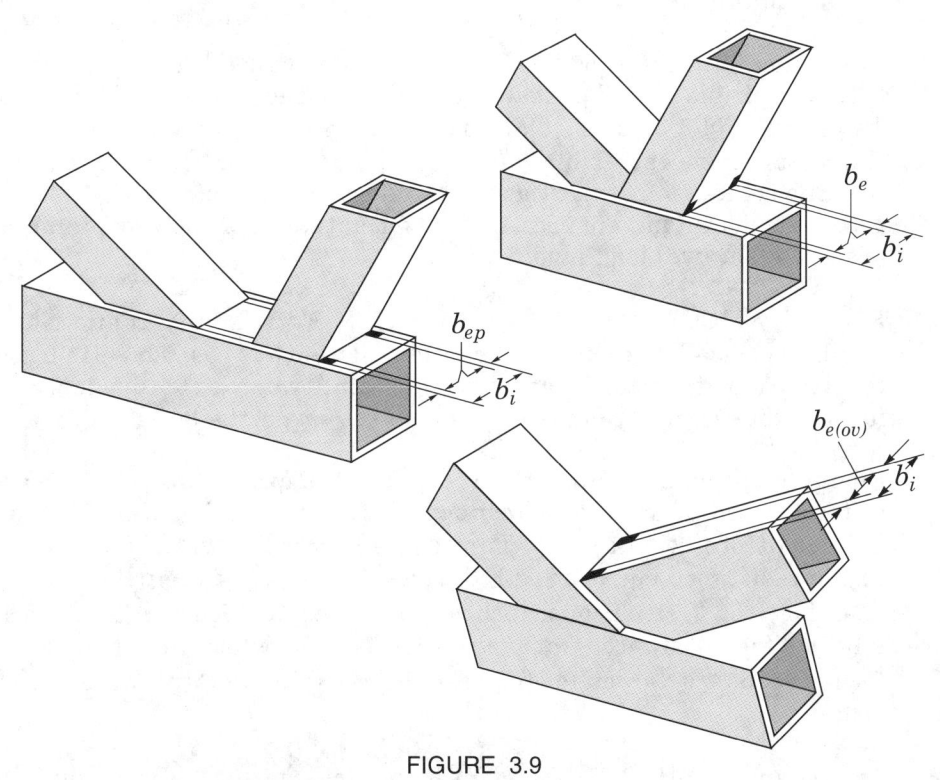

FIGURE 3.9
Physical interpretation of effective width terms
(see Table 3.3 for definitions of b_e, $b_{e(ov)}$ and b_{ep})

3.2.2 T, Y and X Connections

In the same way that an N connection can be considered a particular case of the general K connection, the T connection is a particular case of the Y connection. The basic difference between the two types is that the component of load normal to the chord in T and Y connections is resisted by shear and bending in the chord, whereas with K or N connections the normal component from one web member force is balanced primarily by the similar component in the other.

For X connections, the web member force normal component(s) on one side of the chord are primarily balanced by the web member force component(s) normal to the chord on the <u>other side</u> of the chord member.

3.2.2.1 Design Formulae for T, Y and X Connections

The IIW (1989) limit states design recommendations for HSS T, Y and X connections are summarized in Tables 3.1 and 3.1(a) (for circular chords), Tables 3.2 and 3.2(a) (for square chords), and Tables 3.3 and 3.3(a) (for rectangular chords). As with K and N connections, various observations can be made from the Tables:

Resistance equations in Table 3.1 for circular HSS T, Y and X connections are primarily empirically derived. On the other hand, those in Tables 3.2 and 3.3, for $\beta \leq 0.85$, are based on a yield line mechanism in the relatively more flexible square and rectangular HSS chord face. By limiting connection design capacity under factored loads to the connection yield load, one ensures that deformations will be acceptable at specified (service) load levels.

For full width ($\beta = 1.0$) rectangular HSS T, Y and X connections, flexibility is no longer a problem, and resistance is based on either the tension capacity or the compression instability of the chord side walls, for web tension and compression members respectively.

Compression loaded, full-width X connections for rectangular HSS are differentiated from T or Y connections as their side walls exhibit greater deformation than T connections. Accordingly, the value of F_k in the resistance equation is reduced to $0.8 \sin\theta_1$ of the value that is used for T or Y situations. In both instances, a linear progression is followed for resistances from values for $\beta = 0.85$ (where flexure of chord face governs) to values for $\beta = 1.0$ (where chord side wall failure governs).

All rectangular HSS T, Y and X connections with high web width to chord width ratios ($\beta \geq 0.85$) are also checked for the "effective width" failure modes, and for punching shear of the chord face. For this range of connection width ratios, much of the web member load is carried by the web member side walls parallel to the chord, while the web member walls transverse to the chord have a reduced effectiveness. The expressions for the effective width terms b_e and b_{ep} at the bottom of Table 3.3 are actually derived from plate to rectangular HSS connections, and the coefficients in the expressions have been chosen to provide suitable connection resistances for limit states design.

The upper limit of $\beta = 1 - 1/\gamma$ for checking punching shear is determined by the physical possibility of such a failure, when one considers that the shear has to be between the outer limits of the web width and the inner face of the chord wall.

3.2.3 KT Connections

As shown in Fig. 3.10, KT connections occur in some trusses, and the strength of <u>gap connections</u> can be related to K and N connections by replacing

d_1/d_0 in Table 3.1 with $(d_1 + d_2 + d_3)/3d_0$

$(b_1 + b_2)/2b_0$ in Table 3.2 with $(b_1 + b_2 + b_3)/3b_0$

$(b_1 + b_2 + h_1 + h_2)/4b_0$ in Table 3.3 with $(b_1 + b_2 + b_3 + h_1 + h_2 + h_3)/6b_0$.

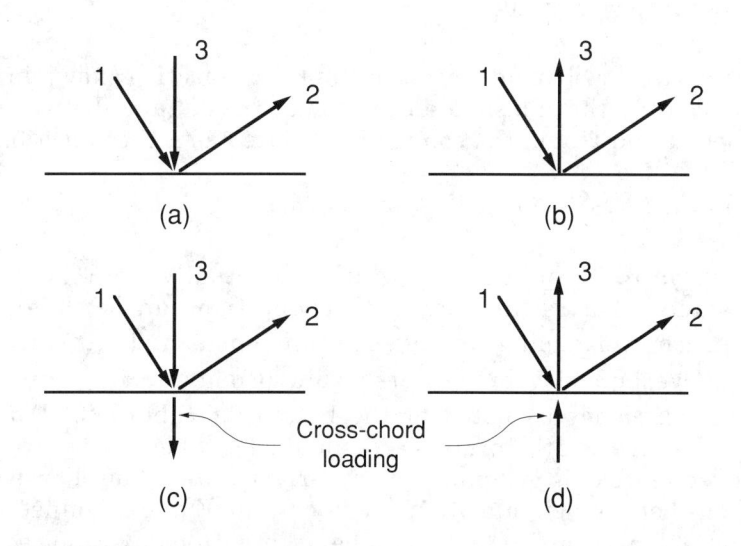

FIGURE 3.10

Four KT connections

In gap KT connections the gap should be taken as the gap between two web members having significant forces acting in the opposite sense.

In the case of KT connections with gap, the force components, normal to the chord, of the two members acting in the same sense are added together to represent the load. The connection resistance component, normal to the chord, of the remaining diagonal is then required to exceed that load. For the examples shown in Fig. 3.10,

$$N_2^* \sin\theta_2 \geq N_1 \sin\theta_1 + N_3 \sin\theta_3 \quad \text{(Fig. 3.10(a))} \quad [3.3]$$

$$N_1^* \sin\theta_1 \geq N_2 \sin\theta_2 + N_3 \sin\theta_3 \quad \text{(Fig. 3.10(b))} \quad [3.4]$$

where N_1^* is calculated from Tables 3.1, 3.2 or 3.3 in which

$$N_2^* \sin\theta_2 = N_1^* \sin\theta_1.$$

When there is a cross-chord load, (for instance, from a purlin or hanger) which acts in the same direction as the load components that were combined above, the connection resistance of the remaining diagonal needs to be examined directly. For the examples shown in Fig. 3.10,

$$N_2^* \sin\theta_2 \geq N_2 \sin\theta_2 \quad \text{(Fig. 3.10(c))} \quad [3.5]$$

$$N_1^* \sin\theta_1 \geq N_1 \sin\theta_1 \quad \text{(Fig. 3.10(d))} \quad [3.6]$$

again, where N_1^* is calculated from Tables 3.1, 3.2 or 3.3 in which

$$N_2^* \sin\theta_2 = N_1^* \sin\theta_1.$$

If the vertical web member in the gap KT connection shown in Fig. 3.10 had no force in it, the gap should be taken as the distance between the toes of members 1 and 2, and the connection treated as a K connection with

$$\beta = (b_1 + b_2 + h_1 + h_2)/4b_0 \quad \text{or} \quad d_1/d_0.$$

For <u>overlap KT connections</u>, which are actually more likely to occur, the resistance of a rectangular HSS connection can be determined by checking each overlapping web member and ensuring that N_i^* (from Table 3.2) $\geq N_i$. (Overlapped web members would also have a restriction on their connection efficiency as noted at the bottom of Table 3.2.) For the web member effective width terms, care should be taken to ensure that the member sequence of overlapping is properly accounted for. For overlap KT connections between circular HSS members, the K and N connection chord plastification check in Table 3.1 can be used with d_1/d_0 replaced by $(d_1 + d_2 + d_3)/3d_0$, and g' estimated in a conservative manner.

Tests have shown that the resistance of a K (or N) connection in the presence of a minor cross-chord loading (i.e., X connection loading) is similar to that given by the resistance formulae for K and N connections. The same is true for a KT connection, providing the force being transferred through the chord member is minor relative to the K (or N) web force components normal to the chord that balance each other. If all the web forces on one side of a connection act in the same sense, or if only one web member is carrying load, the connection should be checked as an X connection using an equivalent web member size.

3.3 Trusses with Wide Flange Chords and HSS Web Members

Some truss configurations use HSS members for the web components of the structure combined with wide flange shapes for the chords. Expressions for the resistance of K, N, T, Y and X connections are related to those presented earlier for trusses with rectangular HSS chords. The IIW (1989) design recommendations are summarized in Table 3.4 and the related limits of validity are listed in Table 3.4(a).

Gap K and N connections are to be checked for chord web stability, chord shear, and web member effective width, whereas overlap connections need only be examined for the effective width of the web.

3.4 Overlapped "Hidden Toe" Welding and Connection Sequence

Welding of the toe of the overlapped member to the chord is particularly important for 100% overlap situations. For partial overlaps, the toe of the overlapped member need not be welded, providing the components (normal to the chord) of the web member forces do not differ by more than about 20% of the lesser one. When these force components do not balance within that limit, the more heavily loaded web member should be the "through member" and its full circumference should be welded to the chord.

Generally, if overlapping HSS members have different widths, the narrower member should land on the wider member. If overlapping HSS members are the same width, the weaker member (defined by wall thickness times yield strength) should be attached to the stronger member, regardless of the load type.

TYPE OF CONNECTION	FACTORED CONNECTION RESISTANCE ($i = 1$ or 2)
T and Y Connections	Basis: CHORD PLASTIFICATION $$N_1^* = \frac{F_{y0}\, t_0^2}{\sin\theta_1} (2.8 + 14.2\,\beta^2)\, \gamma^{0.2}\, f(n')$$
X Connections	Basis: CHORD PLASTIFICATION $$N_1^* = \frac{F_{y0}\, t_0^2}{\sin\theta_1} \left(\frac{5.2}{1 - 0.81\beta} \right) f(n')$$
K and N Gap and Overlap Connections	Basis: CHORD PLASTIFICATION $$N_1^* = \frac{F_{y0}\, t_0^2}{\sin\theta_1} \left(1.8 + 10.2\, \frac{d_1}{d_0} \right) f(\gamma, g')\, f(n')$$ $$N_2^* = N_1^* \frac{\sin\theta_1}{\sin\theta_2}$$

TABLE 3.1 (start....)

Factored resistance of axially loaded welded connections between **circular HSS**

GENERAL CHECK	Basis: PUNCHING SHEAR
Punching shear check for T, Y, and X connections, plus K, N and KT connections with gap	$N_i^* = \dfrac{F_{y0}}{\sqrt{3}} t_0 \pi d_i \dfrac{1 + \sin\theta_i}{2\sin^2\theta_i}$

FUNCTIONS

$f(n') = 1.0$ for $n' \geq 0$ (tension)

$f(n') = 1 + 0.3n' - 0.3n'^2$ for $n' < 0$ (compression)

but ≤ 1.0

$g' = \dfrac{g}{t_0}$ $\gamma = \dfrac{d_0}{2t_0}$

$f(\gamma, g') = \gamma^{0.2}\left(1 + \dfrac{0.024\,\gamma^{1.2}}{\exp(0.5g' - 1.33) + 1}\right)$

TABLE 3.1 (... concluded)

Factored resistance of axially loaded welded connections between **circular HSS**
(See Table 3.1(a) on page 88 for parameter limits of validity)

TYPE OF CONNECTION	FACTORED CONNECTION RESISTANCE ($i = 1$ or 2)
T, Y and X Connections	$\beta \leq 0.85$ Basis: CHORD FACE YIELDING $$N_1^* = \frac{F_{y0}\, t_0^2}{(1-\beta)\sin\theta_1}\left(\frac{2\beta}{\sin\theta_1} + 4(1-\beta)^{0.5}\right) f(n)$$
K and N Gap Connections	$\beta \leq 1.0$ Basis: CHORD FACE PLASTIFICATION $$N_i^* = 8.9\, \frac{F_{y0}\, t_0^2}{\sin\theta_i}\, \frac{b_1 + b_2}{2b_0}\, \gamma^{0.5}\, f(n)$$

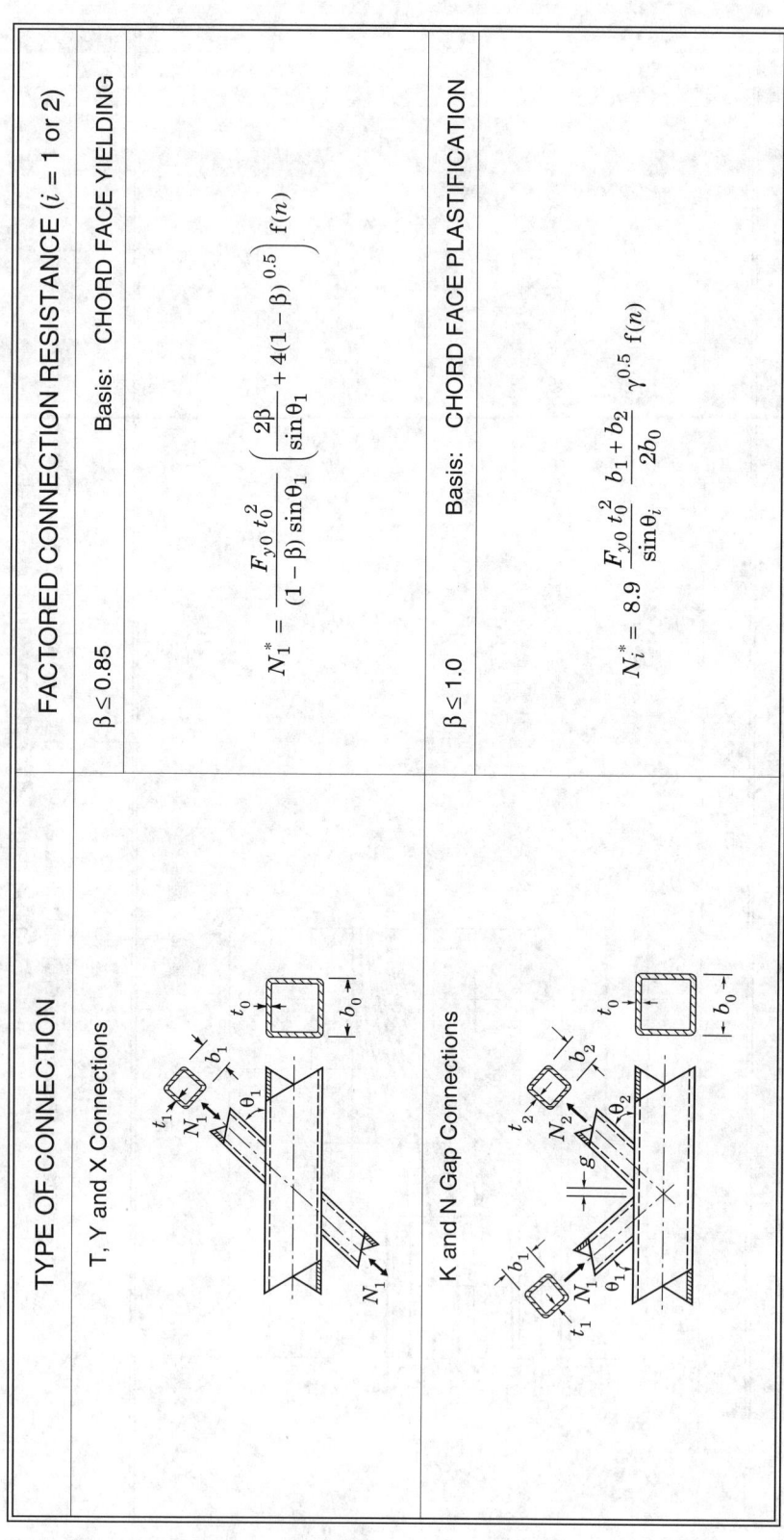

TABLE 3.2 (start ...)

Factored resistance of axially loaded welded connections between square or circular web members and a **square** chord section

K and N Overlap Connections

$25\% \leq O_v \leq 50\%$	Basis: EFFECTIVE WIDTH[1]
	$N_i^* = F_{yi} \cdot t_i \left(\dfrac{O_v}{50} (2h_i - 4t_i) + b_e + b_{e(ov)} \right)$
$50\% \leq O_v < 80\%$	Basis: EFFECTIVE WIDTH[1]
	$N_i^* = F_{yi} \, t_i \, (2h_i - 4t_i + b_e + b_{e(ov)})$
$O_v \geq 80\%$	Basis: EFFECTIVE WIDTH[1]
	$N_i^* = F_{yi} \, t_i \, (2h_i - 4t_i + b_i + b_{e(ov)})$

CIRCULAR BRACES OR WEB MEMBERS

Multiply formulae by $\pi/4$ and replace $b_{1,2}$ and $h_{1,2}$ by $d_{1,2}$

FUNCTIONS

$f(n) = 1.0$ for $n \geq 0$ (tension)

$f(n) = 1.3 + \dfrac{0.4}{\beta} n$ for $n < 0$ (compression) but ≤ 1.0

$b_e = \dfrac{10}{b_0/t_0} \dfrac{F_{y0} t_0}{F_{yi} t_i} b_i \leq b_i$

$b_{e(ov)} = \dfrac{10}{b_j/t_j} \dfrac{F_{yj} t_j}{F_{yi} t_i} b_i \leq b_i$

Subscript i = overlapping web   Subscript j = overlapped web

TABLE 3.2 (... concluded)

Factored resistance of axially loaded welded connections between square or circular web members and a **square** chord section
(See Table 3.2(a) on page 89 for parameter limits of validity)

Note[1]: Effective width computations need only be done for the *overlapping* web member. However, the efficiency (the factored connection resistance divided by the full yield capacity of the web member) of the *overlapped* web member is not to be taken higher than that of the *overlapping* web member.

TYPE OF CONNECTION	FACTORED CONNECTION RESISTANCE (i = 1 or 2)
T, Y and X Connections	Basis: CHORD FACE YIELDING $\beta \leq 0.85$ $N_1^* = \dfrac{F_{y0} t_0^2}{(1-\beta)\sin\theta_1}\left(\dfrac{2\eta}{\sin\theta_1} + 4(1-\beta)^{0.5}\right) f(n)$ Basis: CHORD SIDE WALL FAILURE[1] $\beta = 1.0$ $N_1^* = \dfrac{2F_k t_0}{\sin\theta_1}\left(\dfrac{h_1}{\sin\theta_1} + 5t_0\right)$ $^{(1)}N_1^* = \dfrac{F_{y0}(2h_0 t_0)}{\sqrt{3}\sin\theta_i}$ For $0.85 < \beta \leq 1.0$, use linear interpolation between chord face yielding ($\beta = 0.85$) and chord side wall criteria ($\beta = 1.0$). Basis: EFFECTIVE WIDTH $\beta > 0.85$ $N_1^* = F_{y1} t_1 (2h_1 - 4t_1 + 2b_e)$ Basis: PUNCHING SHEAR $0.85 < \beta \leq (1 - 1/\gamma)$ $N_1^* = \dfrac{F_{y0} t_0}{\sqrt{3}\sin\theta_1}\left(\dfrac{2h_1}{\sin\theta_1} + 2b_{ep}\right)$
K and N Gap Connections	Basis: CHORD FACE PLASTIFICATION $N_i^* = 8.9 \dfrac{F_{y0} t_0^2}{\sin\theta_i} \dfrac{b_1 + b_2 + h_1 + h_2}{4b_0} \gamma^{0.5} f(n)$ Basis: CHORD SHEAR $N_i^* = \dfrac{F_{y0} A_V}{\sqrt{3}\sin\theta_i}$ Also: $N_{0(in\,gap)}^* = (A_0 - A_V)F_{y0} + A_V F_{y0}\left(1 - (V_f/V_p)^2\right)^{0.5}$ Basis: EFFECTIVE WIDTH $N_i^* = F_{yi} t_i (2h_i - 4t_i + b_i + b_e)$ Basis: PUNCHING SHEAR $\beta \leq 1 - 1/\gamma$ $N_i^* = \dfrac{F_{y0} t_0}{\sqrt{3}\sin\theta_i}\left(\dfrac{2h_i}{\sin\theta_i} + b_i + b_{ep}\right)$

TABLE 3.3 (start ...)

Factored resistance of axially loaded welded connections between rectangular, square or circular web members and a **rectangular** chord section

Note[1]: X connections with $\theta_1 < 90°$ must also be checked for chord sidewall shear failure.

K and N Overlap Connections	Similar to connections of square hollow sections (Table 3.2)
CIRCULAR BRACES OR WEB MEMBERS	Multiply chord face plastification formula by $\pi/4$ and replace $b_{1,2}$ and $h_{1,2}$ by $d_{1,2}$
	In addition, check for chord shear if $h_0/b_0 < 1.0$.

FUNCTIONS

Tension: $F_k = F_{y0}$ Compression: $F_k = C_r/\phi A$ for T and Y connections; $F_k = 0.8\sin\theta_1 C_r/\phi A$ for X connections

$C_r/\phi A$ is the unit buckling stress according to CAN/CSA-S16.1-94, using a column slenderness ratio (KL/r) of $3.46\left(\dfrac{h_0}{t_0} - 2\right)\left(\dfrac{1}{\sin\theta_1}\right)^{0.5}$

For Classes C and H use CAN/CSA-S16.1-94, Clause 13.3.1.

$f(n) = 1.0$ for $n \geq 0$ (tension)

$f(n) = 1.3 + \dfrac{0.4}{\beta} n$ for $n < 0$ (compression)

but ≤ 1.0

$V_p = \dfrac{F_{y0} A_V}{\sqrt{3}}$

$\alpha = \left(\dfrac{1}{1 + \dfrac{4g^2}{3t_0^2}}\right)^{0.5}$

For square and rectangular web members $A_V = (2h_0 + \alpha b_0)\, t_0$
For circular web members $A_V = 2h_0 t_0$.

$b_e = \dfrac{10}{b_0/t_0} \dfrac{F_{y0} t_0}{F_{yi} t_i} b_i \leq b_i$

$b_{ep} = \dfrac{10}{b_0/t_0} b_i \leq b_i$

$b_{e(ov)} = \dfrac{10}{b_j/t_j} \dfrac{F_{yj} t_j}{F_{yi} t_i} b_i \leq b_i$

Subscript i = overlapping web
Subscript j = overlapped web

TABLE 3.3   (... concluded)

Factored resistance of axially loaded welded connections
between rectangular, square or circular web members and a **rectangular** chord section
(See Table 3.3(a) on page 90 for parameter limits of validity)

TYPE OF CONNECTION	FACTORED CONNECTION RESISTANCE (i = 1 or 2)
T, Y and X Connections	**Basis: CHORD WEB YIELDING** $$N_1^* = \frac{F_{y0}\, w\, b_m}{\sin\theta_1}$$ **Basis: EFFECTIVE WIDTH** $$N_1^* = 2F_{y1} t_1 b_e$$ No check required for Effective Width if: $$g \leq 20 - 28\beta,$$ $$\beta \leq 1.0 - 0.03\gamma \quad \text{and}$$ $$0.75 \leq \frac{d_1}{d_2}, \frac{b_1}{b_2} \leq 1.33$$
K and N Gap Connections	**Basis: CHORD WEB STABILITY** $$N_i^* = \frac{F_{y0}\, w\, b_m}{\sin\theta_1}$$ **Basis: EFFECTIVE WIDTH** $$N_i^* = 2F_{yi} t_i b_e$$ **Basis: CHORD SHEAR** $$N_i^* = \frac{F_{y0}\, A_V}{\sqrt{3}\,\sin\theta_i}$$ Also: $N_{0(in\,gap)}^* = (A_0 - A_V)F_{y0} + A_V F_{y0}\left(1 - (V_f/V_p)^2\right)^{0.5}$

TABLE 3.4 (start ...)

Factored resistance of axially loaded welded connections between rectangular, square or circular web members and an **I-shaped** chord section

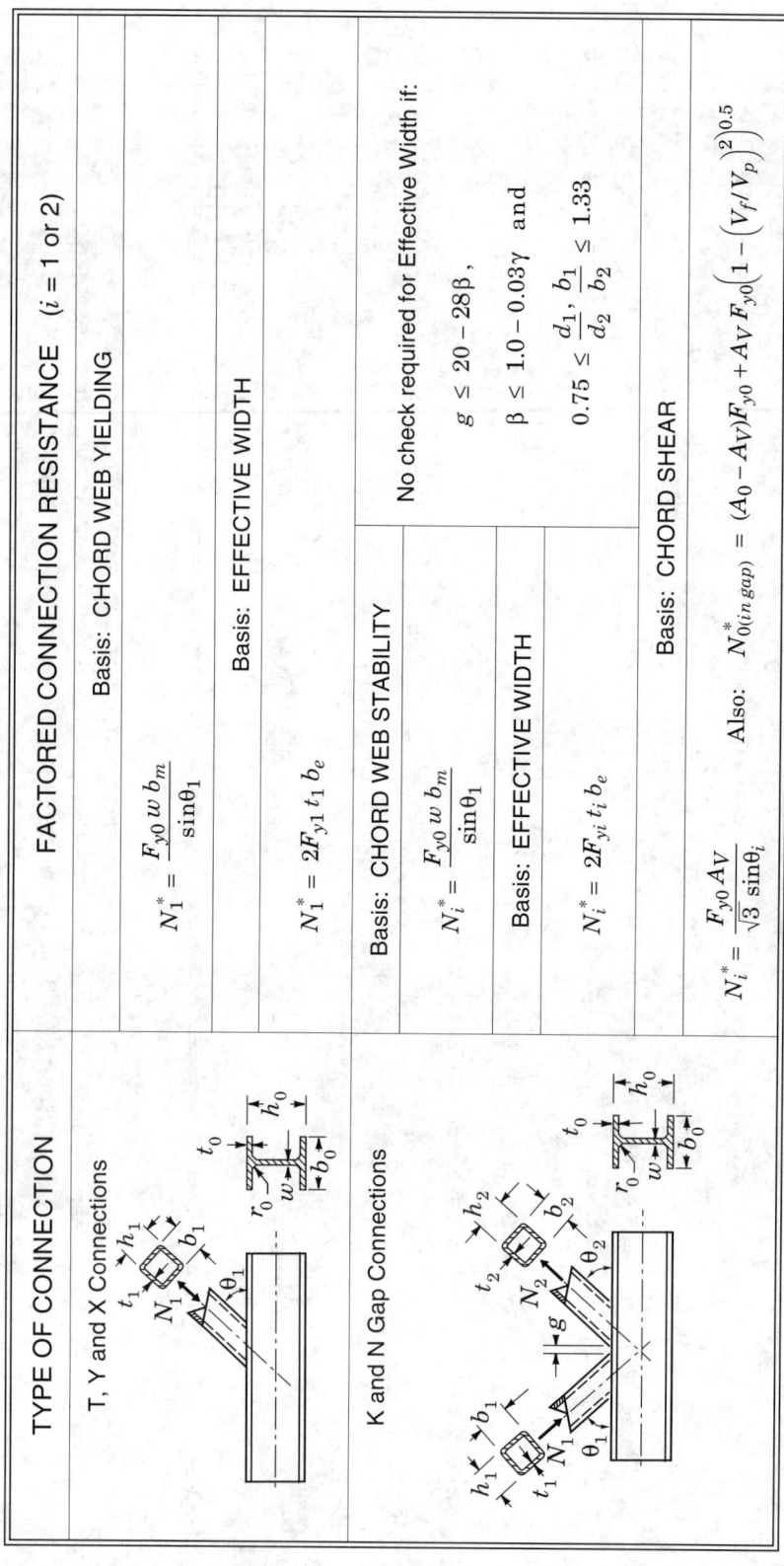

K and N Overlap Connections

	EFFECTIVE WIDTH[1]
$25\% \leq O_v < 50\%$	$N_i^* = F_{yi} t_i \left(\dfrac{O_v}{50} (2h_i - 4t_i) + b_e + b_{e(ov)} \right)$
	EFFECTIVE WIDTH[1]
$50\% \leq O_v < 80\%$	$N_i^* = F_{yi} t_i (2h_i - 4t_i + b_e + b_{e(ov)})$
	EFFECTIVE WIDTH[1]
$O_v \geq 80\%$	$N_i^* = F_{yi} t_i (2h_i - 4t_i + b_i + b_{e(ov)})$

FUNCTIONS

$$V_p = \dfrac{F_{y0} A_V}{\sqrt{3}} \qquad V_f = (N_i \sin\theta_i)_{max} \qquad \alpha = \left(\dfrac{1}{1 + \dfrac{4g^2}{3t_0^2}} \right)^{0.5}$$

For square and rectangular web members, $A_V = A_0 - (2 - \alpha) b_0 t_0 + (w + 2r_0) t_0$
For circular web members, $A_V = A_0 - 2b_0 t_0 + (w + 2r_0) t_0$

Rect. webs $\begin{cases} b_m = \dfrac{h_i}{\sin\theta_i} + 5(t_0 + r_0) \\ b_m \leq 2t_i + 10(t_0 + r_0) \end{cases}$

Cir. webs $\begin{cases} b_m = \dfrac{d_i}{\sin\theta_i} + 5(t_0 + r_0) \end{cases}$

$$b_{e(ov)} = \dfrac{10}{b_j/t_j} \dfrac{F_{yj} t_j}{F_{yi} t_i} b_i \leq b_i$$

Subscript i = overlapping web
Subscript j = overlapped web

$$b_e = w + 2r_0 + 7 \dfrac{F_{y0}}{F_{yi}} t_0 \leq b_i$$

TABLE 3.4   (... concluded)

Factored resistance of axially loaded welded connections
between rectangular, square or circular web members and an **I-shaped** chord section
(See Table 3.4(a) on page 91 for parameter limits of validity)

Note[1]: Effective width computations need only be done for the *overlapping* web member. However, the efficiency (the factored connection resistance divided by the full yield capacity of the web member) of the *overlapped* web member is not to be taken higher than that of the *overlapping* web member.

CONNECTION PARAMETERS (i = 1 or 2; j = overlapped web member)

$0.2 < \dfrac{d_i}{d_0} \leq 1.0$	$\dfrac{d_i}{t_i} \leq 50$	$\gamma \leq 25$ $\gamma \leq 20$ (X connections)	$-0.55 \leq \dfrac{e}{d_0} \leq 0.25$	$\dfrac{t_i}{t_j} \leq 1.0$	$O_v \geq 25\%$ $g \geq t_1 + t_2$

Connection efficiency for the compression web member ($N_1^* / (A_1 F_{y1})$) must be limited to the values given below, because of the possibility of premature local buckling of the compression web member.

Limits of d_1/t_1 up to which the connection factored resistance (efficiency) is not decreased

For yield strengths up to:	No reduction up to d_1/t_1 limit of:	Compression Web Member Efficiency ($N_1^* / (A_1 F_{y1})$) Limit				
		For F_{y1} up to:	For d_1/t_1 of:			
			30	35	40	45	50
F_{y1} = 235 MPa (34 ksi)	$d_1/t_1 \leq 43$	235 MPa	1.0	1.0	1.0	0.98	0.93
F_{y1} = 275 MPa (40 ksi)	$d_1/t_1 \leq 37$	275 MPa	1.0	1.0	0.96	0.88	0.86
F_{y1} = 355 MPa (52 ksi)	$d_1/t_1 \leq 28$	355 MPa	0.98	0.88	0.82	0.78	0.76

TABLE 3.1(a)

Range of validity for Table 3.1

(Factored resistance of axially loaded welded connections between **circular HSS**)

CONNECTION PARAMETERS ($i = 1$ or 2; $j =$ overlapped web member)

TYPE OF CONNECTION	b_i/b_0	b_i/t_i Compression	b_i/t_i Tension	b_0/t_0	b_1/b_2 t_i/t_j and b_i/b_j	Gap or Overlap	Eccentricity
T, Y and X	(1) $0.25 \leq \beta \leq 0.85$	(2) Minimum section is CSA-S16.1, Class 2 ≤ 35		(1) $10 \leq \dfrac{b_0}{t_0} \leq 35$			
K and N Gap	$\geq 0.1 + 0.01 \dfrac{b_0}{t_0}$ $\beta \geq 0.35$	(2) Minimum section is CSA-S16.1, Class 2 ≤ 35	≤ 35	(1) $15 \leq \dfrac{b_0}{t_0} \leq 35$	(1) smaller $b_i \geq 0.63$ the larger b_i	(3) $0.5(1-\beta) \leq \dfrac{g}{b_0}$ and $g \geq t_1 + t_2$	$-0.55 \leq \dfrac{e}{h_0} \leq 0.25$
K and N Overlap	≥ 0.25	(2) Minimum section is CSA-S16.1, Class 1		$\dfrac{b_0}{t_0} \leq 40$	$\dfrac{t_i}{t_j} \leq 1.0$ $\dfrac{b_i}{b_j} \geq 0.75$	$25\% \leq O_v \leq 100\%$	
For Circular Braces or Web Members	$0.4 \leq \dfrac{d_i}{b_0} \leq 0.8$	(2) Minimum section is CSA-S16.1, Class 1	$\dfrac{d_2}{t_2} \leq 50$			Limitations as above for $d_i = b_i$	

TABLE 3.2(a)

Range of validity for Table 3.2 (Factored resistance of axially loaded welded connections between square or circular web members and a **square** chord section)

Note[1]: Outside this range of validity, other failure criteria may be governing; e.g., punching shear, effective width, side wall failure, chord shear or local buckling. If these particular limits of validity are violated, the connection may still be checked as one having a rectangular chord using Table 3.3, provided the limits of validity in Table 3.3(a) are still met.

Note[2]: CAN/CSA-S16.1-94 definition for rectangular Class 1 sections is $\dfrac{b}{t} \leq \dfrac{420}{\sqrt{F_y}}$; for Class 2 sections is $\dfrac{b}{t} \leq \dfrac{525}{\sqrt{F_y}}$ where b is the width of the *flat* surface of the HSS section. For circular sections, Class 1 is $\dfrac{d}{t} \leq \dfrac{13\,000}{F_y}$.

Note[3]: If g/b_0 is large, treat as two Y connections. Maximum gap size for K connection action will be controlled by e/h_0 limit.

CONNECTION PARAMETERS (i = 1 or 2; j = overlapped web member)

TYPE OF CONNECTION	b_i/b_0 h_i/b_0	$b_i/t_i, h_i/t_i, d_i/t_i,$ Compression	$b_i/t_i, h_i/t_i, d_i/t_i,$ Tension	h_i/b_i	b_0/t_0 h_0/t_0	Gap or Overlap b_i/b_j t_i/t_j	Eccentricity
T, Y and X	≥ 0.25	(1) Minimum section is CSA-S16.1, Class 2 ≤ 35	≤ 35	$0.5 \leq \dfrac{h_i}{b_i} \leq 2$	≤ 35		
K and N Gap	$\geq 0.1 + 0.01\dfrac{b_0}{t_0}$ $\beta \geq 0.35$				≤ 35	(2) $0.5(1-\beta) \leq \dfrac{g}{b_0}$ and $g \geq t_1 + t_2$	$-0.55 \leq \dfrac{e}{h_0} \leq 0.25$
K and N Overlap	≥ 0.25	(1) Minimum section is CSA-S16.1, Class 1			≤ 40	$25\% \leq O_v \leq 100\%$ $\dfrac{b_i}{b_j} \geq 0.75$ $\dfrac{t_i}{t_j} \leq 1.0$	
For Circular Braces or Web Members	$0.4 \leq \dfrac{d_i}{b_0} \leq 0.8$	(1) Minimum section is CSA-S16.1, Class 1	≤ 50			Limitations as above for $d_i = b_i$	

TABLE 3.3(a)

Range of validity for Table 3.3

(Factored resistance of axially loaded welded connections between rectangular, square or circular web members and a **rectangular** chord section)

Note$^{(1)}$: CAN/CSA-S16.1-94 definition for rectangular Class 1 sections is $\dfrac{b}{t}$ or $\dfrac{h}{t} \leq \dfrac{420}{\sqrt{F_y}}$; for Class 2 sections is $\dfrac{b}{t}$ or $\dfrac{h}{t} \leq \dfrac{525}{\sqrt{F_y}}$ where b (or h) is the width of the *flat* surface of the HSS section. For circular sections, Class 1 is $\dfrac{d}{t} \leq \dfrac{13\,000}{F_y}$.

Note$^{(2)}$: If g/b_0 is large, treat as two Y connections. Maximum gap size for K connection action will be controlled by e/h_0 limit.

CONNECTION PARAMETERS (i = 1 or 2; j = overlapped web member)

TYPE OF CONNECTION	$\dfrac{h_w}{w}$	$\dfrac{b_0}{t_0}$	β	b_i/t_i, h_i/t_i, d_i/t_i Compression	b_i/t_i, h_i/t_i, d_i/t_i Tension	$\dfrac{h_i}{b_i}$	$\dfrac{b_i}{b_j}$	Gap	Eccentricity
X	$\dfrac{h_w}{w} \leq 1.2\sqrt{\dfrac{E}{F_{y0}}}$ $h_w \leq 400$ mm								
T and Y		≤ 20	≤ 1.0	(1) Minimum section is CSA-S16.1, Class 1	$\dfrac{h_i}{t_i}, \dfrac{b_i}{t_i} \leq 35$ $\dfrac{d_i}{t_i} \leq 50$	$0.5 \leq \dfrac{h_1}{b_1} \leq 2.0$			$-0.55 \leq \dfrac{e}{h_0} \leq 0.25$
K and N Gap	$\dfrac{h_w}{w} \leq 1.5\sqrt{\dfrac{E}{F_{y0}}}$ $h_w \leq 400$ mm					$\dfrac{h_i}{b_i} = 1.0$		$g \geq t_1 + t_2$	
K and N Overlap						$0.5 \leq \dfrac{h_i}{b_i} \leq 2.0$	$\dfrac{b_i}{b_j} \geq 0.75$		

TABLE 3.4(a)

Range of validity for Table 3.4

(Factored resistance of axially loaded welded connections between rectangular, square or circular web members and an **I-shaped** chord section)

Note[1]: CAN/CSA-S16.1-94 definition for rectangular Class 1 sections is $\dfrac{b}{t}$ or $\dfrac{h}{t} \leq \dfrac{420}{\sqrt{F_y}}$ where b (or h) is the width of the *flat* surface of the HSS section. For circular sections, Class 1 is $\dfrac{d}{t} \leq \dfrac{13\,000}{F_y}$.

3.5 Bending Moments combined with Axial Forces in Web Members

A comprehensive examination of moment connections is presented in Chapter 6.

<u>Primary</u> bending moments will be produced at the ends of web members (in addition to axial forces) if a rigid-jointed truss analysis is performed, with or without external loads applied to the web members, but such an analysis is generally not recommended. Web member primary bending moments can be avoided at the connections if pin-ended web conditions (as described in Section 2.2.2) are used for the analysis of trusses, even when external off-node chord loads are present. In such cases, all primary moments produced by both transverse chord loads and noding eccentricities will be distributed to the chord members.

Design recommendations given herein are applicable only within the eccentricity limits in Fig. 3.1, and within these limits primary bending moments from the nodal eccentricity can be ignored when computing connection resistances. If the eccentricity limits are exceeded, the resulting primary bending moment at the node must be taken into account. Since the connection eccentricity then exceeds the range that has been justified by experimental and analytical research, the connection capacity should be verified by laboratory proof tests and/or careful finite element analysis.

If a web member transmits both a bending moment and an axial force, the following conservative interaction formula is recommended:

$$\frac{N_i}{A_i} + \frac{M_{ip}}{S_i} \leq \frac{N_i^*}{A_i} \qquad [3.7]$$

where M_{ip} is the bending moment to be transmitted by the web member at the connection. In addition, the bending stress produced in the web member must not exceed the axial stress

$$\left| \frac{M_{ip}}{S_i} \right| \leq \left| \frac{N_i}{A_i} \right| \qquad [3.8]$$

<u>Secondary</u> bending moments resulting from induced deformations in the structural system do not influence connection strengths (as mentioned in Chapter 2), if the connections have sufficient deformation capacity. That will be the case for connections conforming to the validity ranges in Tables 3.1(a), 3.2(a), 3.3(a) and 3.4(a).

3.6 Limits of Validity

The expressions for resistances in Tables 3.1 to 3.4 are valid only within specific ranges of various dimensional parameters for the connected members. Such ranges have been published by IIW (1989), and are expressed in the tables in mathematical terms. For convenience, they are presented in writing in the following sections.

3.6.1 Circular HSS Chords and Web Members

The parameters and their ranges of validity for circular member HSS connections (Table 3.1a) follow.

- Diameter of web members must be between 0.2 and 1.0 times the chord diameter.
- Diameter to wall thickness ratio of a member must not exceed 50 (40 for chord members in X connections).
- Wall thickness of the overlapping member must not exceed the wall thickness of the overlapped member (for equal yield strength materials).
- Nodal eccentricity limits are − 0.55 and + 0.25 of the chord diameter.
- Minimum overlap is 25%.
- Minimum gap is the sum of the wall thicknesses of the members on either side of the gap.
- With increasing diameter to wall thickness ratio of a web compression member (beyond 28 for 350 MPa yield material), there is an increasing reduction in the permitted connection resistance relative to the yield of the member.

3.6.2 Square HSS Chords with Square or Circular HSS Web Members

The parameters and their ranges of validity for square member HSS K and N connections, gap and overlap (Table 3.2a) are given below.

For gap connections:

- Average width of the two web members must be at least 0.35 times the chord width.
- Width of the web members, relative to the chord width, must be at least a hundredth of the chord width-to-thickness ratio, plus 0.1.
- Compression web members must be either CAN/CSA-S16.1-94 Class 1 or Class 2 sections.
- Width to thickness ratio of tension web members must not exceed 35.
- Width to thickness ratio of the chord must be between 15 and 35.

- » Width of the smaller web member must be at least 0.63 times the width of the other web member (otherwise resort to the more general rectangular HSS Table 3.3).
- » A limit for minimum gap is defined as a function of chord and web member widths. If the maximum gap for a K or N connection (as allowed by the eccentricity limit) is exceeded, one should treat it as a pair of T / Y connections.
- » Absolute minimum gap is the sum of the wall thicknesses of the two web members.
- » Nodal eccentricity limits are − 0.55 and + 0.25 of the chord height.

For overlap connections:

- » Width of the web members must be at least 0.25 times the chord width.
- » Compression web members must be CAN/CSA-S16.1-94 Class 1 sections.
- » Width to thickness ratio of tension web members must not exceed 35.
- » Width to thickness ratio of the chord must not exceed 40.
- » Wall thickness of the overlapping member must not exceed the wall thickness of the overlapped member (for equal yield strength materials).
- » Width of the overlapping web member must be at least 0.75 times the width of the overlapped web member.
- » Minimum overlap is 25%.
- » Nodal eccentricity limits are the same as for gap connections.

The parameters and their ranges of validity for square member HSS T, Y and X connections are as follows.

- » Width of web members must be between 0.25 and 0.85 times the width of the chord (otherwise, resort to the more general rectangular HSS Table 3.3).
- » Compression web members must be either CAN/CSA-S16.1-94 Class 1 or Class 2 sections.
- » Width to thickness ratio of tension web members must not exceed 35.
- » Width to thickness ratio of the chord must be between 10 and 35.

When circular HSS members are welded to a square HSS chord, the following constraints apply to the circular web members.

- » Diameter of web members must be between 0.4 and 0.8 times the width of the chord.

- » Compression web members must be CAN/CSA-S16.1-94 Class 1 sections.
- » Diameter to thickness ratio of tension web members must not exceed 50.
- » For K or N gap connections, diameter of the smaller web member must be at least 0.63 times the diameter of the other web member (otherwise resort to the more general rectangular HSS Table 3.3).
- » For K and N gap connections, a limit for minimum gap is defined as a function of chord and web member widths. (If the maximum gap for a K or N connection is exceeded, as allowed by the eccentricity limit, one should treat it as a pair of T / Y connections.)
- » Absolute minimum gap for K and N connections is the sum of the wall thicknesses of the two web members.
- » Wall thickness of the overlapping member cannot exceed the wall thickness of the overlapped member (for equal yield strength materials).
- » Diameter of the overlapping web member must be at least 0.75 times the diameter of the overlapped web member (K or N overlap connections).
- » Minimum overlap is 25%.

3.6.3 Rectangular HSS Chords with Rectangular, Square or Circular HSS Web Members

The parameters and their ranges of validity, for rectangular HSS chords with rectangular or square HSS web member K and N connections, gap and overlap, (Table 3.3a) are given below.

For gap connections:

- » Average of the widths plus the heights of the two web members (four dimensions) must be at least 0.35 times the chord width.
- » Width or height of web members, relative to the chord width, must be at least a hundredth of the chord width-to-thickness ratio, plus 0.1.
- » Compression web members must be either CAN/CSA-S16.1-94 Class 1 or Class 2 sections.
- » Width to thickness ratio and height to thickness ratio of tension web members must not exceed 35.
- » Height to width ratio of all members must be between 0.5 and 2.0.
- » Width to thickness ratio, and the height to thickness ratio, of the chord must not exceed 35.

- » A limit for minimum gap is defined as a function of web member widths and heights, and chord width. If the maximum gap for a K or N connection (as allowed by the eccentricity limit) is exceeded, one should treat it as a pair of T / Y connections.
- » Absolute minimum gap is the sum of the wall thicknesses of the two web members.
- » Nodal eccentricity limits are −0.55 and +0.25 of the chord height.

For overlap connections:

- » Width and height of web members must be at least 0.25 times the chord width.
- » Compression web members must be CAN/CSA-S16.1-94 Class 1 sections.
- » Width to thickness ratio, and the height to thickness ratio, of tension web members must not exceed 35.
- » Height to width ratio of all members must be between 0.5 and 2.0.
- » Width to thickness ratio and height to thickness ratio of the chord must not exceed 40.
- » Wall thickness of the overlapping member must not exceed the wall thickness of the overlapped member (for equal yield strength materials).
- » Width of the overlapping web member must be at least 0.75 times the width of the overlapped web member.
- » Minimum overlap is 25%.
- » Nodal eccentricity limits are the same as for gap connections.

The parameters and their ranges of validity for rectangular HSS chords with rectangular or square HSS web member T, Y and X connections are as follows.

- » Width and height of the web members must be at least 0.25 times the chord width.
- » Compression web members must be either CAN/CSA-S16.1-94 Class 1 or Class 2 sections.
- » width to thickness ratio and height to thickness ratio of tension web members must not exceed 35.
- » Height to width ratio of all members must be between 0.5 and 2.0.
- » Width to thickness ratio and height to thickness ratio of the chord must be not exceed 35.

When circular HSS members are welded to a rectangular HSS chord, the following constraints apply to the circular web members.

- » Diameter of web members must be between 0.4 and 0.8 times the width of the chord.
- » Compression web members must be CAN/CSA-S16.1-94 Class 1 sections.
- » Diameter to thickness ratio of tension web members must not exceed 50.
- » For K and N gap connections, a limit for minimum gap is defined as a function of chord and web member widths. (If the maximum gap for a K or N connection is exceeded, as allowed by the eccentricity limit, one should treat it as a pair of T / Y connections.)
- » Absolute minimum gap is the sum of the wall thicknesses of the two web members.
- » Diameter of the overlapping web member must be at least 0.75 times the diameter of the overlapped web member (K or N overlap connections).
- » Wall thickness of the overlapping member must not exceed the wall thickness of the overlapped member (for equal yield strength materials).
- » Minimum overlap is 25%.

3.6.4 Wide Flange Chords with Rectangular, Square or Circular HSS Web Members

The parameters and their ranges of validity for wide flange chords with rectangular, square or circular web member K and N connections, gap and overlap (Table 3.4a) are given below.

For gap connections:

- » Height (between root fillets) to thickness ratio of chord webs must not exceed $671/\sqrt{F_{y0}}$.
- » Web heights between root fillets must not exceed 400 mm.
- » Full width to thickness ratio of chord flanges must not exceed 20.
- » Web members may be the full width of the chord.
- » Compression web members must be CAN/CSA-S16.1-94 Class 1 sections.
- » Width to thickness ratio of tension web members must not exceed 35 (50 if circular).
- » Web members must be either square or circular sections.
- » Nodal eccentricity limits are −0.55 and +0.25 of the chord height.
- » Absolute minimum gap is the sum of the wall thicknesses of the two web members.

For overlap connections:

> Parameters and validity ranges are the same as for gap connections except that rectangular web sections are permitted, provided height to width ratios are between 0.5 and 2.0.

> Width of the overlapping web member must be at least 0.75 times the width of the overlapped web member.

> Wall thickness of the overlapping member cannot exceed the wall thickness of the overlapped member (for equal yield strength materials).

> Minimum overlap is 25%.

> Nodal eccentricity limits are the same as for gap connections.

The parameters and their ranges of validity for wide flange chords with rectangular, square or circular web member T, Y and X connections are as follows.

> Height (between root fillets) to thickness ratio of chord webs must not exceed $671/\sqrt{F_{y0}}$ (T and Y connections).

> Height (between root fillets) to thickness ratio of chord webs must not exceed $537/\sqrt{F_{y0}}$ (X connections).

> Web heights between root fillets must not exceed 400 mm.

> Full width to thickness ratio of chord flanges must not exceed 20.

> Web members may be the full width of the chord.

> Compression web members must be CAN/CSA-S16.1-94 Class 1 sections.

> Width to thickness ratio of tension web members must not exceed 35 (50 if circular).

> Height to width ratio of rectangular web members must be between 0.5 and 2.0.

3.7 Design Charts and Tables

Several series of design charts have been developed to facilitate the application of formulations published by CIDECT, AWS (American Welding Society) and IIW.

Packer (1986) presented Fig. 3.11 which allows a quick evaluation of whether a K or N gap connection can be configured from proposed square web sections on a rectangular or square chord section. The chart indicates the maximum value of β (average width of web members relative to the chord width) which can be accommodated while remaining within the allowable bounds of nodal eccentricity. Input parameters are chord to web angle and chord aspect ratio h_0/b_0 on the horizontal axis. If the intersection

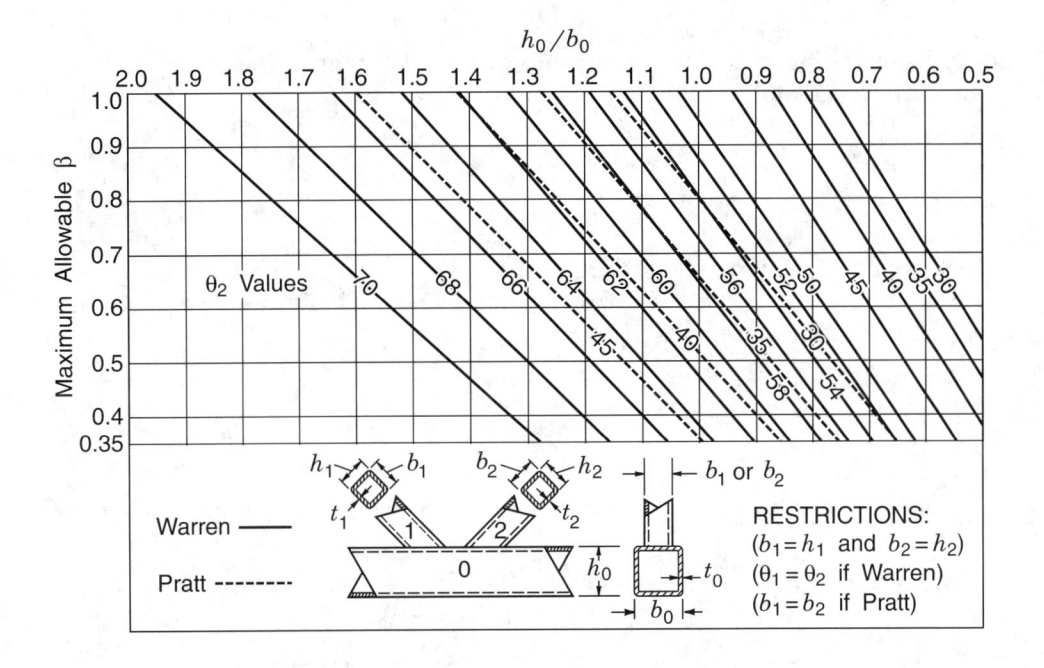

FIGURE 3.11

Maximum β based on allowable eccentricity limits, chord aspect ratio and inclination of web members

of these two parameters lies above the graph, the maximum allowable β is 1.0.

3.7.1 Design Charts

Reusink and Wardenier (1989a and 1989b) have produced sets of consistent design charts for the preliminary design of K, N, T, Y, and X connections which are based on IIW (1989) and therefore Eurocode 3 Annex K (European Committee for Standardization 1992) recommendations. The first set (Figs. 3.12 to 3.18) are for *circular* HSS and are derived from the formulae presented in Table 3.1. The second set (Figs. 3.19 to 3.26) are for *square* HSS, and they relate to formulae in Table 3.2.

The concept of "connection efficiency" (defined as the connection factored resistance divided by the full section yield load of the particular web member) is employed for these charts. That is, connection efficiency is equal to $N_i^*/(A_i F_{yi})$. The efficiencies given by the charts for all but the overlapped K connections (and the T, Y, X connection effective width checks) are termed C_K, C_T or C_X depending on the type of connection. These efficiencies need to be multiplied by three factors to obtain the final connection efficiency in each case.

The first factor which corrects for different strengths between the chord and the web member is

$$\frac{F_{y0}\, t_0}{F_{yi}\, t_i}$$

In Canada, however, all HSS is generally ordered to the requirements of CAN/CSA-G40.21, Grade 350W which has a specified minimum yield strength of 350 MPa; therefore, this factor reduces to t_0/t_i.

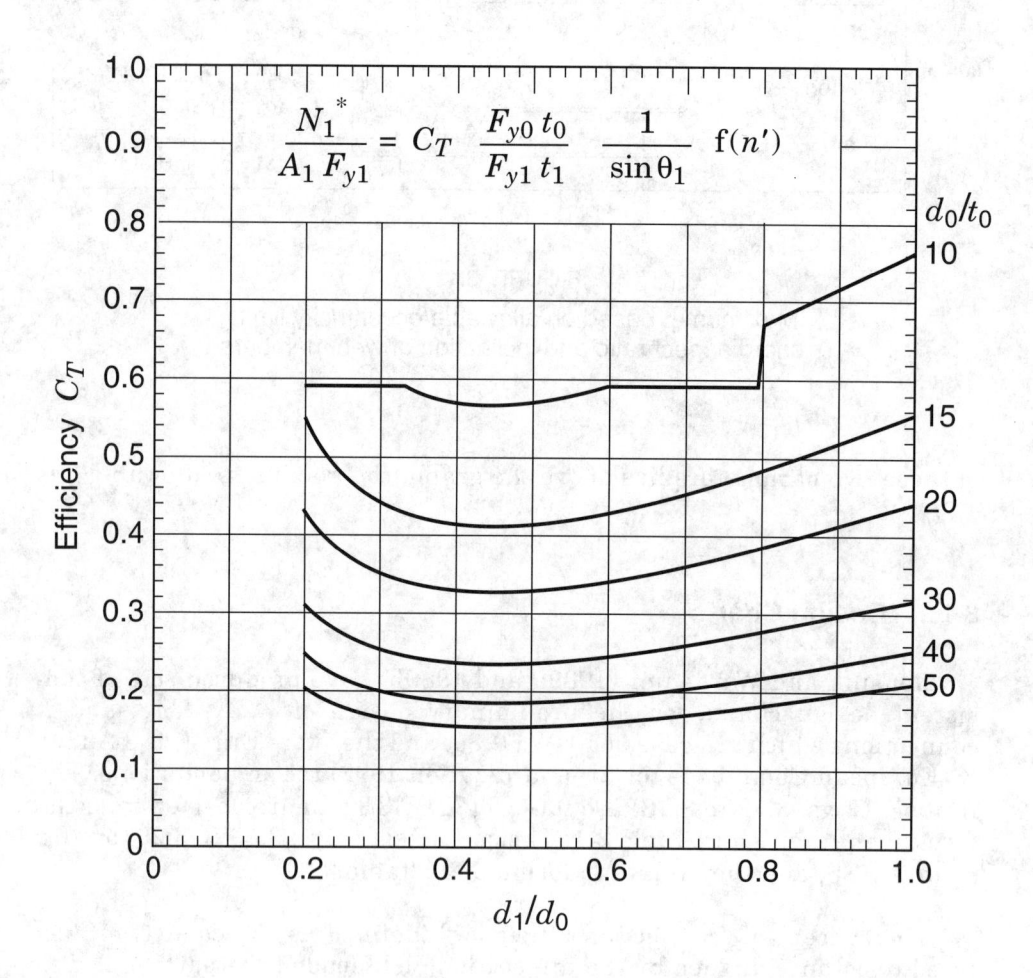

$$\frac{N_1^*}{A_1\, F_{y1}} = C_T\, \frac{F_{y0}\, t_0}{F_{y1}\, t_1}\, \frac{1}{\sin\theta_1}\, f(n')$$

FIGURE 3.12

Branch member efficiency for **circular HSS** T and Y connections

The second factor which adjusts for the angle between the web member and the chord is $1/\sin\theta_i$ for both circular HSS and square HSS connections. For overlap square HSS connections it is 1.0.

The third factor which corrects for the influence of compression chord longitudinal stresses on the connection efficiency is $f(n')$ for circular HSS or $f(n)$ for square HSS. The function $f(n')$ is defined in Table 3.1 and plotted in Fig. 3.18, while $f(n)$ is defined in Table 3.2 and plotted in Fig. 3.24. The function $f(n)$ is 1.0 for overlap square HSS connections.

In the design charts a lower bound of the various failure modes was generally drawn to result in C_K, C_T and C_X functions which depend only on the type of connection. Simplifying conservative assumptions, and nar-

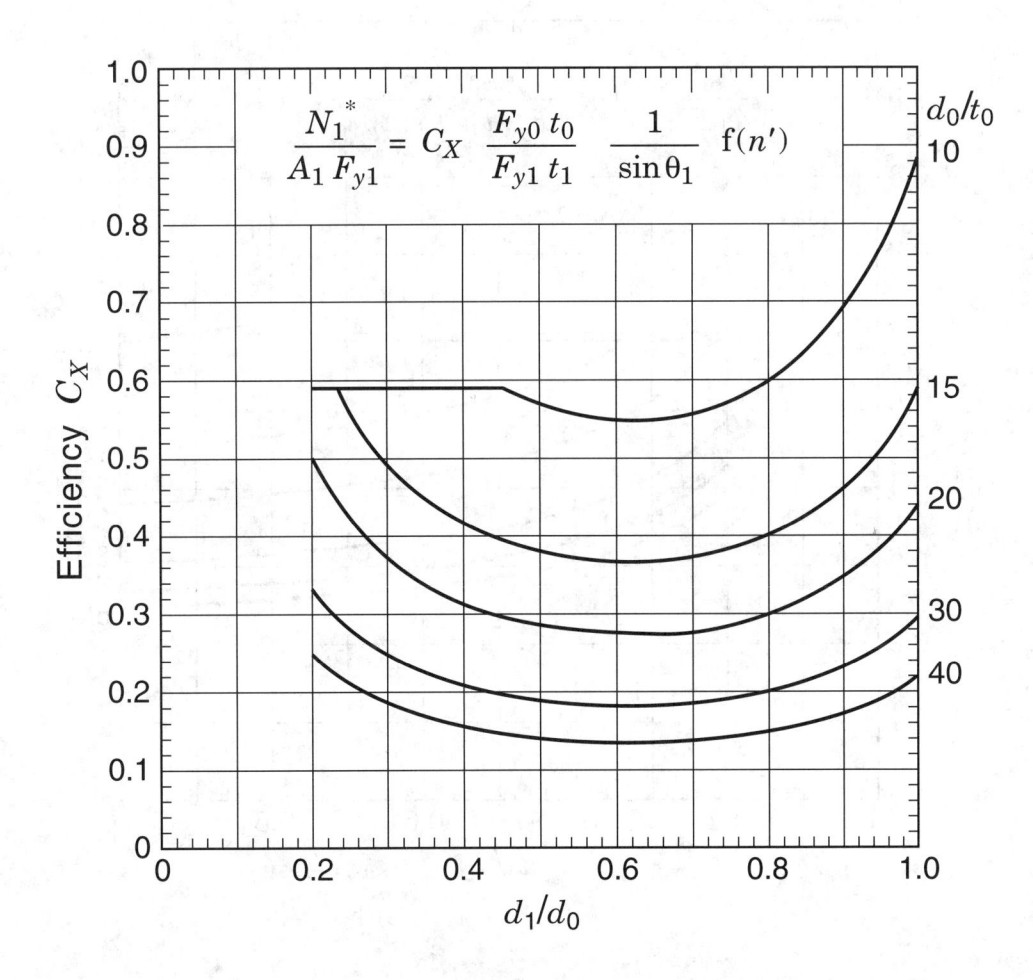

$$\frac{N_1^*}{A_1 F_{y1}} = C_X \frac{F_{y0} t_0}{F_{y1} t_1} \frac{1}{\sin\theta_1} f(n')$$

FIGURE 3.13

Branch member efficiency for **circular HSS** X connections

rower parameter ranges of validity were sometimes necessary in the process. One should be aware that these design charts can be very conservative for web members having a low value of the angle θ.

Comparing the formulae for connection resistance from Tables 3.1 and 3.2 with the related design charts can be interesting. Expressions from the tables are formulated to give connection resistance directly, while the charts are structured to give connection efficiency which can be converted into connection resistance by multiplying that efficiency by $A_i F_{yi}$. The formulae from the tables generally do not include cross sectional area terms for the members since the connection resistances are basically a function

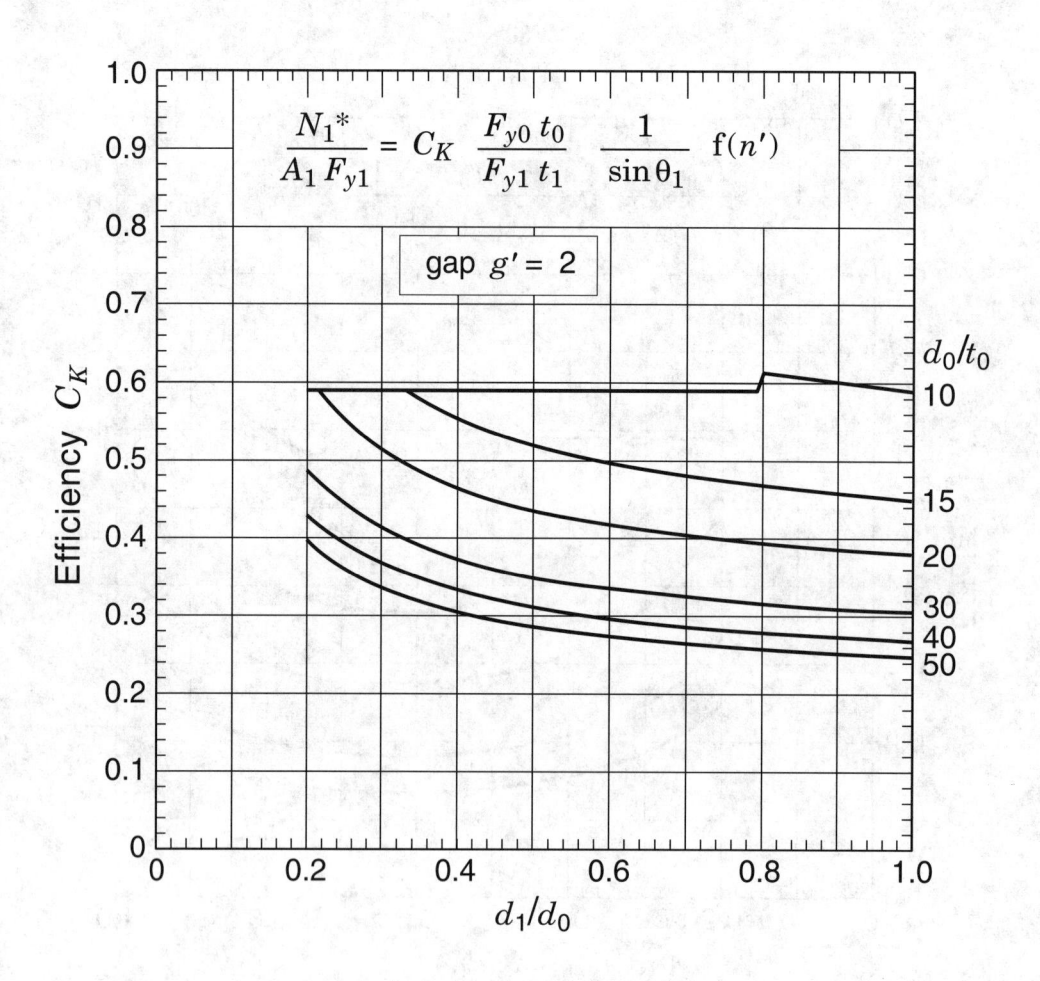

$$\frac{N_1^*}{A_1 F_{y1}} = C_K \frac{F_{y0} t_0}{F_{y1} t_1} \frac{1}{\sin \theta_1} f(n')$$

FIGURE 3.14

Web member efficiency for **circular HSS**
K and N connections with gap $g' = 2$

of their relative geometries and their wall thicknesses, whereas the charts specifically use cross sectional areas when converting from efficiencies to resistances.

A particular situation arises regarding gap K or N connections for circular and square HSS. The formulae in Tables 3.1 and 3.2 provide connection resistances that are the same for both compression and tension web members (for equal θ's) since the action is basically a push-pull combination of loads. In the event of a cross-chord load at the K or N connection, one web member may be considerably lighter than the other, and the calculated connection resistance for that member may exceed the

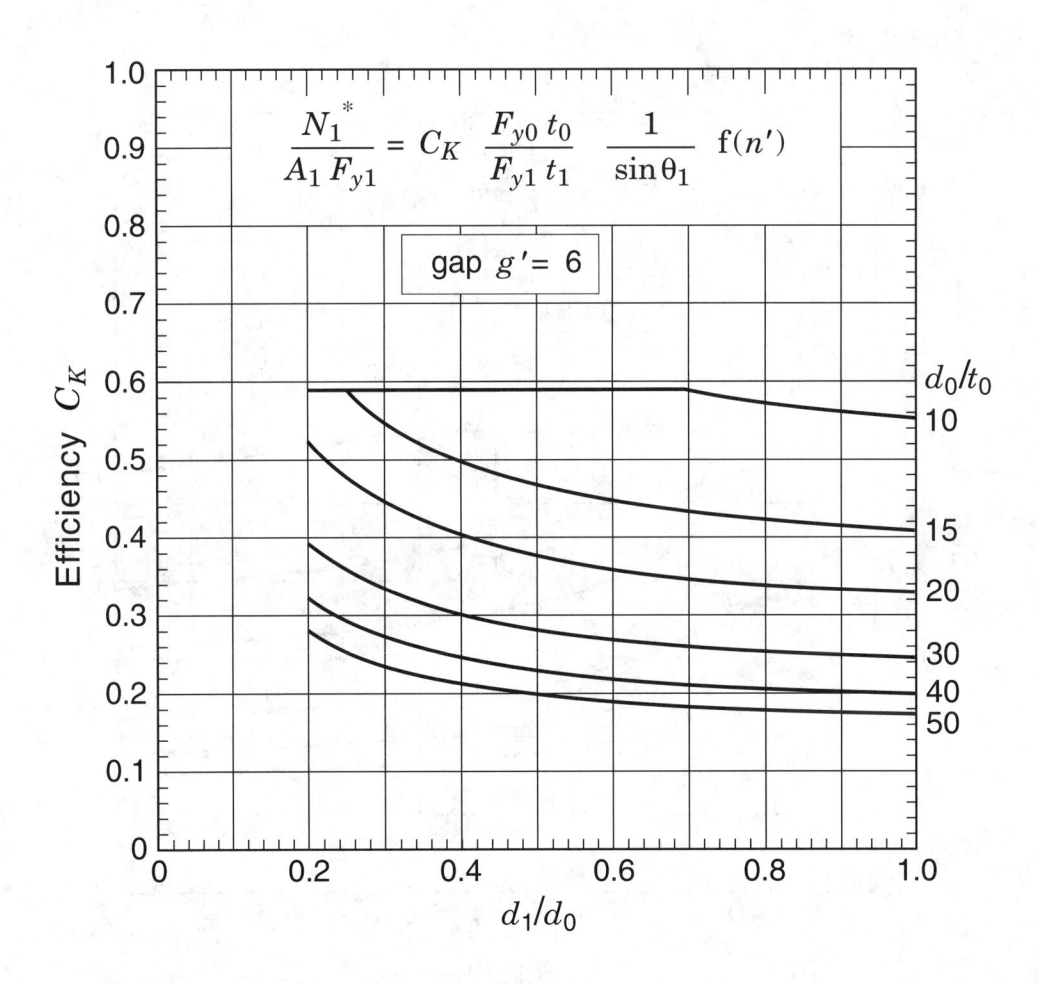

$$\frac{N_1^*}{A_1 F_{y1}} = C_K \frac{F_{y0} t_0}{F_{y1} t_1} \frac{1}{\sin\theta_1} f(n')$$

gap $g' = 6$

FIGURE 3.15

Web member efficiency for **circular HSS** K and N connections with gap $g' = 6$

resistance of the member itself. In other words, the member governs, not the connection.

The corresponding design chart for square HSS (Fig. 3.23) can be used to calculate the connection resistances of the two web members independently, and it will tend to give the same value to both (within the accuracy of graphical aids). This means that it too may give connection resistances for some web members which exceed the resistance of the member itself. Again, the member governs rather than the connection.

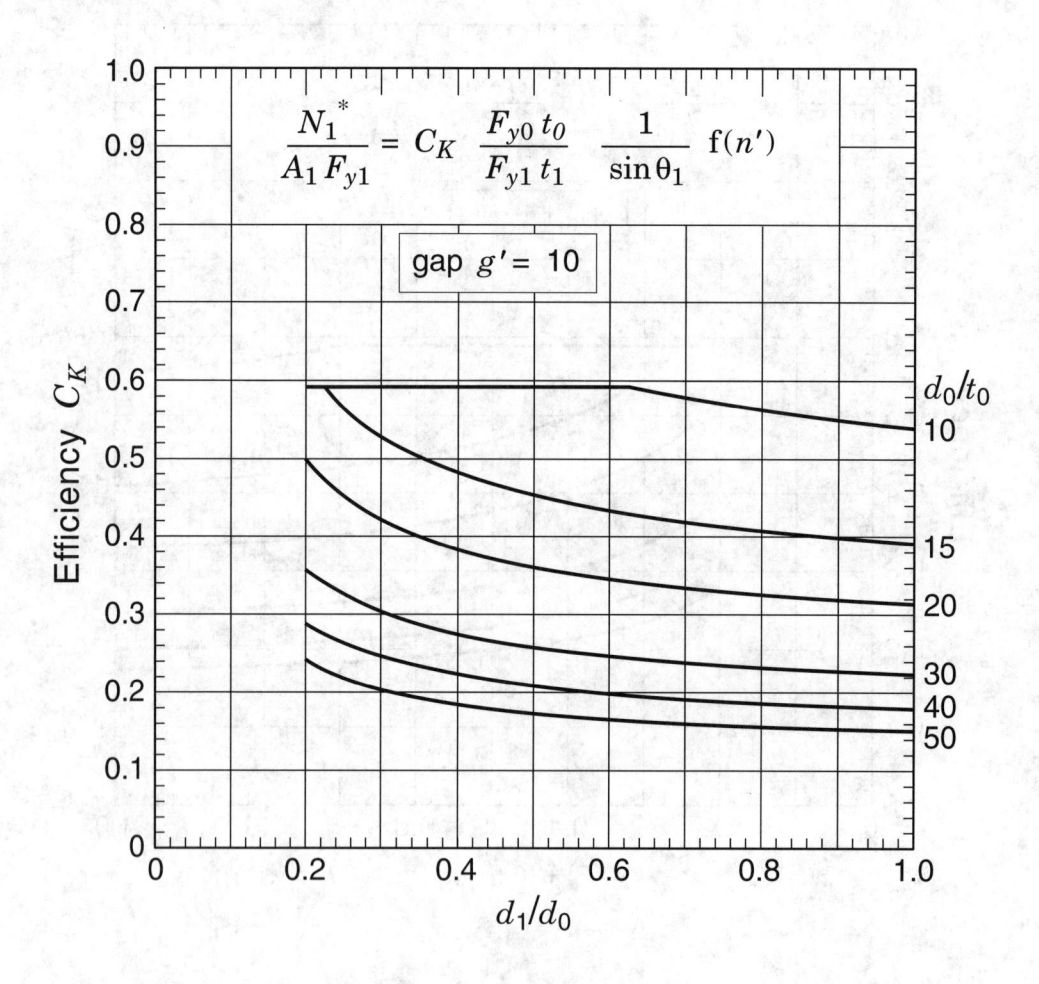

$$\frac{N_1^*}{A_1 F_{y1}} = C_K \frac{F_{y0} t_0}{F_{y1} t_1} \frac{1}{\sin \theta_1} f(n')$$

gap $g' = 10$

FIGURE 3.16

Web member efficiency for **circular HSS** K and N connections with gap $g' = 10$

The charts for circular HSS, Figs. 3.12 to 3.17, are entered along the horizontal d_1/d_0 axis, read up to the appropriate d_0/t_0 curve, then across to the vertical axis to obtain the uncorrected connection efficiency. The final efficiency is then determined by the equation on the chart, with $f(n')$ obtained from Fig. 3.18. The full parameter range of validity given in Table 3.1(a) applies. Discontinuities in the curves for $d_0/t_0 = 10$ are due to the upper boundary limit for punching shear of $\beta = 1 - 1/\gamma$.

Figures 3.14, 3.15 and 3.16, for gap K and N connections, provide efficiency values which can be linearly interpolated over a wide range of

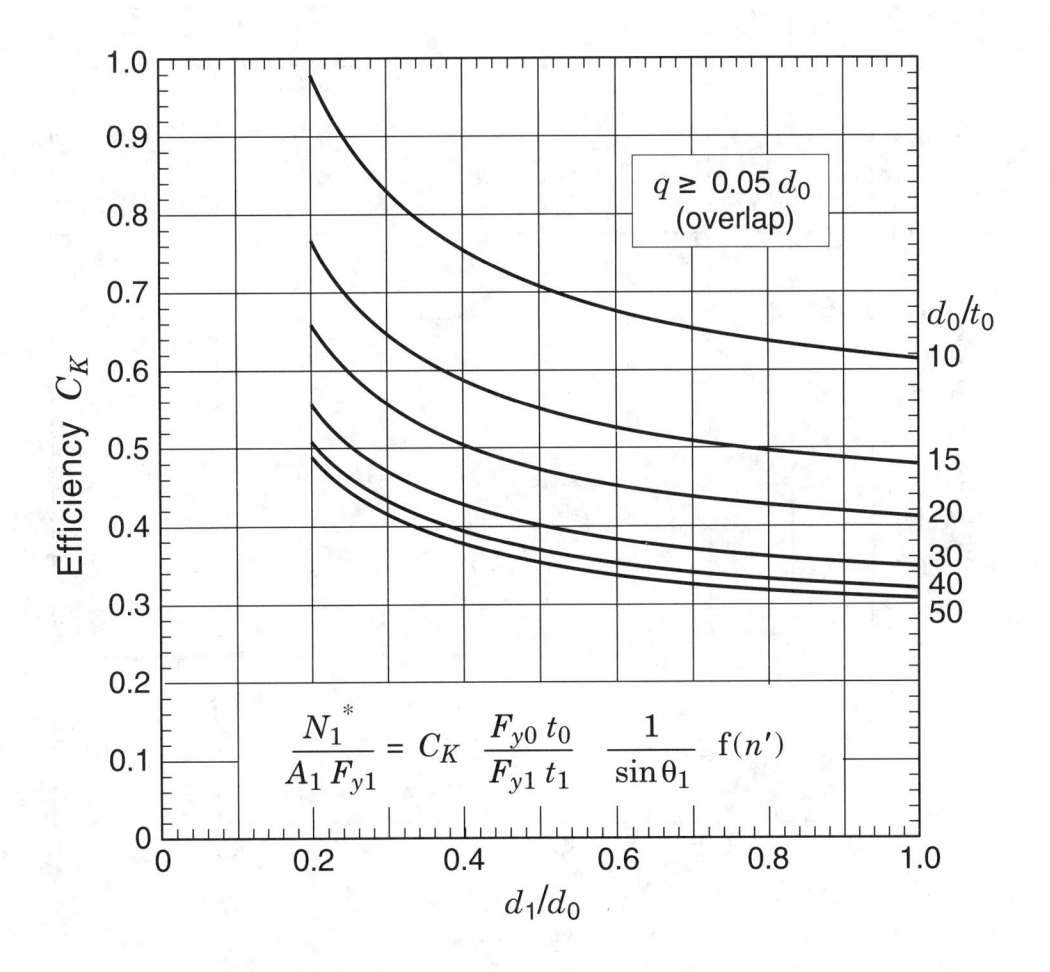

FIGURE 3.17

Web member efficiency for **circular HSS** K and N overlap connections with overlap q at least 5% of the chord diameter

gaps. Fig. 3.17 does the same for all overlapped connections within the validity limit of $O_v \geq 25\%$.

The charts for square HSS, Figs. 3.19 to 3.26, are used in a similar manner. Four charts are presented for T, Y and X connections (Figs. 3.19 to 3.22). The first applies to all three types of connection when they are loaded in tension; the second applies to T and Y connections, when loaded in compression; the third to X connections, when loaded in compression; and the fourth is an effective width failure mode check, necessary only when β exceeds 0.85. The first three charts are identical for β values up to 0.85.

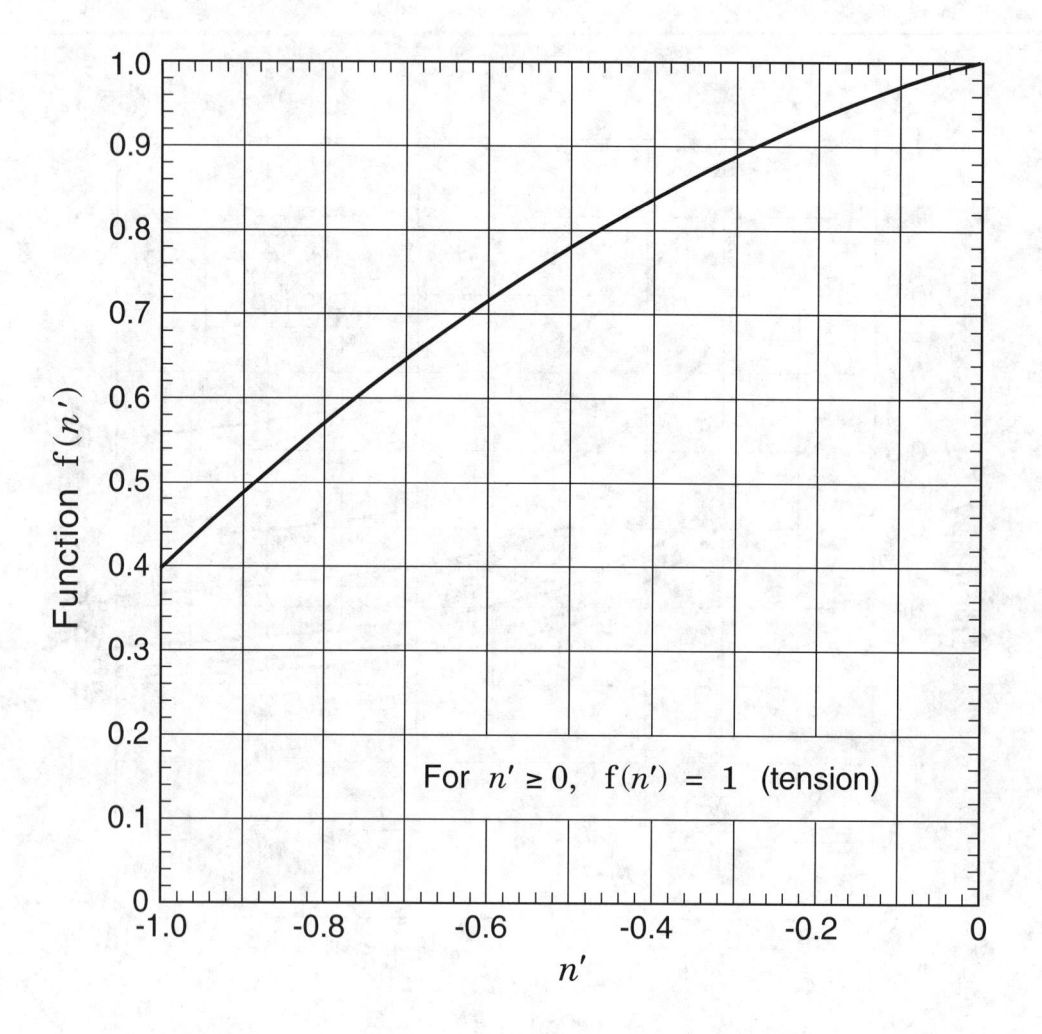

FIGURE 3.18

Function f(n') which describes the influence of chord "prestress" on the total efficiency of **circular HSS** connections

When β exceeds 0.85 the behaviour of the chord side walls is different for the three situations, and three charts become necessary. They show linear interpolations between known resistance values at β = 0.85 and at β = 1.0.

Figure 3.23 gives values of the efficiency coefficient C_K for gap K and N connections. Values of C_K need to be multiplied by the same three factors discussed for Figs. 3.12 to 3.17 in order to obtain the connection efficiency which can then be converted to connection resistance as mentioned previously. The third of these factors f(n) is provided in Fig. 3.24 as a function of β and n.

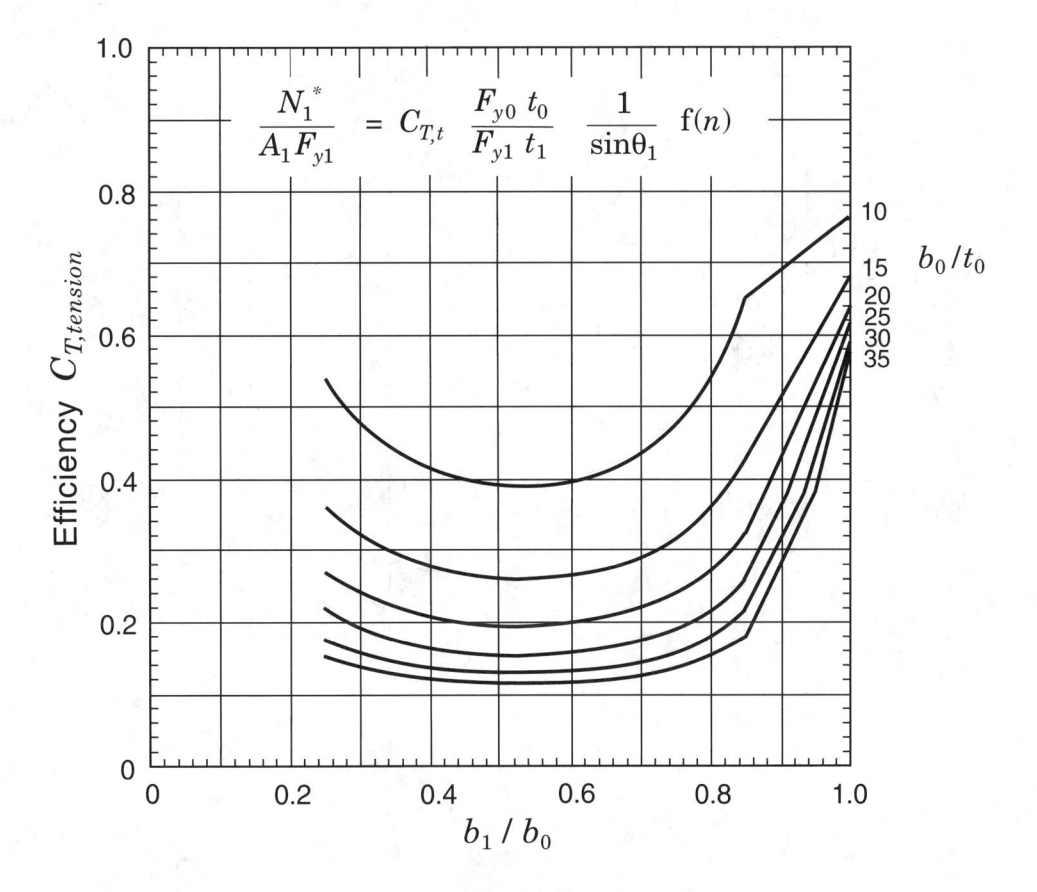

$$\frac{N_1^*}{A_1 F_{y1}} = C_{T,t} \frac{F_{y0} t_0}{F_{y1} t_1} \frac{1}{\sin\theta_1} f(n)$$

FIGURE 3.19

Branch member efficiency for **square HSS** T, Y and X connections with the branch member loaded in tension

The range of overlap for square K and N connections is from 50% to 100% rather than from 25% as in Table 3.2. This avoids the more complex lower range where resistance varies constantly with amount of overlap. In the 100% overlap chart, Fig. 3.26, the subscript *j* applies to the *overlapped* member while subscript *i* refers to the *overlapping* member. Unlike circular HSS overlapped connections, only the *overlapping* member need be checked for square HSS connections; however, there is a check on the connection efficiency for the *overlapped* web members, as noted in Table 3.2.

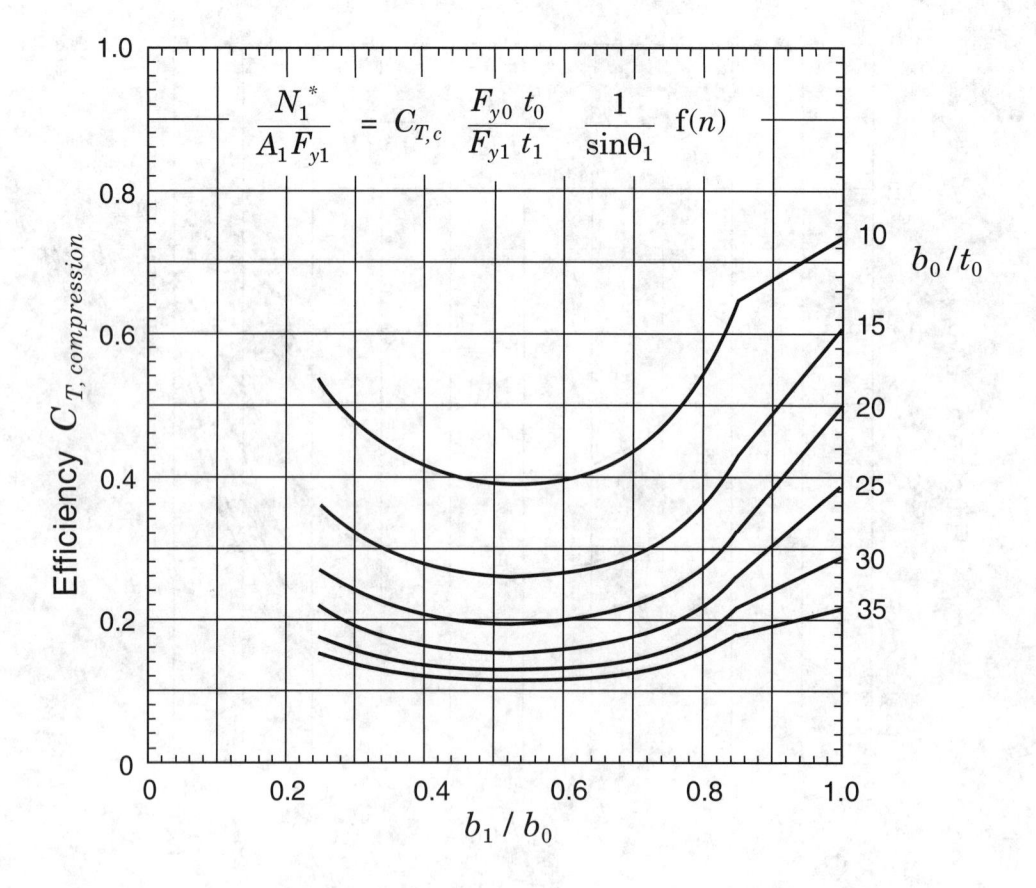

FIGURE 3.20

Branch member efficiency for **square HSS** T and Y connections with the branch member loaded in compression

Figure 3.25 for partially overlapped connections requires entering twice, once for the overlap on the chord (when 0 and i terms are used) and a second time for the overlap on the other web member (when j and i terms are used). The two part-efficiencies are then added together for the total efficiency. Validity ranges are the same as for Table 3.2 except that overlap starts at 50%. It can be seen from the charts that efficiencies of K and N overlap connections always exceed 0.8 for full overlap (Fig. 3.26) and 0.6 for partial overlap (Fig. 3.25). Therefore, fully overlapped connections are usually stronger than partially overlapped ones.

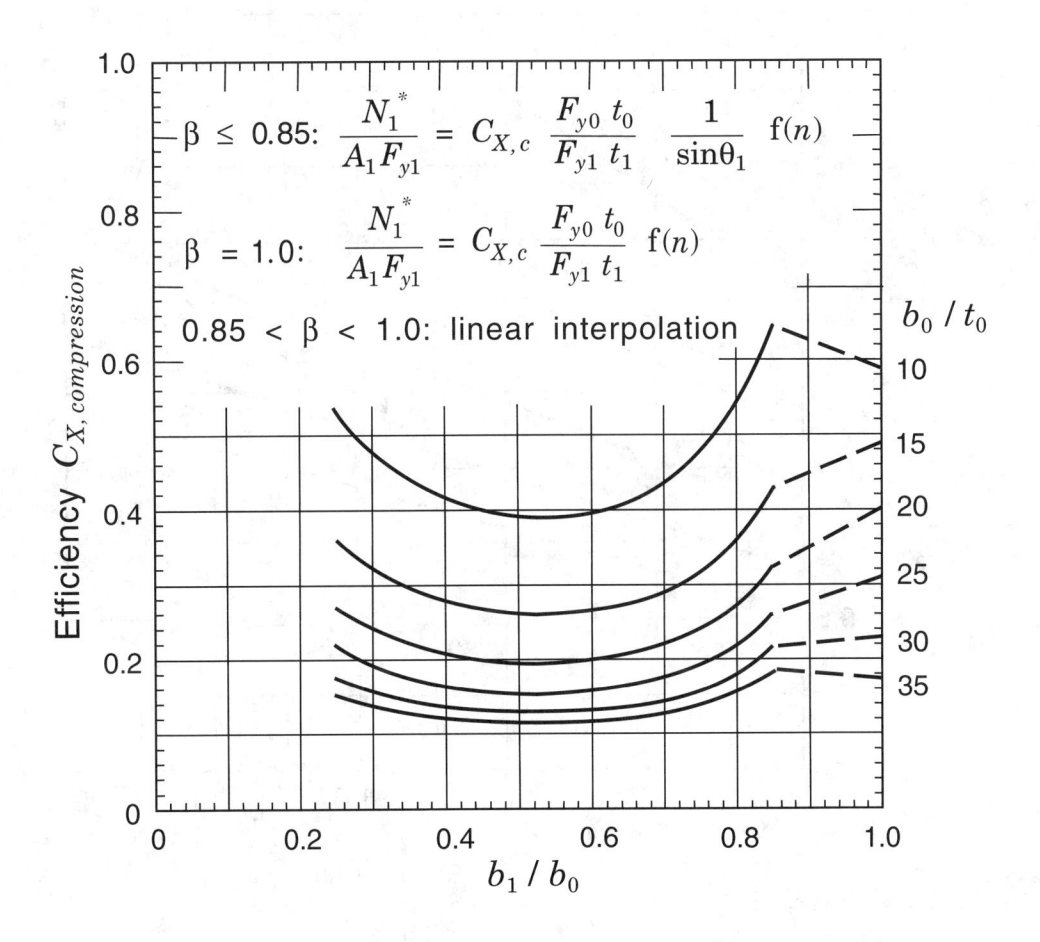

$\beta \leq 0.85$: $\dfrac{N_1^*}{A_1 F_{y1}} = C_{X,c} \dfrac{F_{y0} t_0}{F_{y1} t_1} \dfrac{1}{\sin\theta_1} f(n)$

$\beta = 1.0$: $\dfrac{N_1^*}{A_1 F_{y1}} = C_{X,c} \dfrac{F_{y0} t_0}{F_{y1} t_1} f(n)$

$0.85 < \beta < 1.0$: linear interpolation

FIGURE 3.21

Branch member efficiency for **square HSS** X connections with the branch member loaded in compression

3.7.2 Design Tables

Another alternative is to make use of "pre-engineered" connection tables. The factored resistances of some popular, standard, welded truss connections are given in Tables 3.5 to 3.10 (Packer *et al.* 1996). Three connection shapes are covered: 90° T connections, K gap connections and K 100% overlap connections, with the members subjected to predominantly axial loads. These three connection configurations have resistances tabulated for round-to-round members and square-to-square members. The steel grade assumed is 350W, with a minimum guaranteed yield strength of 350 MPa.

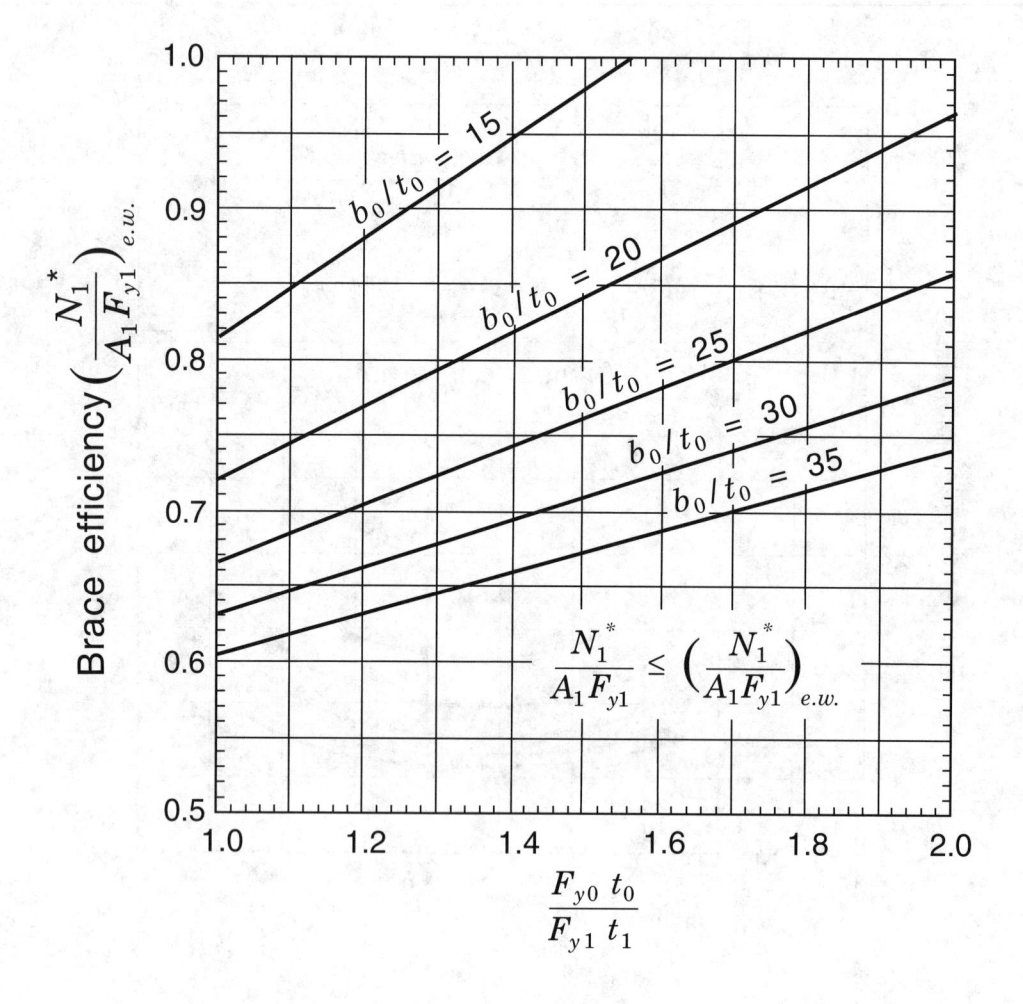

FIGURE 3.22

Total efficiency check for effective width, **square HSS** T, Y and X connections (for use when $\beta > 0.85$)

The K connections are for a specific web member angle (45°) and a particular gap size (g) or amount of overlap (O_v), whereas in practice a huge number of possible parameter combinations is possible. These tables, however, will enable the designer

(i) to get a very quick estimate of a connection factored resistance, even for a slightly different connection, and

(ii) to confirm that manual or computer-coded calculations for connection resistance formulae are being performed correctly.

Blank spaces in these tables indicate that either

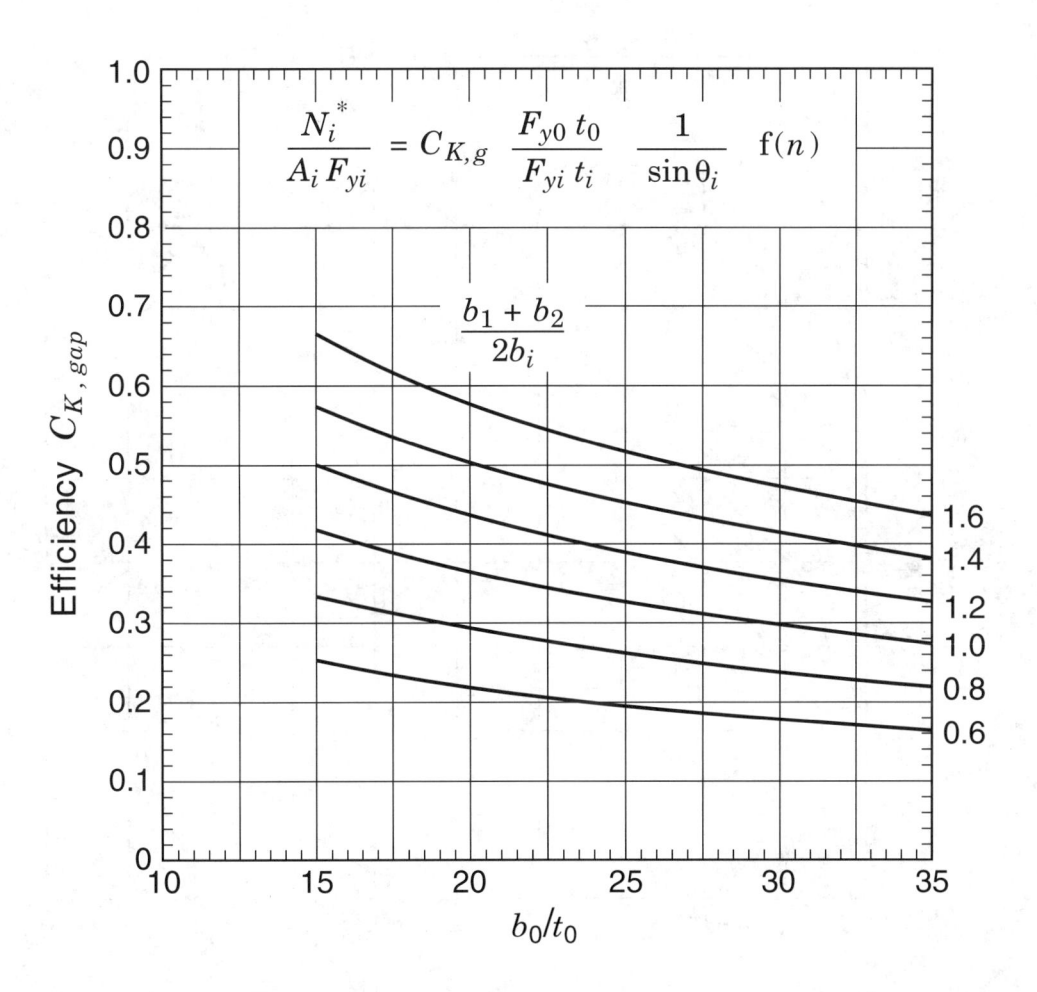

FIGURE 3.23
Web member efficiency for **square HSS** K and N gap connections

(i) a particular combination of members is outside the range of validity of the design formulae available, or

(ii) the connection is impractical (for example the web member width is greater than the chord width), or

(iii) the connection is not recommended (for example web member widths equal to the chord member width, for square and rectangular HSS connections).

Where such blank spaces arise the combination of members may still be possible, and recourse to the more definitive Tables 3.1, 3.2 and 3.3 in this Chapter is recommended. In some tables, for instance those for K gap connections, one should realize that the use of a particular parameter size

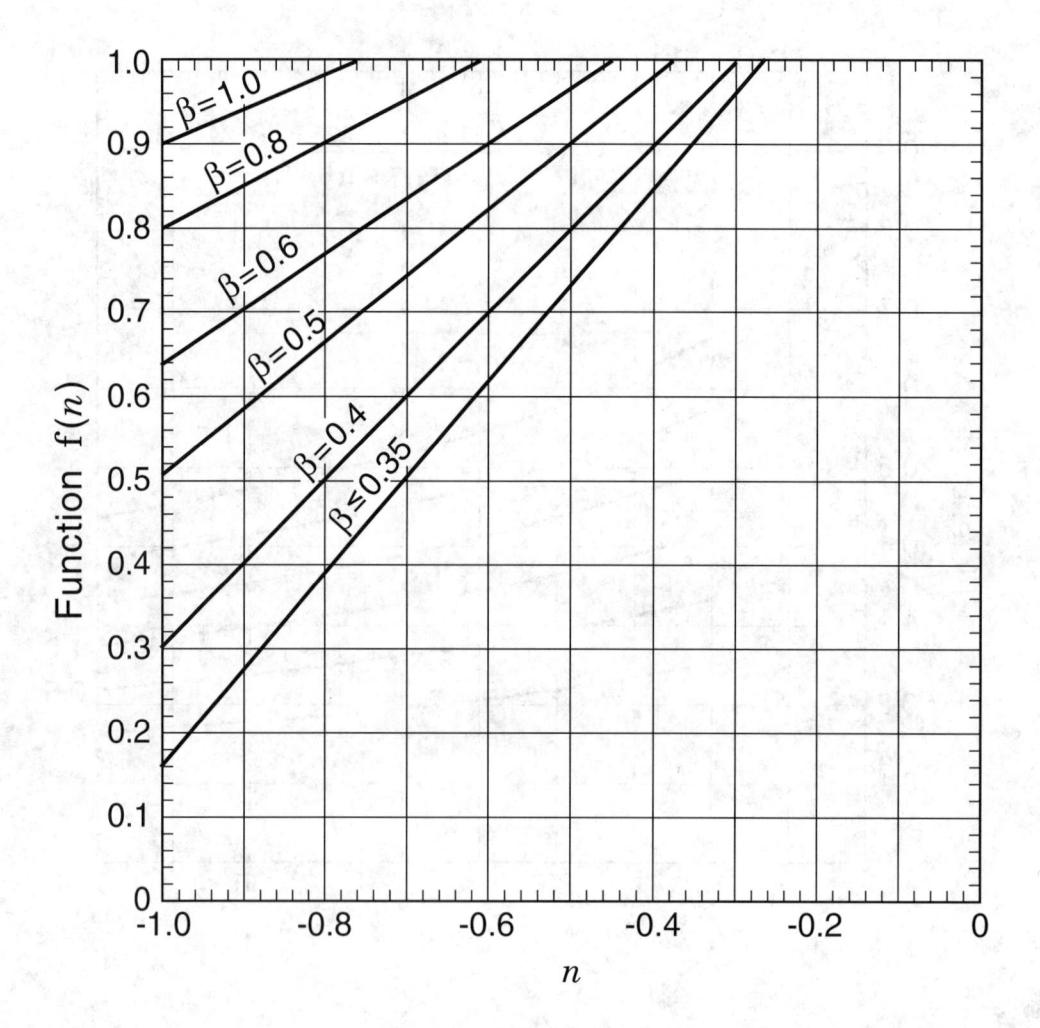

FIGURE 3.24

Function $f(n)$ which describes the influence of chord stress on the total efficiency of **square HSS** T, Y, X, K and N gap connections

(such as $g = 30$ mm) has severely restricted the number of possible connection combinations.

The connection factored resistances tabulated usually need to be reduced by a correction factor $f(n')$ or $f(n)$ if the chord member is loaded in compression.

For circular HSS: $\quad f(n') = 1 + 0.3\,n' - 0.3\,n'^2 \quad$ but ≤ 1.0

For square HSS: $\quad f(n) = 1.3 + \dfrac{0.4}{\beta}\,n \quad$ but ≤ 1.0

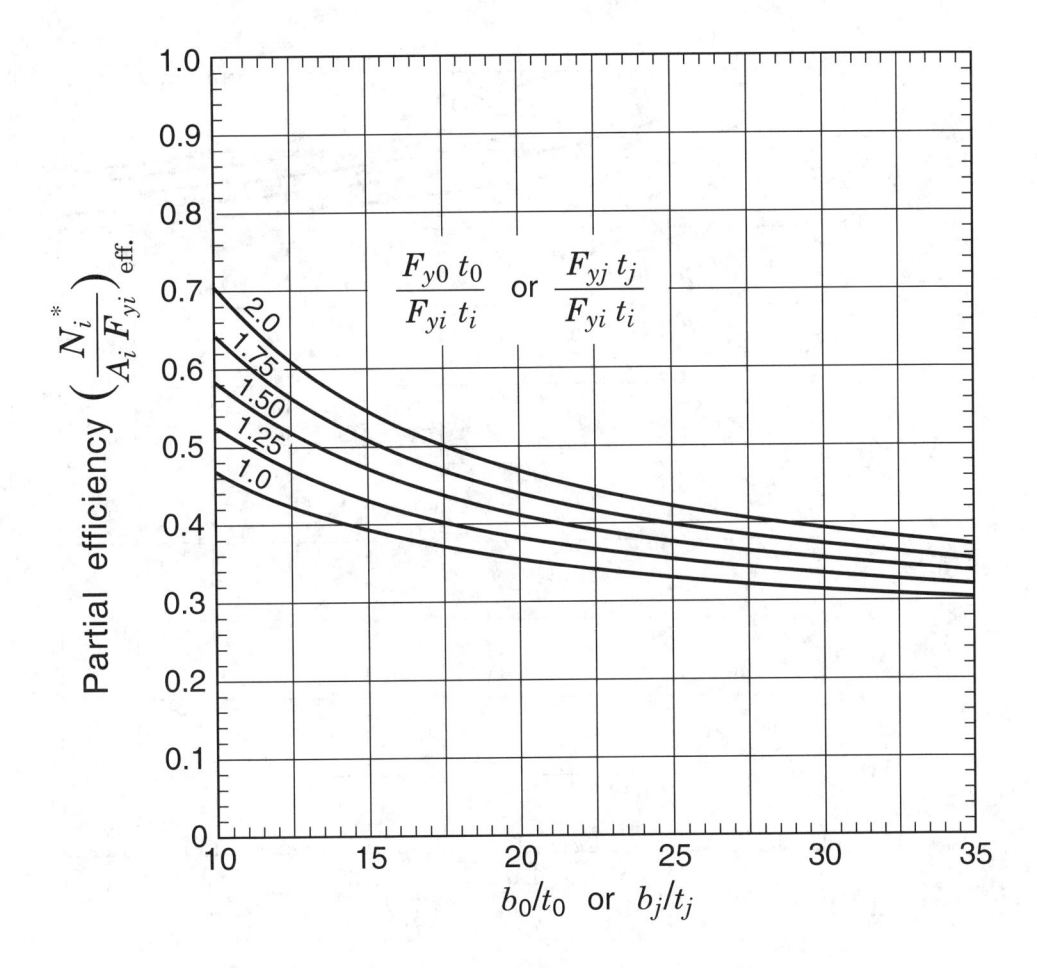

FIGURE 3.25

Web member efficiency for **square HSS** K and N overlap connections having $50\% \leq O_v < 80\%$

For axial compression load in the chord, n and n' will be negative numbers. The variable n is the axial force in the chord (the larger for either side of the connection) divided by the chord member squash load (area times yield strength), and n' is the additional axial force in a truss chord at a panel point, other than that required to maintain equilibrium with web member forces (or the "prestress force"), divided by the chord member squash load.

Another valuable design aid is the computer program CIDJOINT (Parik et al. 1994). This Windows program performs factored resistance checks for planar, welded and bolted, truss-type, statically-loaded connections made from circular, square and rectangular HSS members, and even has a member database of sections produced in accordance with CAN/CSA-G40.21-M92. (CSA 1992).

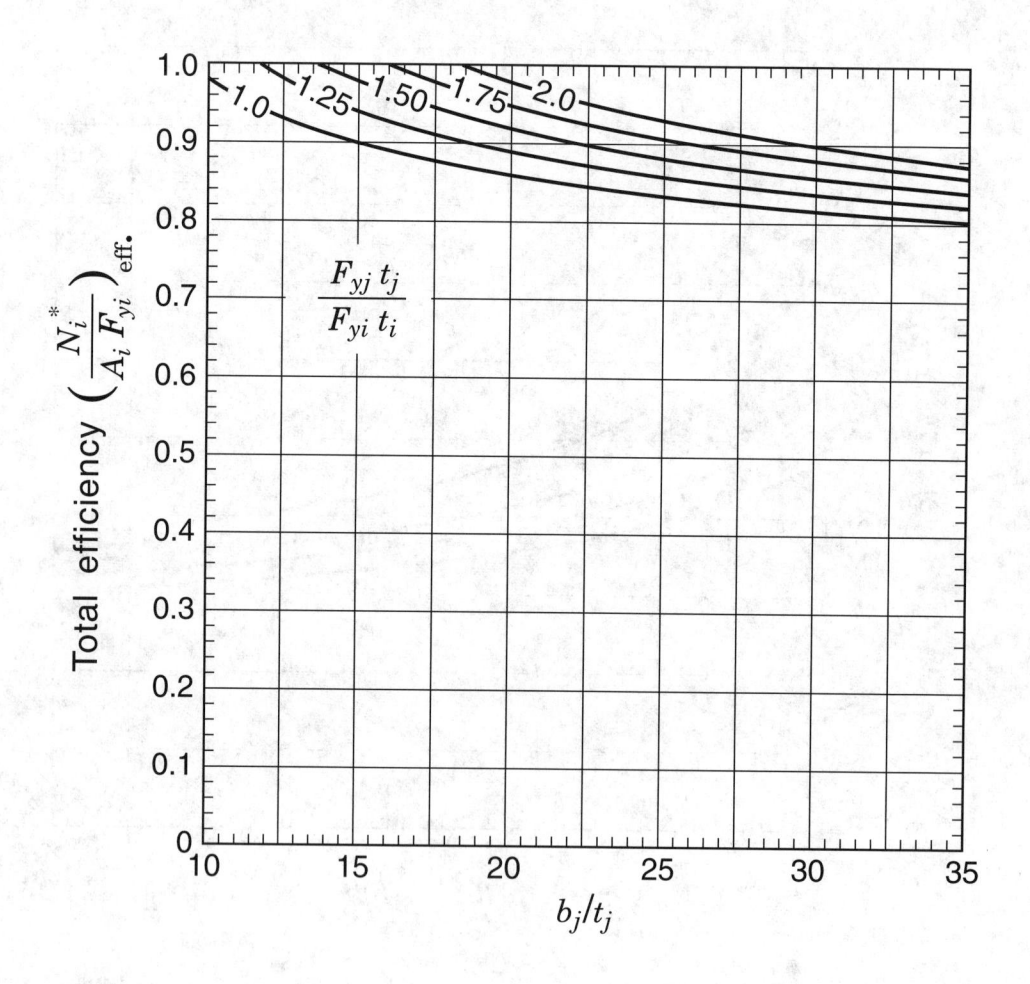

FIGURE 3.26

Web member efficiency for **square HSS** K and N overlap connections having $80\% \leq O_v \leq 100\%$

Chord		Factored Connection Resistances (N_1^*) in kN for Web Width (d_1 in mm) of:								
d_0 (mm)	t_0 (mm)	60	89	114	168	219	273	324	406	508	610
60	3.2	94								
60	3.8	131								
60	4.8	183								
60	6.4	243								
89	3.8	78	141							
89	4.8	117	212							
89	6.4	194	354							
89	8.0	291	449							
114	4.8	89	150	223						
114	6.4	148	250	372						
114	8.0	222	375	558						
168	4.8	66	96	132	241					
168	6.4	110	160	221	402					
168	8.0	164	240	331	603					
168	9.5	227	333	459	835					
219	4.8	58	77	99	167	254				
219	6.4	97	128	166	278	424				
219	8.0	145	192	248	417	635				
219	9.5	201	266	344	578	881				
219	11	264	350	453	760	1160				
273	6.4	91	112	138	213	311	443			
273	8.0	136	168	206	319	466	664			
273	9.5	189	233	286	443	646	920			
273	11	249	307	376	583	850	1210			
273	13	317	391	479	742	1080	1540			
324	8.0		156	184	268	375	521	687		
324	9.5		217	255	371	520	722	952		
324	11		285	336	488	685	950	1250		
324	13		363	428	622	872	1210	1600		
406	9.5		204	230	307	406	540	694	996	
406	11		268	302	403	535	711	913	1310	
406	13		342	385	514	681	906	1160	1670	
508	11			284	351	439	557	692	957	1370
508	13			361	447	559	709	881	1220	1750
610	13				413	494	602	726	969	1350	1810

CORRECTION FACTORS:
1. If there is compressive load in chord, multiply by reduction factor $f(n')$.
2. If the web member is in compression, the maximum permitted connection resistance, N_1^*, is limited to:

For d_1/t_1 of: 30   35   40   45   50
N_1^* max. (kN): $0.343 A_1$ $0.308 A_1$ $0.298 A_1$ $0.273 A_1$ $0.266 A_1$

where A_1 is the web member cross-sectional area in mm².

TABLE 3.5

Factored resistances for T connections between **circular HSS**
(F_y = 350 MPa)

Chord		Factored Connection Resistances (N_1^* or N_2^*) in kN for Web Width (d_1 in mm) of:								
d_0 (mm)	t_0 (mm)	60	89	114	168	219	273	324	406	508	610
60	3.2	128								
60	3.8	168								
60	4.8	239								
60	6.4	378								
89	3.8	149	205							
89	4.8	208	286							
89	6.4	320	441							
89	8.0	458	630							
114	4.8	196	266	327						
114	6.4	297	403	496						
114	8.0	419	568	699						
168	4.8	192	252	306	421					
168	6.4	280	369	447	615					
168	8.0	385	507	615	846					
168	9.5	503	663	804	1110					
219	4.8	198	255	305	413	515				
219	6.4	281	363	434	588	733				
219	8.0	379	489	586	793	988				
219	9.5	489	631	755	1020	1270				
219	11	609	787	942	1280	1590				
273	6.4	292	369	436	581	719	864			
273	8.0	387	489	578	771	953	1150			
273	9.5	491	621	735	980	1210	1460			
273	11	606	767	908	1210	1500	1800			
273	13	734	929	1100	1470	1810	2180			
324	8.0		498	583	767	941	1130	1300		
324	9.5		626	734	965	1180	1420	1630		
324	11		766	898	1180	1450	1730	2000		
324	13		921	1080	1420	1740	2080	2410		
406	9.5		649	751	969	1180	1390	1600	1930	
406	11		785	908	1170	1420	1690	1940	2340	
406	13		934	1080	1400	1700	2010	2300	2780	
508	11			945	1200	1430	1680	1920	2300	2770
508	13			1110	1410	1690	1980	2260	2710	3270
610	13				1450	1720	2000	2270	2700	3240	3780

CORRECTION FACTORS:
1. If there is compressive load in chord, multiply by reduction factor $f(n')$.
2. For the compression web member, the maximum permitted connection resistance, N_1^*, is limited to:

   For d_1/t_1 of: 30   35   40   45   50
   N_1^* max. (kN): $0.343 A_1$ $0.308 A_1$ $0.298 A_1$ $0.273 A_1$ $0.266 A_1$
   where A_1 is the web member cross-sectional area in mm².

NOTE: The thickness ratio between web members is limited as follows: $t_i/t_j \leq 1.0$, where "j" refers to the overlapped member.

TABLE 3.6

Factored resistances for K overlap connections between **circular HSS**
$O_v = 100\%$, and $\theta_1 = \theta_2 = 45°$
($F_y = 350$ MPa)

Chord		Factored Connection Resistances (N_1^* or N_2^*) in kN for Web Width (d_1 in mm) of:									
d_0 (mm)	t_0 (mm)	60	89	114	168	219	273	324	406	508	610
60	3.2										
60	3.8										
60	4.8										
60	6.4										
89	3.8										
89	4.8										
89	6.4										
89	8.0										
114	4.8	142									
114	6.4	242									
114	8.0	365									
168	4.8	121	160	193							
168	6.4	208	274	332							
168	8.0	314	414	502							
168	9.5	434	572	693							
219	4.8	112	144	172	233						
219	6.4	194	250	299	405						
219	8.0	293	378	452	612						
219	9.5	404	521	624	844						
219	11	526	679	812	1100						
273	6.4		236	279	372	460					
273	8.0		358	423	564	698					
273	9.5		493	583	777	961					
273	11		640	757	1010	1250					
273	13		804	951	1270	1570					
324	8.0		348	408	537	659					
324	9.5		479	562	739	906					
324	11		621	728	958	1180					
324	13		778	912	1200	1470					
406	9.5			547	706	856	1020	1170			
406	11			706	911	1110	1310	1510			
406	13			881	1140	1380	1640	1880			
508	11				890	1070	1250	1430	1710		
508	13				1110	1320	1560	1770	2130		
610	13					1310	1520	1720	2050	2460	

CORRECTION FACTORS:
1. If there is compressive load in chord, multiply by reduction factor $f(n')$
2. For the compression web member, the maximum permitted connection resistance, N_1^*, is limited to:
   For d_1/t_1 of: 30   35   40   45   50
   N_1^* max. (kN): $0.343 A_1$ $0.308 A_1$ $0.298 A_1$ $0.273 A_1$ $0.266 A_1$
   where A_1 is the web member cross-sectional area in mm².

TABLE 3.7

Factored resistances for K gap connections between **circular HSS**
$g = 30$ mm and $\theta_1 = \theta_2 = 45°$
($F_y = 350$ MPa)

Chord		Factored Connection Resistances (N_1^*) in kN for Web Width (b_1 in mm) of:									
b_0 (mm)	t_0 (mm)	51	64	76	89	102	127	152	203	254	305
51	3.2										
51	3.8										
51	4.8										
64	3.2	60									
64	3.8	86									
64	4.8	136									
64	6.4	239									
76	3.2	39	70								
76	3.8	56	101								
76	4.8	87	158								
76	6.4	154	279								
89	3.2	31	44								
89	3.8	45	63								
89	4.8	70	100								
89	6.4	124	176								
102	3.2	27	35	49							
102	3.8	39	50	70							
102	4.8	61	78	111							
102	6.4	108	138	196							
102	8.0	169	217	307							
102	9.5	243	312	441							
127	4.8	52	61	75	96	137					
127	6.4	92	108	132	169	242					
127	8.0	144	169	206	265	380					
127	9.5	207	243	296	380	546					
152	4.8	47	53	61	72	88	160				
152	6.4	83	94	108	127	156	283				
152	8.0	131	148	170	200	245	443				
152	9.5	188	212	244	287	351	636				
152	13	333	377	433	510	624	1130				
203	6.4	75	81	88	97	109	139	197			
203	8.0	117	127	139	152	170	219	308			
203	9.5	168	182	199	219	244	314	443			
203	13	298	324	354	389	434	558	787			
254	8.0		117	125	134	144	169	206	374		
254	9.5		168	179	192	207	243	295	537		
254	13		298	318	341	368	432	525	953		
305	9.5			168	177	188	212	243	346	628	
305	13			298	315	334	376	431	615	1120	

CORRECTION FACTOR: If there is compressive load in chord, multiply by reduction factor $f(n)$

NOTE: The width to thickness ratio for web members must be ≤ 35

TABLE 3.8

Factored resistances for T connections between **square HSS** (F_y = 350 MPa)

Webs		Factored Connection Resistance
b_1, b_2 (mm)	t_1, t_2 (mm)	(N_1^* or N_2^*) in kN
51	3.2	191
51	3.8	234
51	4.8	303
64	3.2	233
64	3.8	285
64	4.8	367
64	6.4	508
76	3.2	276
76	3.8	335
76	4.8	430
76	6.4	593
89	3.2	318
89	3.8	386
89	4.8	494
89	6.4	677
102	3.2	362
102	3.8	439
102	4.8	560
102	6.4	765
102	8.0	984
102	9.5	1210
127	4.8	685
127	6.4	931
127	8.0	1190
127	9.5	1460
152	4.8	811
152	6.4	1100
152	8.0	1400
152	9.5	1710
152	13	2370
203	6.4	1440
203	8.0	1830
203	9.5	2220
203	13	3050
254	8.0	2250
254	9.5	2730
254	13	3730
305	9.5	3240
305	13	4410

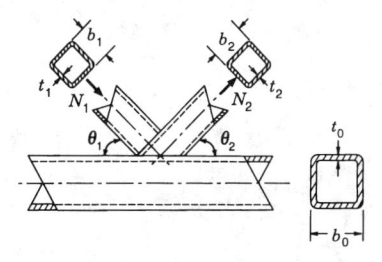

NOTES: (1) The width to thickness ratio for web members must be ≤ 35. Also, compression webs must be CSA-S16.1 Class 1 (plastic design) sections.

(2) The width to thickness ratio for the chord member must be ≤ 40.

(3) The width ratio between web members and chord must be ≥ 0.25.

TABLE 3.9

Factored resistances for K overlap connections between **square HSS**
$O_v = 100\%$ and $\theta_1 = \theta_2 = 45°$
($F_y = 350$ MPa)

Chord		Factored Connection Resistances (N_1^* or N_2^*) in kN for Web Width (b_1 in mm) of:									
b_0 (mm)	t_0 (mm)	51	64	76	89	102	127	152	203	254	305
51	3.2										
64	3.2										
64	3.8										
76	3.8	103									
76	3.8	135									
76	4.8	189									
89	3.2	95	119								
89	3.8	125	156								
89	4.8	175	219								
102	3.2	89	111	133							
102	3.8	117	146	174							
102	4.8	164	205	245							
102	6.4	251	313	375							
127	4.8			220	257	293					
127	6.4			336	394	449					
127	8.0			471	551	629					
152	4.8					268	335				
152	6.4					410	513				
152	8.0					575	719				
152	9.5					754	944				
203	6.4							533			
203	8.0							747			
203	9.5							981			
203	13							1510			
254	8.0								889		
254	9.5								1170		
254	13								1800		
305	9.5									1330	
305	13									2050	

CORRECTION FACTOR: If there is compressive load in chord, multiply by reduction factor $f(n)$
NOTE: The width to thickness ratio for web members must be ≤ 35

TABLE 3.10

Factored resistances for K gap connections between **square HSS**
$g = 30$ mm and $\theta_1 = \theta_2 = 45°$
($F_y = 350$ MPa)

3.8 Chord Splices

If chord splices are required, conventional practice is to use welding in the shops when fabricating assemblies for shipping. Field connections are usually high strength bolted, rather than welded, the former being easier to control in varying construction environments. For welded splices, fabricators generally make either full penetration welds onto backing bars tack welded against the back side of the joint, or partial penetration welds without the use of backing bars. The former can develop the full strength of structural sections, while the latter develop less strength, depending on the size of the deposited weld. Welding processes and techniques are presented in more detail in Chapter 8 and bolted splices are discussed in Chapter 7.

3.9 Design of Reinforced Connections

Instances may occur when a truss connection has inadequate resistance, and a designer must resort to some form of connection reinforcement. Such a situation can arise when HSS members are ordered on the basis of minimum mass, without connection resistance being confirmed. Labour costs associated with connection reinforcement are significant, and the resulting structure may lose its aesthetic appeal.

A detailed structural steelwork costing survey by Watson *et al.* (1996) has highlighted the higher number of labour hours, per connection, for cutting, handling, fitting and welding operations when connection plates are used rather than tube-to-tube direct connections. However, if only one or a few connections of a truss are inadequate, the reinforcement of these connections may be an acceptable solution.

3.9.1 With Stiffening Plates

The most common method of strengthening HSS connections is to weld a stiffening plate or plates to the HSS chord or "through" member. It is particularly applicable to gap K connections with rectangular chord members, although an unstiffened overlap connection is generally preferable from the viewpoints of economy and fatigue. However, a gap connection with a stiffening plate eliminates the necessity for double cuts on the web members, and may prove more attractive to the fabricator. The addition of a flat plate welded to the connecting face of the chord member greatly reduces local deformations of the connection, and consequently the overall truss deformation is lessened. It also permits a more uniform stress distribution in the web members.

3.9.1.1 K and N Connections with Rectangular (or Square) HSS Chords

The type of reinforcement required depends upon the governing failure mode that causes the inadequate connection capacity. Two types of plate reinforcement are shown in Fig. 3.27: (a) welded to the chord connecting face, and (b) welded to the chord side walls. Both would be applicable to connections with rectangular HSS chord members and either circular or rectangular HSS web members. An alternative to stiffening a connection with plates is to insert a length of chord with thicker walls at the connection, the length of which would be the same as L_p given below. (This is equivalent to the use of a "joint can" in offshore steel structures.)

The capacity of rectangular HSS gap K connections is typically controlled by the criteria for either chord face plastification or chord shear, as summarized in Tables 3.2 and 3.3. When chord face plastification controls, the connection capacity can be increased by using flange plate reinforcement as shown in Fig. 3.27(a). This will usually occur when $\beta < 1.0$ and the

(a) Flange plate reinforcement

(b) Side plate reinforcement

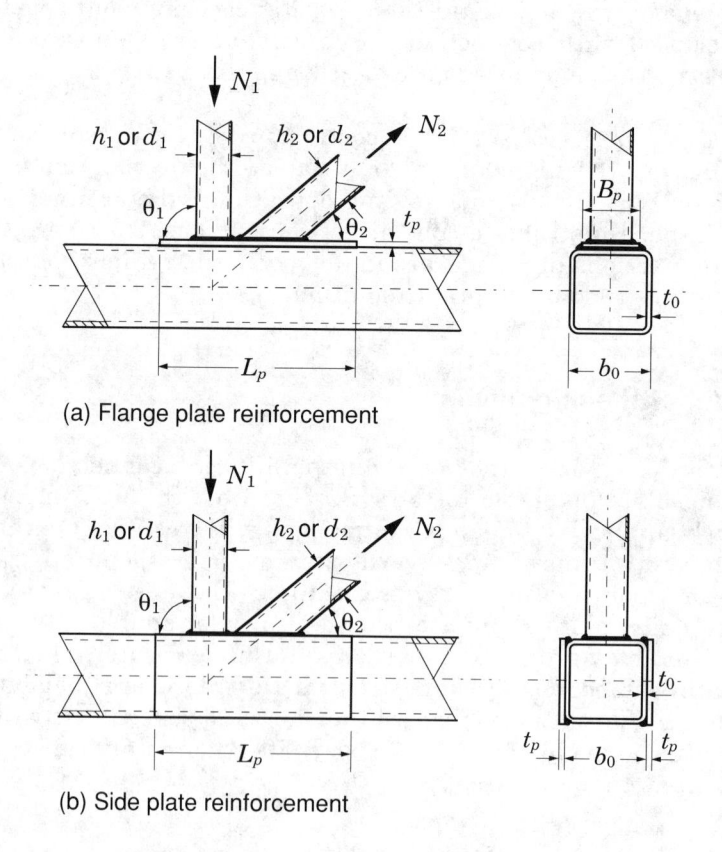

FIGURE 3.27
Pratt truss connections with plate stiffening

members are square. When chord shear controls, the connection capacity can be increased by reinforcing with a pair of side plates as shown in Fig. 3.27(b). This failure mode will usually govern when β = 1.0 and $h_0 < b_0$.

Design guidance for K connections *stiffened with a flange plate*, as shown in Fig. 3.27(a), was first given by Shinouda (1967). However, this method was based on an elastic deformation requirement of the connection plate under specified (service) loads. A more logical limit states approach recommended for calculating the necessary stiffening plate thickness for gap K connections is to use the connection resistance expressions in Table 3.2 (for square or circular web members to square chord members), and Table 3.3 (for rectangular members) by considering t_p as the chord face thickness and neglecting t_0 (Wardenier 1982). The plate yield stress should be used.

It is suggested that proportioning of the stiffening plate be based on the principle of developing the capacity of the web members ($A_i F_{yi}$). Dutta and Würker (1988) consider that this will be achieved providing $t_p \geq 2t_1$ and $2t_2$. Careful attention should be paid to the stiffening plate-to-chord welds which should have a weld throat size at least equal to the wall thickness of the adjacent web member (Dutta and Würker 1988). The stiffening plate should have a minimum length L_p, (see Fig. 3.27(a)), such that

$$L_p \geq 1.5 \left(\frac{h_1}{\sin\theta_1} + g + \frac{h_2}{\sin\theta_2} \right) \qquad [3.9]$$

A minimum gap between the web members, just sufficient to permit welding of the web members independently to the plate is suggested (Stelco 1981). All-round welding is generally required to connect the stiffening plate to the chord member, and to prevent corrosion on the two inner surfaces. It may also be advisable to drill a small hole in the stiffening plate under a web member to allow entrapped air to escape prior to closing the weld. This will prevent the expanding heated air from causing voids in the closing weld (Stelco 1981).

If the capacity of a gap K connection is inadequate and chord shear is the governing failure mode, then as mentioned before one should stiffen *with side plate reinforcement*, as shown in Fig. 3.27(b). A recommended procedure for calculating the necessary stiffening plate thickness is to use the chord shear resistance expression in Table 3.3, by calculating A_V as $2h_0 (t_0 + t_p)$. The stiffening plates should again have a minimum length L_p (see Fig. 3.27(b)), given by [3.9] and have the same depth as the chord member.

In order to avoid partial overlapping of one web member onto another in a K connection, fabricators may elect to weld each web member to a vertical stiffener as shown in Fig. 3.28(a). Another variation of this concept is to use the reinforcement shown in Fig. 3.28(b). For both of these connections, $t_p \geq 2t_1$ and $2t_2$ is recommended (Dutta and Würker 1988). The K connection shown in Fig. 3.28(c) is <u>not</u> acceptable, as it does not develop

(a) Acceptable

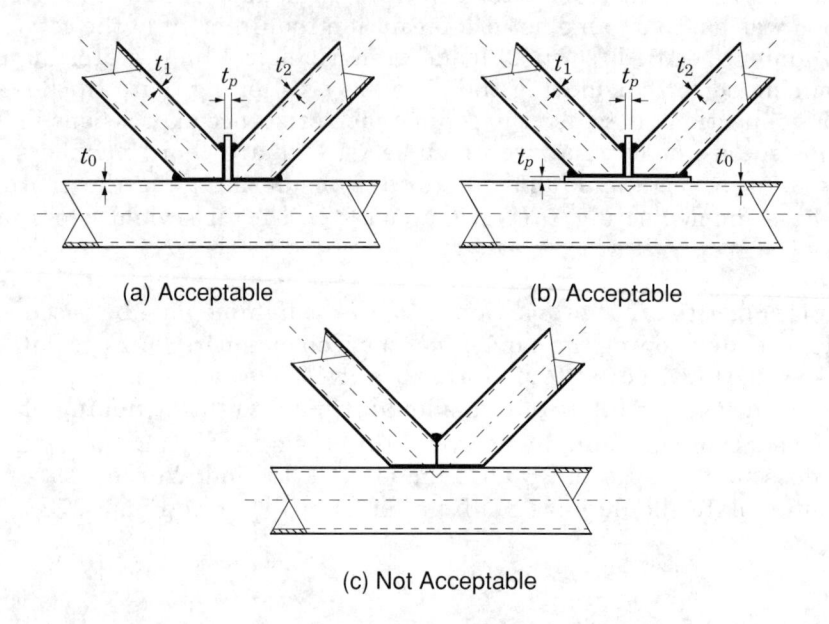

(b) Acceptable

(c) Not Acceptable

FIGURE 3.28

Some acceptable and unacceptable non-standard K connections

the strength of an overlapped K connection where one web member is welded only to the chord face. Also, it is difficult to create and ensure an effective saddle weld between the two web members.

3.9.1.2 K and KK Connections with Circular HSS Chords

For circular-to-circular HSS K connections, stiffening plates shaped to saddle the chord member have been used, particularly where cross-chord external loads are applied (Fig. 3.29), but no design guidelines exist. The authors tentatively suggest that the connection resistance expression for a K connection in Table 3.1 be used, by again considering t_p as the chord wall thickness and neglecting t_0. As before, a reasonable check would be to ensure that $t_p \geq 2t_1$ and $2t_2$.

There are benefits to using circular HSS members with relatively large diameter-to-thickness ratios in three dimensional truss systems, but this may produce relatively weak multiplanar (or KK) gap connections. It is

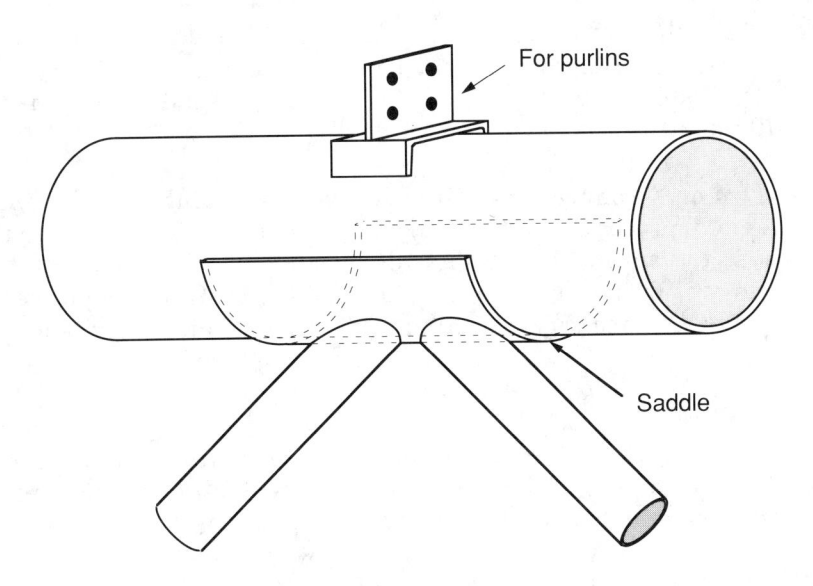

FIGURE 3.29
Circular HSS K connection with saddle reinforcement

extremely difficult to fabricate three dimensional circular HSS overlap KK connections, so a preferable way of increasing the strength of gap KK connections is to use diaphragm stiffeners. Makino *et al.* (1993) have performed tests on KK connections stiffened by

(a) cutting the chord through the longitudinal gap and welding each chord to this circular plate stiffener, and

(b) using the above stiffener plus two further internal ring stiffeners located beneath the "footprint" of a pair of braces.

The ultimate capacity of the stiffened KK connections with <u>one</u> diaphragm was found to be predictable by utilizing modified ultimate capacity equations for unstiffened KK connections. The only modification required was to reduce the longitudinal gap dimension g in the direction of the chord to the distance between the diaphragm and the compression brace. Presumably, as this design technique has been verified for multiplanar gap KK stiffened connections, this reduction in the longitudinal gap could also be applied to predicting the capacities of planar, gap, stiffened K connections.

3.9.1.3 T, Y and X Connections with Rectangular (or Square) HSS Chords

Under tension or compression web loading, the capacity of a rectangular HSS T, Y or X connection is typically controlled by either chord face yielding or chord side wall failure, as summarized in Tables 3.2 and 3.3. When chord face yielding controls, the connection capacity can be increased by using flange plate reinforcement similar to the connection shown in Fig.

3.27(a). This will usually occur when $\beta \leq 0.85$. When chord side wall failure controls, the connection capacity can be increased by reinforcing with a pair of side plates similar to the connection shown in Fig. 3.27(b). This failure mode will usually govern when $\beta \approx 1.0$.

For T, Y or X connections stiffened *with side plate reinforcement*, a recommended procedure for calculating the necessary stiffening plate thickness is to use the chord side wall resistance expression in Table 3.3, by replacing t_0 with $(t_0 + t_p)$ for the side walls. The stiffening plates should have a length L_p, (see Fig. 3.27(b)), such that for T and Y connections

$$L_p \geq 1.5 \, (h_1/\sin\theta_1) \qquad [3.10]$$

For T, Y and X connections stiffened *with a flange plate* there is a difference in behaviour of the stiffening plate depending on the sense of the load in the web member. With a tension load in the web member, the plate tends to lift off the chord member and behave as a plate clamped (welded) along its four edges. The strength of the connection depends on the plate geometry and properties, and not on the chord connecting face. Thus, for *tension web loading*, if one applies yield line theory to the plate-reinforced rectangular HSS T, Y or X connection, the connection factored resistance can be reasonably estimated by:

$$N_i^* = \frac{F_{yp} \, t_p^2}{(1 - \beta_p) \sin\theta_1} \left[\frac{2\eta_p}{\sin\theta_1} + 4 \, (1 - \beta_p)^{0.5} \right] \quad \text{for } \beta_p \leq 0.85 \qquad [3.11]$$

where

t_p = thickness of the stiffening plate
F_{yp} = yield strength of the stiffening plate
β_p = width ratio of branch member to plate = b_1/B_p
η_p = branch member depth to plate width ratio = h_1/B_p
B_p = width of plate.

The similarity between [3.11] and the factored resistance of *unstiffened* T, Y and X connections, based on chord face yielding as given by Table 3.3, is clearly evident. In order to develop the yield line pattern in the stiffening plate implicit in [3.11], the length of the plate L_p should be at least

$$L_p \geq \frac{h_1}{\sin\theta_1} + \sqrt{B_p \, (B_p - b_1)} \qquad [3.12]$$

Also, the plate width B_p should be such that a good transfer of loading to the side walls is achieved, i.e., B_p should be nearly equal to the flat width of the chord face (see Fig. 3.27(a)).

For T, Y and X connections stiffened *with a flange plate*, and under *compression web loading*, the plate and connecting chord face can be expected to act integrally with each other. This type of connection has been studied by Korol *et al.* (1982), also using yield line theory. Hence for $\beta_p \leq 0.85$, (a reasonable upper limit for application of yield line analysis also employed for unreinforced connections), the following plate design recommendations are made to obtain a full strength connection:

1. B_p = nearly the flat width of the chord face
2. $L_p \geq 2b_0$ (Increase proportionately if $\theta_1 \neq 90°$ to allow for greater web member "footprint".)
3. $t_p \geq 4t_1 - t_0$.

The application of these guidelines for compression-loaded T, Y and X connections should ensure that the connection capacity exceeds the web member capacity, provided chord side wall failure by web crippling is avoided (Korol *et al.* 1982).

3.9.2 With Concrete Filling

A less visible alternative to adding stiffening plates to the exterior of an HSS is for the fabricator to fill a length of the critical chord member (or members) with concrete or grout. With short span trusses, all of the chord member could be filled, but with long span trusses, which can be assembled with bolted flange plates in the chord to facilitate transportation and erection, only a part of the chord needs to be filled with concrete or grout, as shown in Fig. 3.30.

Filling the chord members of a truss has two main disadvantages: (1) the concrete increases the dead weight of the structure; and (2) it involves a secondary material with associated costs. On the other hand, the strength of *certain* connections is likely to increase, and if the members are completely filled there are further benefits of enhanced member capacity (due

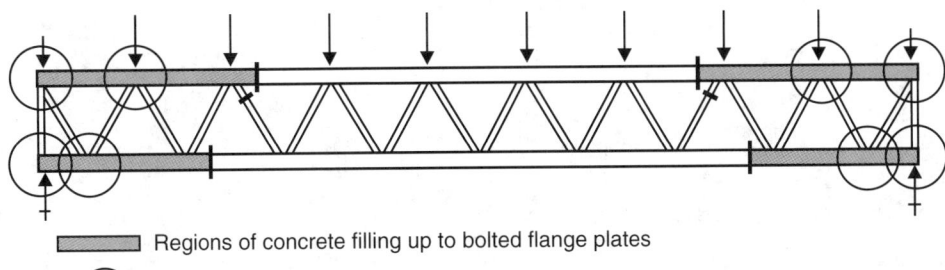

▬ Regions of concrete filling up to bolted flange plates

○ Critical connections that benefit from reinforcement by concrete filling

FIGURE 3.30

Partial filling of HSS chord members in critical connection regions to composite action), increased truss stiffness, and improved fire endurance.

Packer (1995) has performed experimental research on a variety of concrete-filled rectangular HSS connections, resulting in the design recommendations below. In particular, concrete filling of hollow sections greatly enhances their performance under transverse compression. The hollow section provides confinement for the concrete, which allows it to reach bearing capacities greater than its crushing strength as determined by cylinder compression tests (Packer and Fear 1991). Examples of connections under transverse compression, and hence that are likely to particularly benefit from concrete filling, include: truss reaction points, truss connections at which there is a significant external concentrated load, and beam-to-column moment connections, as illustrated in Fig. 3.31.

For the connection resistances recommended (Packer 1995), only the chord (or "through" member) must be filled with concrete or grout.

3.9.2.1 Compression-Loaded Rectangular (or Square) HSS X Connections

The resistance of this connection type is limited by the bearing strength of the concrete in confined compression and can be taken as:

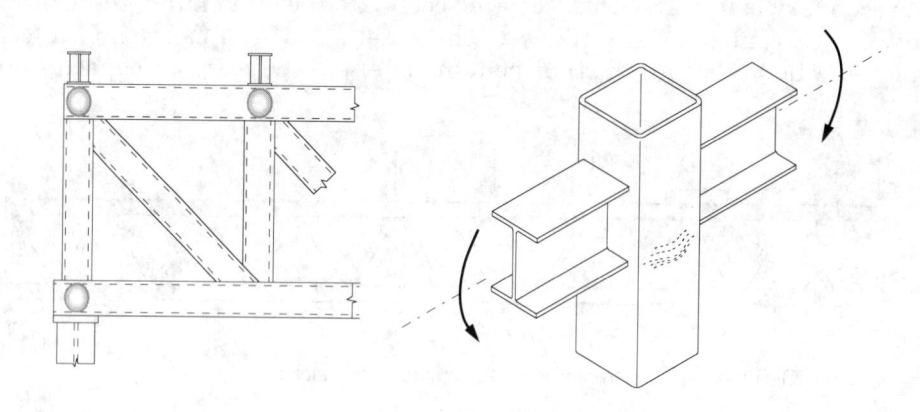

FIGURE 3.31
Instances where concrete filling may significantly improve connection resistance

$$N_1^* = \phi_c f_c' \frac{A_1}{\sin\theta_1} \sqrt{\frac{A_2}{A_1}}$$

[3.13]

where

ϕ_c = resistance factor for concrete in bearing = 0.6
f_c' = 28 day crushing strength of concrete by cylinder tests
A_1 = bearing area over which the transverse load is applied
A_2 = dispersed bearing area.

Here, A_2 should be determined by dispersion of the bearing load at a slope of 2:1 longitudinally along the chord member, as shown in Fig. 3.32. The value of A_2 may be limited by the length of concrete, and $\sqrt{A_2/A_1}$ cannot be taken greater than 3.3. The following are also recommended for general design applications:

$$\frac{h_0}{b_0} \leq 1.4 \quad \text{(limit of experimental verification)}$$

$$L_c \geq \frac{h_1}{\sin\theta_1} + 2h_0$$

where L_c = length of concrete in HSS chord member.

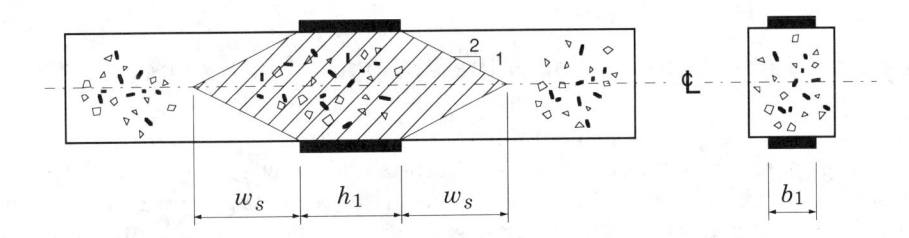

$A_1 = h_1 b_1$
$A_2 = (h_1 + 2w_s) b_1$

FIGURE 3.32

Applied load area (A_1) and dispersed load area (A_2), for a concrete-filled rectangular HSS in transverse compression

3.9.2.2 Compression-Loaded Rectangular (or Square) HSS T and Y Connections

Since the load in this case is being resisted by shear forces in the chord rather than being transferred through the chord, the dispersed bearing area A₂ should in this case be calculated assuming a stress distribution longitudinally at a slope of 2:1 through the <u>entire</u> depth of the chord, rather than to an (A₂ /A₁) limit. Thus, with respect to Fig. 3.32, the dispersed bearing area (A₂) would be adjusted (for an inclined branch) to the following:

$$A_2 = \left(\frac{h_1}{\sin\theta_1} + 4\,h_0\right) b_1$$

Similarly the limit of validity for L_c would need to be adjusted to the following:

$$L_c \geq \frac{h_1}{\sin\theta_1} + 4\,h_0$$

The resistance for these connections can then be calculated using [3.13].

3.9.2.3 Tension-Loaded Rectangular (or Square) HSS T, Y and X Connections

In tests, none of the concrete-filled connections exhibited a decrease in connection yield or ultimate strength, relative to their unfilled counterparts, by more than a few percent. The concrete-filled connections still had large connection deformations, so their design should also be based on the connection yield load. Thus it is recommended that the design capacity of these connections be calculated using existing design rules for unfilled rectangular HSS connections (see Tables 3.2 and 3.3).

3.9.2.4 Gap K Connections with Rectangular (or Square) HSS

For the range of connection parameters studied experimentally (Packer 1995), gap K connections with concrete-filled chords were found to have superior connection yield strengths and ultimate strengths relative to their unfilled counterparts. Also, concrete filling of such connections has been found to produce a significant change in connection failure mode. This is shown in Fig. 3.33. It was also found that the strength of a gap K connection with concrete in the chord is unaffected by a moderate amount of shrinkage of the concrete away from the chord walls, thereby making this filling procedure an option for any fabricator.

It is recommended that the connection resistance be calculated separately for the compression web member and the tension web member. For the compression web member, which presses on a relatively rigid founda-

FIGURE 3.33
Failure modes for gap K connections with the chord unfilled (K3) and filled (K3C) with concrete

tion of concrete, the connection strength would appear to be limited by bearing failure of the concrete. Hence, calculations should be performed as for a compression-loaded Y connection (see Section 3.9.2.2). For the tension web member, the concrete filling only permits two possible failure modes, which must be checked: (i) "effective width" or premature failure of one of the web members, and (ii) "punching shear" of the chord face around one of the web members. These are a subset of the four possible failure modes experienced with unfilled gap K connections (Table 3.3), as presented in Table 3.11.

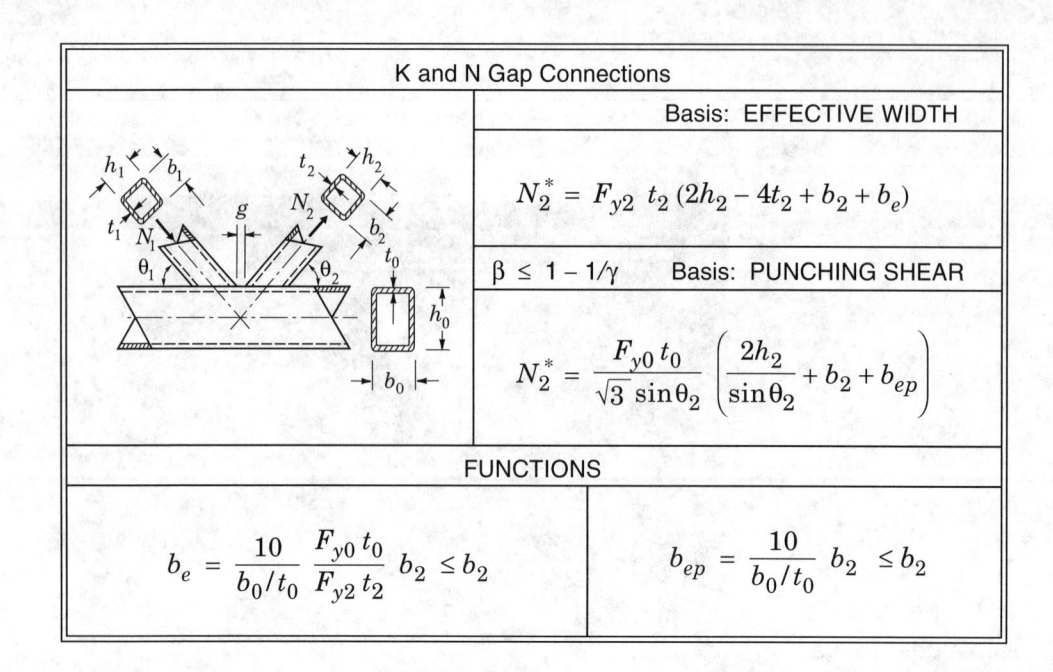

K and N Gap Connections	
	Basis: EFFECTIVE WIDTH
	$N_2^* = F_{y2}\, t_2\, (2h_2 - 4t_2 + b_2 + b_e)$
	$\beta \leq 1 - 1/\gamma$ Basis: PUNCHING SHEAR
	$N_2^* = \dfrac{F_{y0}\, t_0}{\sqrt{3}\, \sin\theta_2} \left(\dfrac{2h_2}{\sin\theta_2} + b_2 + b_{ep} \right)$
FUNCTIONS	
$b_e = \dfrac{10}{b_0/t_0} \dfrac{F_{y0}\, t_0}{F_{y2}\, t_2}\, b_2 \leq b_2$	$b_{ep} = \dfrac{10}{b_0/t_0}\, b_2 \leq b_2$

TABLE 3.11
Failure mode checks for the tension web member in concrete-filled gap
K connections between rectangular HSS (a subset of Table 3.3)

3.9.2.5 Design Example

Check that the square HSS gap K connection shown in Fig. 3.34 is adequate under the forces shown. The bending moments shown are primary moments produced by transverse loads on the chord between panel points, and these do have an effect on connection resistance. The primary moments produced in the chord member due to noding eccentricity are not shown as they are already considered in the connection resistance expressions (provided the noding eccentricity falls within specified limits of validity).

Select a gap g of 25 mm.

The square HSS members chosen (in CAN/CSA-G40.21, Grade 350W) are:

    Chord member (member 0):   HSS 152x152x6.4

    Compression web member (member 1):   HSS 127x127x4.8   (Class 2)

    Tension web member (member 2):   HSS 102x102x4.8.

$$\beta = \dfrac{b_1 + b_2}{2b_0}, \text{ since the members are square}$$

$$= (127 + 102)\, /\, 2\,(152) = 0.753$$

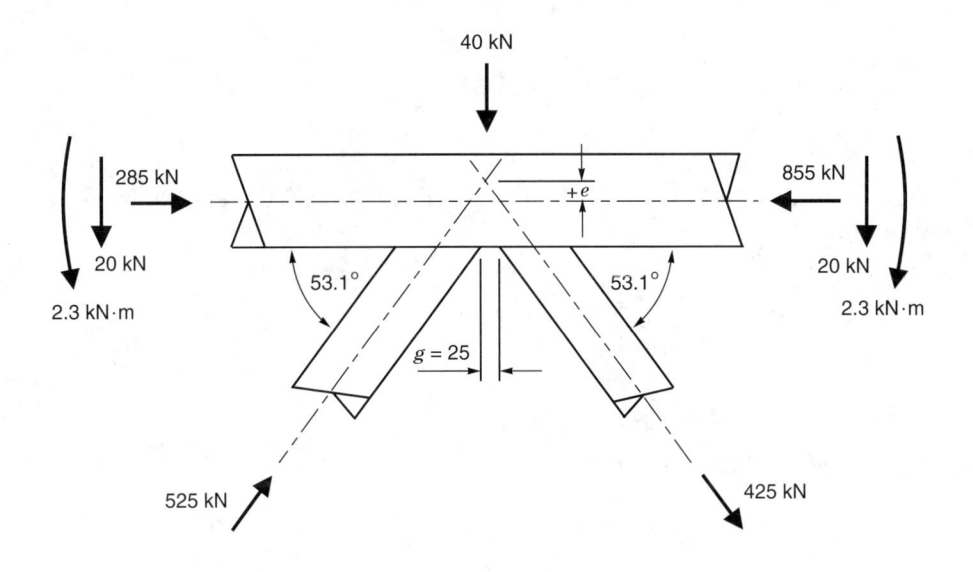

FIGURE 3.34

Forces acting on a square HSS gap K connection for the design example

Check as an Unreinforced Connection

First check that the connection parameters fall within the acceptable range of validity, as given by Table 3.2(a). It will be found that all parameters are OK. Calculate the noding eccentricity e, as shown in Section 3.1, which is +36.0 mm.

Maximum allowed positive e is $0.25\, h_0 = 38.0$ mm $\quad > 36.0 \quad \therefore$ OK

According to Table 3.2, for square HSS connections with $\beta \leq 1.0$, which we have in this case, the connection only needs to be checked for the failure mode of "Chord Face Plastification". First, calculate the reduction factor $f(n)$.

$$n = \frac{N_0}{A_0\, F_{y0}} + \frac{M_0}{S_0\, F_{y0}}$$

$$= -855/(3\,610\,(0.350)) - 2.3/(166\,(0.350))$$

$$= -0.716$$

Note: The moment of 2.3 kN·m was taken as negative as it causes compression on the connecting face of the chord.

$$f(n) = 1.3 + \left(\frac{0.4}{\beta}\right)n$$

$$= 1.3 + (0.4/0.753)(-0.716)$$

$$= 0.920$$

Now, since $\theta_1 = \theta_2 = 53.1°$, the connection factored resistance from Table 3.2 is

$$N_1^*, N_2^* = 8.9 \frac{F_{y0} t_0^2}{\sin\theta_{1,2}} \frac{b_1 + b_2}{2b_0} \gamma^{0.5} f(n)$$

$$= 8.9 \frac{0.350(6.35)^2}{0.8} \frac{127 + 102}{2(152)} \left(\frac{152}{2(6.35)}\right)^{0.5} 0.920$$

$$= 376 \text{ kN} \quad < 525 \text{ and } 425 \quad \therefore \text{ no good}$$

The connection resistance of both web members is inadequate. It is interesting to note that the next chord size up (HSS 152x152x8.0 with a 23% greater mass) would prove adequate.

Check as a Concrete-filled Connection

Now the chord member only will be filled with a 40 MPa concrete or grout.

Check connection resistance of compression web member:

$$N_1^* = \phi_c f_c' \frac{A_1}{\sin\theta_1} \sqrt{\frac{A_2}{A_1}} \qquad ([3.13])$$

$$A_1 = \frac{127}{\sin 53.1°} \, 127 = 20\,170 \text{ mm}^2$$

$$A_2 = \left(\frac{127}{\sin 53.1°} + 4(152)\right) 127 = 97\,390 \text{ mm}^2$$

$$\therefore \left(\frac{A_2}{A_1}\right)^{0.5} = 2.20 \leq 3.3$$

$$\therefore N_1^* = 0.6(0.040)(20\,170/0.80)\,2.20$$

$$= 1\,330 \text{ kN} \quad \gg 525 \quad \therefore \text{ OK}$$

Check connection resistance of tension web member:

For the "effective width" failure mode from Table 3.11,

$$N_2^* = F_{y2} t_2 (2h_2 - 4t_2 + b_2 + b_e)$$

$$b_e = \frac{10}{b_0/t_0} \frac{F_{y0} t_0}{F_{y2} t_2} b_2$$

$$= (10/(152/6.35))(6.35/4.78)\, 102 = 56.6 \text{ mm}$$

$$\therefore N_2^* = 0.350\,(4.78)\,(204 - 19.1 + 102 + 56.6)$$

$$= 575 \text{ kN} \geq 425 \quad \therefore \text{ OK}$$

For the chord "punching shear" failure mode from Table 3.11,

Confirm applicability

$$\beta = 0.75 \leq 1 - \frac{1}{\gamma} = 1 - \frac{1}{152/(2\,(6.35))} = 0.92 \quad \therefore \text{ applicable}$$

$$N_2^* = \frac{F_{y0} t_0}{\sqrt{3} \sin\theta_2} \left(\frac{2h_2}{\sin\theta_2} + b_2 + b_{ep} \right)$$

$$b_{ep} = \frac{10}{b_0/t_0} b_2$$

$$= (10/(152/6.35))\, 102 = 42.6 \text{ mm}$$

$$\therefore N_2^* = \frac{0.350\,(6.35)}{1.732\,(0.8)} \left(\frac{2\,(102)}{0.8} + 102 + 42.6 \right)$$

$$= 641 \text{ kN} > 425 \quad \therefore \text{ OK}$$

Therefore, the connection is now acceptable once the chord is filled with concrete, and a much thicker chord member has been avoided.

3.10 Cranked Chord Connections

Although not a reinforced truss connection, "cranked-chord" K connections arise in certain Pratt or Warren trusses such as the one shown in Fig. 3.35, and are characterized by a change in direction of the chord member at the connection noding point. The crank is achieved by cutting and groove welding two common sections together at the appropriate angle. Intersection of the three member centrelines is usually made coincident. The uniqueness of this cranked-chord K connection lies both in its lack of a

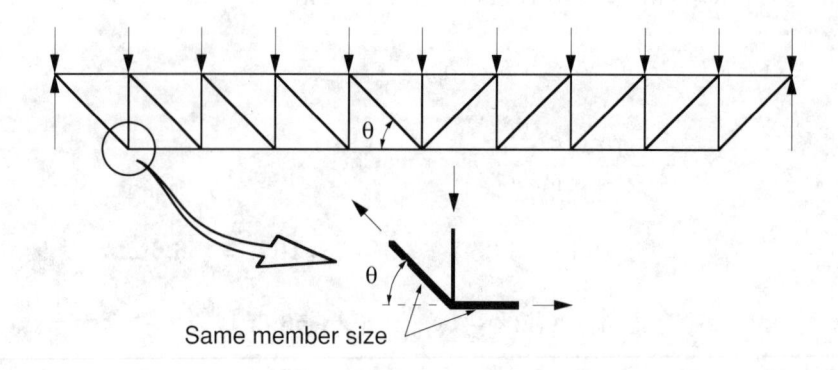

FIGURE 3.35
Example of a cranked-chord K connection in a Pratt truss

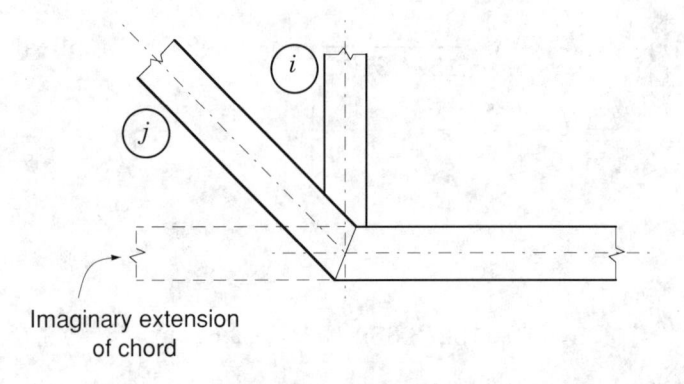

FIGURE 3.36
Cranked-chord connection represented as an overlapped K or N connection

straight chord member and the role of the chord member as an "equal width branch member".

An experimental research program with square and rectangular members has revealed that unstiffened, welded, cranked-chord HSS connections behave generally in a manner *dissimilar* to HSS T or Y connections, despite their *similar* appearance. (They all have a single branch member welded to a uniform-size chord member). Rather, cranked-chord HSS connections have been shown to behave as *overlapped* K or N connections, and their capacity can be predicted using the recommendations given in Table 3.2, subject to the limits of application in Table 3.2(a) (Packer 1991).

Figure 3.36 illustrates a cranked-chord connection interpreted as an overlapped K connection, wherein the chord member is given an imaginary extension and the cranked-chord member is considered to be the *overlapped* web member.

REFERENCES

CSA. 1992. Structural quality steels, CAN/CSA-G40.21-M92. Canadian Standards Association, Rexdale, Ontario.

CSA. 1994. Limit states design of steel structures, CAN/CSA-S16.1-94. Canadian Standards Association, Rexdale, Ontario.

DUTTA, D., and WÜRKER, K. 1988. Handbuch hohlprofile in stahlkonstruktionen. Verlag TÜV Rheinland GmbH, Köln, Federal Republic of Germany.

EUROPEAN COMMITTEE for STANDARDIZATION. 1992. Eurocode 3: design of steel structures. Part 1.1—general rules and rules for buildings. ENV 1993-1-1:1992E, British Standards Institution, London, England.

IIW. 1989. Design recommendations for hollow section joints—predominantly statically loaded, 2nd. ed. International Institute of Welding Subcommission XV-E, IIW Doc. XV-701-89, IIW Annual Assembly, Helsinki, Finland.

KOROL, R.M., MITRI, H., and MIRZA, F.A. 1982. Plate reinforced square hollow section T-joints of unequal width. Canadian Journal of Civil Engineering, **9**(2): 143–148.

KUROBANE, Y. 1981. New developments and practices in tubular joint design. IIW Doc. XV-488-81 + Addendum, IIW Annual Assembly, Oporto, Portugal.

KUROBANE, Y., MAKINO, Y., and MITSUI, Y. 1980. Re-analysis of ultimate strength data for truss connections in circular hollow sections. IIW Doc. XV-461-80, Kumamoto University, Japan.

MAKINO, Y., KUROBANE, Y., and PAUL, J.C. 1993. Ultimate behaviour of diaphragm-stiffened tubular KK-joints. Proceedings, 5th. International Symposium on Tubular Structures, Nottingham, England, pp. 465–472.

PACKER, J.A. 1986. Design examples for HSS trusses. Canadian Journal of Civil Engineering, **13**(4): 460–473.

PACKER, J.A. 1991. Cranked-chord HSS connections. Journal of Structural Engineering, American Society of Civil Engineers, **117**(8): 2224–2240.

PACKER, J.A. 1995. Concrete-filled HSS connections. Journal of Structural Engineering, American Society of Civil Engineers, **121**(3): 458–467.

PACKER, J.A., and FEAR, C.E. 1991. Concrete-filled rectangular hollow section X and T connections. Proceedings, 4th. International Symposium on Tubular Structures, Delft, The Netherlands, pp. 382–391.

PACKER, J.A., FRATER, G.S., and KITIPORNCHAI, S. 1996. Resistance tables for welded hollow structural section truss connections. Proceedings, International Conference on Tubular Structures, Vancouver, B.C., pp. 32–47.

PACKER, J.A., WARDENIER, J., KUROBANE, Y., DUTTA, D., and YEOMANS, N. 1992. Design guide for rectangular hollow section (RHS) joints under predominantly static loading. CIDECT (ed.) and Verlag TÜV Rheinland GmbH, Köln, Federal Republic of Germany.

PARIK, J., DUTTA, D., and YEOMANS, N. 1994. User guide for PC-program CIDJOINT for hollow section joints under predominantly static loading. CIDECT (ed.) and Ing.-Software Dlubal GmbH, Tiefenbach, Federal Republic of Germany.

REUSINK, J.H., and WARDENIER, J. 1989a. Simplified design charts for axially loaded joints of circular hollow sections. Proceedings, 3rd.International Symposium on Tubular Structures, Lappeenranta, Finland, pp. 154–161.

REUSINK, J.H., and WARDENIER, J. 1989b. Simplified design charts for axially loaded joints of square hollow sections. Proceedings, 3rd. International Symposium on Tubular Structures, Lappeenranta, Finland, pp. 54–61.

SHINOUDA, M.R. 1967. Stiffened tubular joints. Ph.D. Thesis, University of Sheffield, England.

STELCO. 1981. Hollow structural sections—design manual for connections, 2nd. ed. Stelco Inc., Hamilton, Ontario.

WARDENIER, J. 1982. Hollow section joints. Delft University Press, Delft, The Netherlands.

WARDENIER, J., and STARK, J. W. B. 1978. The static strength of welded lattice girder joints in structural hollow sections: Parts 1 to 10. CIDECT Final Report 5Q/78/4, Delft University of Technology, The Netherlands.

WARDENIER, J., KUROBANE, Y., PACKER, J.A., DUTTA, D., and YEOMANS, N. 1991. Design guide for circular hollow section (CHS) joints under predominantly static loading. CIDECT (ed.) and Verlag TÜV Rheinland GmbH, Köln, Federal Republic of Germany.

WATSON, K.B., DALLAS, S., van der KREEK, N., and MAIN, T. 1996. Costing of steelwork from feasibility through to completion. Steel Construction, Australian Institute of Steel Construction, **30**(2): 2–47.

4

NON-STANDARD TRUSS DESIGN

4.1 Trusses with Double HSS Chords

Chapter 1 included a discussion of the aesthetic and economic advantages available to the designer who builds with Hollow Structural Sections. These advantages include the greater visual appeal of clear uncluttered structures, the structural efficiencies and increased torsional stiffness of closed compression members, and the lower finishing costs resulting from reduced surface areas. Unfortunately, since the largest available section in Canada has been an HSS 305x305x13, spans of HSS roof trusses have been restricted to about 30 or 40 metres. The need for buildings such as auditoria and sports complexes with clear spans of 60 metres or more has generated interest in HSS trusses which employ two sections for each chord.

Immediate advantages of double chord HSS trusses include not only their greater capacity, but also more efficient and stiffer connections compared to some single chord trusses. Enhanced lateral stiffness can reduce bracing requirements and also facilitate handling and erection of the trusses.

4.1.1 Types of Double Chord Trusses

A number of arrangements are possible for framing webs with double chords. Principal configurations are illustrated in Fig. 4.1 where (a) and (b) show connections for welded and bolted versions of trusses with double chords separated so that web members fit between them. This configuration requires that all web members have the same width, but variation in size is possible by changing the wall thickness or the member depth. A separated

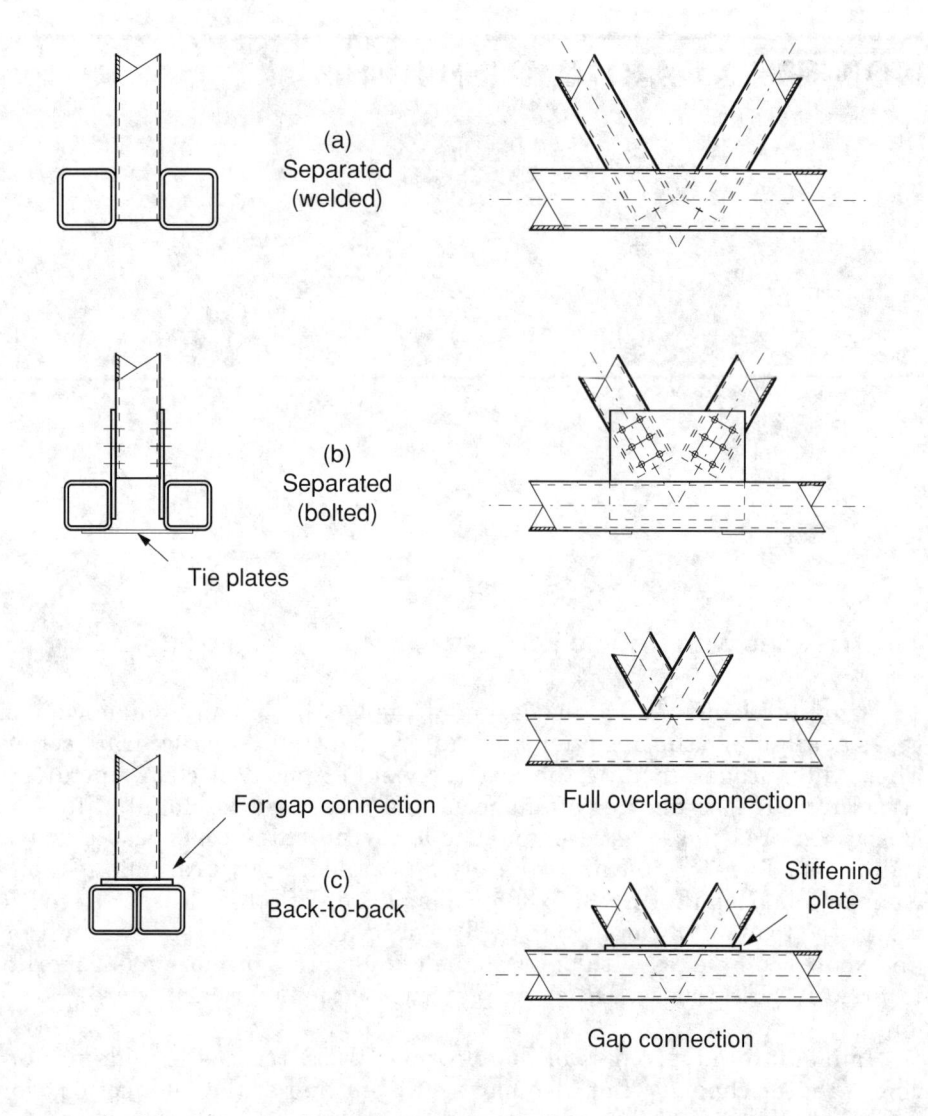

FIGURE 4.1
Types of HSS double chord truss connections

double chord truss and connection are shown in Fig. 4.2. Back-to-back chords (i.e., twin HSS sections in full contact), with the web members centred on them, are illustrated in Fig. 4.1(c). The first detail is a 100% overlap connection and the second is a gap connection on a stiffener plate. As might be expected, these alternatives have various advantages and disadvantages.

Welded separated chords provide easy and economical fabrication, along with enhanced lateral stability. But the effectiveness of vertical shear transfer from one diagonal to the other at panel points may be limited to the shear capacity of the inner walls of the HSS chord members.

Bolted separated chords, necessary for trusses too large to ship as a single unit, give superior structural performance but are the most expensive type.

Back-to-back chords with overlap connections produce stiff, efficient trusses, but are more expensive than the welded separated type. Gapped connections for back-to-back chords provide easier fitting and welding, but the trusses are less stiff and more expensive (because of the stiffener plate employed with this gap arrangement) than those with overlap connections.

Luft *et al.* (1991) reported the results of a comparative analysis regarding the relative material and fabrication costs of single chord, back-to-back double chord and separated double chord trusses. Nineteen configurations of rectangular HSS Warren trusses were attempted for each of the three truss types, with spans of 20 m, 40 m and 60 m, with light, medium and heavy loads, with span to depth ratios of 10, 16, 20, 24, 27, 32 and 40, and with web member slopes from 32° to 63°. Connection arrangements were limited to gapped or stiffened-gapped for single chord, and stiffened-gapped for back-to-back double chord trusses.

Not all configurations were feasible for all three truss types. Viable single chord trusses could only be established for shorter, more lightly loaded, deeper spans, while the separated double chord trusses had the greatest applicability of the three types.

Single chord trusses were the most economical and sometimes the lightest for the short spans, and they were generally between the two types of double chord trusses in both cost and mass for medium spans. Back-to-back double chord trusses were often the heaviest and always the most expensive for all spans. However, they were feasible for almost the complete range of truss configurations. (One needs to keep in mind that back-to-back double chord trusses having overlap web connections are both more economical and somewhat stiffer than the gap type with stiffener plate considered in this study.) Separated double chord trusses were the heaviest, but not the most expensive, for most short spans. They were usually both the lightest and the least expensive for medium, and particularly for long spans.

The cost studies indicated that, for short spans, single chord trusses were around 20% less expensive than back-to-back double chord trusses, and for long spans separated double chord trusses were about 10% more economical than the back-to-back type considered.

FIGURE 4.2
Separated double chord truss, Hamilton Convention Centre, Ontario

FIGURE 4.3
Back-to-back double chord trusses with overlap connections
for Ballantyne Pier Warehouse, Vancouver, B.C.
(Photograph courtesy of J.A. Rapson, P.Eng., Sandwell Inc., Vancouver)

4.1.2 Experimental Studies on Double Chord Trusses

Research on double chord trusses is limited. Korol and Chidiac (1980) investigated a range of isolated K connections using HSS 152x152x6.4 chord and HSS 127x127 web sections to confirm the most viable options. Subsequently, Korol et al. (1983) tested five double chord Warren trusses, each 15 metres long and 1.7 metres deep, fabricated from the same section sizes used for the isolated connections.

Figure 4.1 shows the connection arrangements selected, and Fig. 4.4 illustrates the truss geometry. All trusses had web members inclined at 63° to the chords. Two trusses employed welded separated chords, one with square cut web members and the other with web members cut at a slight bevel (40° from vertical, as fitted) to permit reduced eccentricity upon assembly. All trusses were uniformly loaded at their top panel points until they failed.

4.1.3 Design Guidance for Double Chord Trusses

Much of the following information is derived from Korol's summary (1983) of research referred to in the previous section.

Korol recommends that double chord truss analysis be performed on the assumption of pin jointed behaviour, and that 0.9 be used for K factors with the compression chord. A value of 0.75 may be used for K with web members, as for single chord trusses. Axial and shear forces generated from the analysis may then be used for connection design.

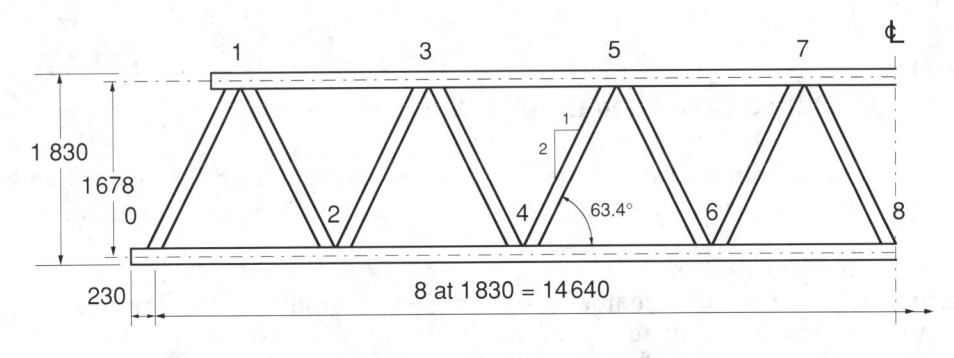

NOTE: 1) Chords were HSS152x152; webs were HSS127x127
2) Thickness of chord members: = 6.4 mm
   Thickness of web members: 0-1 = 9.5 mm
   1-2 to 4-5 = 6.4 mm
   5-6 to 7-8 = 4.8 mm

FIGURE 4.4

Dimensions of load-tested double chord trusses

4.1.3.1 Welded Separated Double Chord Trusses

The truss ultimate capacity can depend upon the chord resistance in shear between the web members at outer panel points of the truss. If the chords end near a panel point, that chord resistance is improved by capping the chord sections to provide greater restraint against distortion of the chord cross sections.

At a panel point of a tested truss, shear resistance of the four webs of the square HSS chord sections was only 72% of their theoretical plastic shear resistance. This can be stated as the inner webs being 100% effective in shear while the outer webs were only 44% effective. In general, it would seem prudent to count on the outer webs of *square* HSS being no more than 30% effective in shear. This is equivalent to considering that the four webs of the two chord sections are 65% effective in resisting shear. Korol (1983) believes that the outer webs of *rectangular* HSS will be close to 100% effective in shear if the sections are oriented so that $h_0 = 2b_0$.

Strength of the panel points is not highly sensitive to eccentricity of the member noding. It seems that a larger connection gap results in increased transfer of shear loads from the inner to the outer webs of the chord members, which is an advantage. However, this benefit is offset by the greater local moment resulting from the increased eccentricity.

Studies by Lau and Dawe (1982) report that if the web member ends are cut on a bevel so that the ends are closer to being perpendicular to the chord, a more uniform stress pattern (caused by the transfer of load between web members) is produced along the inner webs of the chord members. This also reduces the connection eccentricity, which should provide a truss with less deflection. An optimum arrangement may be ends bevel cut and a gap sufficiently large to ensure good access for sound welding.

4.1.3.2 Bolted Separated Double Chord Trusses

The bolted separated double chord truss performed well in tests, exhibiting high stiffness and strength.

Tie plates between the chord HSS sections on the side of the connection opposite to the web members (see Fig. 4.1) significantly increase the stiffness of the truss by maintaining alignment of the sections, with one tie acting in compression and the other in tension. However, connections without tie plates were almost as strong, just more flexible than those with them.

4.1.3.3 Back-to-Back Double Chord Trusses

T or Y Connections

It is suggested by the authors that guidance for T or Y connections on a back-to-back double chord can be obtained from Table 3.4 for HSS web members used with an I-shaped chord. Structurally, there would seem to be similar stiffness provided by the web of an I-shaped chord and the central side walls of the double HSS chord. If the flare bevel recess between the chord sections is welded flush in the vicinity of the T or Y connection, the web member can be welded across the top face of the chord. (Stiffnesses offered by the I-shaped chord and the double HSS chord do differ of course at locations other than the central region of the chords—the tips of the I-shaped chord are free, but the outer walls of the double HSS chord provide restraint.) Reinforcing plates on the top surface of the chord will likely be necessary for large web loads, particularly tension.

Overlap K or N Connections

Back-to-back double chord trusses with overlap connections are somewhat stiffer than those with gap connections.

The use of 100% overlap connections avoids additional effective width complexities which would result from partially overlapping web members landing on the surface of a double chord with, in effect, a middle longitudinal stiffener.

Gap K or N Connections

Stiffening plates on the surface of the chord sections are necessary. A plate thickness of about three times the chord wall thickness was necessary in tests for the connection strength to develop the web member strength.

4.1.4 Connection Design for Separated Double Chord Trusses

The resistance of HSS sections in the gap region of separated double chord truss connections results from a complex interplay of forces and geometry, with numerous variables to consider. A number of researchers suggested analytical approaches to the problem during the 1980s, but a satisfactory definitive solution for design office use was not produced.

Early formulations accommodated axial, shear and moment forces in an interactive format that utilized different portions of the cross section for different types of force. They assumed that inner and outer webs of the chord members were equally effective for shear, at least for square sections and those whose depth exceeded their width. Results of the full scale testing described above prompted new testing and analytical work (Korol and Mitri

1985) that incorporated varying chord wall effectiveness, but apparently did not account for bending moments acting in the connections.

Complex formulations that considered all the forces along with a varying degree of chord outer wall effectiveness were derived in the early 1990s. These were incorporated in a computer program named DCTRUSS developed for CIDECT as project #5AT, which may still be available. The program is an interactive package intended to aid the practising engineer in the analysis and design of separated double chord rectangular HSS Warren trusses (Luft 1991).

To manually calculate the resistance of separated double chord K connections at this time, it is suggested that the equations for factored resistance based on chord shear from Table 3.3 for K connections be adjusted to reflect the fact that the outer webs of the chord sections are not always fully effective in shear. To the extent that the relatively thin-walled square chord members of the trusses tested by Korol (1983) represent low outer sidewall participation, while thicker-walled rectangular (with height greater than width) members represent more outer sidewall participation, then the 65% effectiveness suggested in Section 4.1.3.1. approaches a lower bound. It is therefore suggested that the term A_V in the chord shear equations be conservatively defined (with $\alpha = 0$) for this application as

$$A_V = 4\,(0.65\,h_0\,t_0).$$

The equations in question are

$$N_i^* = \frac{F_{y0}\,A_V}{\sqrt{3}\,\sin\theta_i} \qquad [4.1]$$

and

$$N_{0(in\,gap)}^* \leq (2A_0 - A_V)\,F_{y0} + A_V\,F_{y0}\,[1 - (V_f/V_p)^2]^{0.5} \qquad [4.2]$$

where

$$V_p = \frac{F_{y0}\,A_V}{\sqrt{3}}$$

Values of ϕ are built into [4.1] and [4.2].

Welds need to be proportioned for the web member loads.

4.2 Trusses with Cropped or Flattened Web Members

Statically loaded trusses of moderate size are sometimes constructed with circular HSS web members whose ends have been cropped or flattened as shown in Fig. 4.5. Web slenderness ratios should be calculated with $K = 1.0$. In the case of full or partial flattening, Wardenier et al. (1991) recommend that the length of flattened tube be minimized and the taper not be shorter than 1:4. Flattening will reduce the compressive strength of the tube if the ratio of diameter to thickness is more than 25. Combining the

A   B   C   D

Cropping (A)   Full flattening (B,C)   Partial flattening (D)

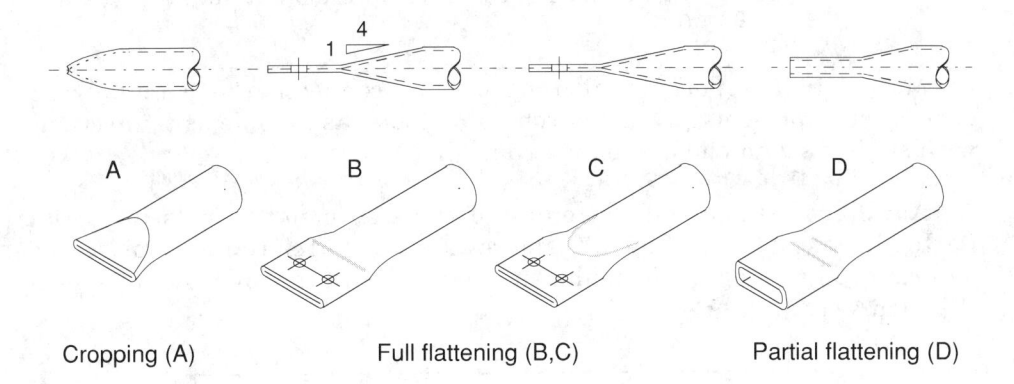

FIGURE 4.5
Various types of flattening for circular HSS web and bracing members

flattening operation with shearing the member to length can provide some savings in fabrication.

4.2.1 Circular HSS Chords

Wardenier *et al.* (1991) published connection resistance formulations with graphical assistance for cropped overlapped N connections (as shown in Fig. 4.6) for a limited range of tube sizes:

Dimensions tested

$114 \leq d_0 \leq 169$
$42 \leq d_1 \leq 90$
$3 \leq t_0 \leq 8$
$3 \leq t_1 \leq 4.6$
$F_y \leq 400$ MPa

Parameters tested

$14 \leq d_0/t_0 \leq 57$
$0.35 \leq d_1/d_0 \leq 0.8$
$d_1/d_2 = 1.0$ and $t_1/t_2 = 1.0$
$\theta_1 = 90°$ and $\theta_2 = 45°$
overlap $\leq 75\%$.

Connection resistance is given by

$$N_1^* = 0.8 \left[10.5 + 40.6 \left(\frac{d_1}{d_0}\right)^2 - 172 \left(\frac{t_0}{d_0}\right) \right] \left(\frac{t_1}{d_1}\right) \left(\frac{d_0}{t_0}\right) t_0^2 F_{y0} \, f_2(n') \qquad [4.3]$$

149

where

$f_2(n') = 1.0$ when the chord is in tension or

$f_2(n') = 1 + 0.2n'$ when the chord is in compression (n' is negative), and $0 \geq n' \geq -0.8$.

Morris (1985) reported that cropped web N connections on circular HSS chords are approximately 12% stronger and 45% as flexible as connections with similar size members using square HSS chords.

An alternative to full flattening of the web member ends is partial flattening, as shown in Fig. 4.7. The resistance of such connections can be determined by using the formulae in Table 3.1 with the following revisions (Wardenier et al. 1991):

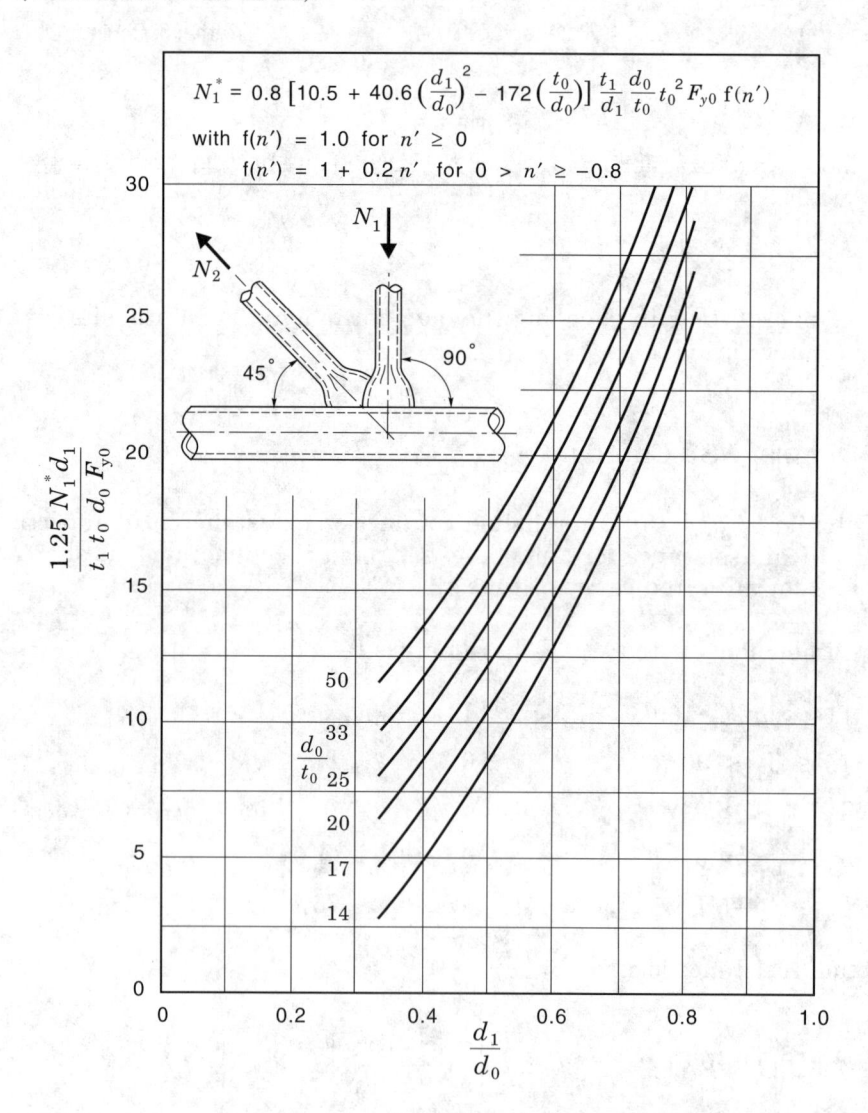

$$N_1^* = 0.8 \left[10.5 + 40.6\left(\frac{d_1}{d_0}\right)^2 - 172\left(\frac{t_0}{d_0}\right)\right] \frac{t_1}{d_1} \frac{d_0}{t_0} t_0^2 F_{y0} f(n')$$

with $f(n') = 1.0$ for $n' \geq 0$
$f(n') = 1 + 0.2n'$ for $0 > n' \geq -0.8$

FIGURE 4.6
Design recommendations for cropped end connections to circular HSS chords

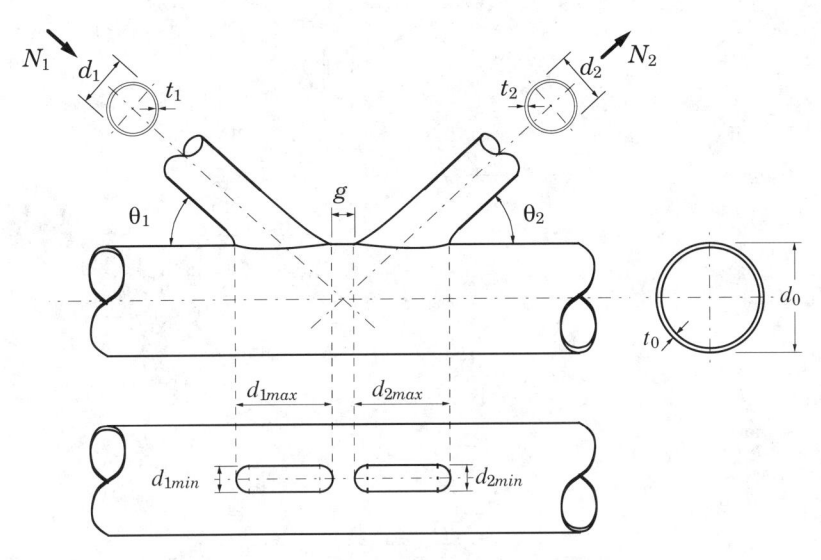

FIGURE 4.7
Web members with semi-flattened ends connected to a circular HSS chord

For T and X connections in compression,

　　use d_{1min} rather than d_1

For gapped K connections,

　　use $(d_1 + d_{1min})/2$ rather than d_1.

The 0.8 factor at the beginning of [4.3] is the appropriate value of ϕ.

4.2.2 Square HSS Chords

Pratt Connections

Morris (1985) reported the results of an experimental program which included cropped web members framing into square HSS chord sections in the form of N connections as shown in Fig. 4.8.

Dimensions tested	Parameters tested
$102 \leq b_0 \leq 152$	$13 \leq b_0/t_0 \leq 32$
$42 \leq d_1 \leq 73$	$0.32 \leq d_1/b_0 \leq 0.72$
$5 \leq t_0 \leq 8$	$d_1/d_2 = 1.0$　and　$t_1/t_2 = 1.0$
$3 \leq t_1 \leq 6$	$\theta_1 = 90°$ and $\theta_2 = 45°$
$F_y \leq 400$ MPa	overlap $\leq 75\%$.

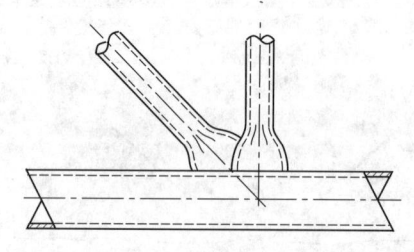

(a) Square chord longitudinal cropping

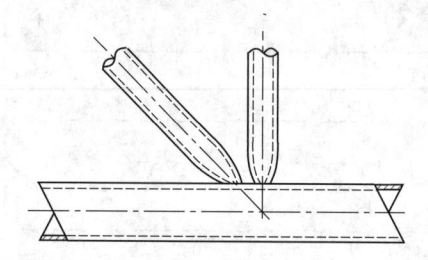

(b) Square chord transverse cropping

FIGURE 4.8
Web members with cropped ends connected to a square HSS chord

Connection resistance is given by

$$N_1^* = 0.8 \left[0.504 + 6.10 \left(\frac{d_1}{b_0}\right)^3 - 43.3 \left(\frac{d_1}{b_0}\right)^2 \left(\frac{t_0}{b_0}\right) \right] t_0 b_0 F_{y0} \, f_2(n') \qquad [4.4]$$

where $f_2(n')$ is the same as for circular HSS chords.

It was reported that, based upon limited experimental evidence, the direction of web cropping, parallel to the chord or perpendicular to it, appears to make little difference to the ultimate strength of the connection, nor does the amount of overlap (Morris 1985). The contribution of joint deformation to truss deflection at specified loads is up to 15% of that due to member deformation, provided b_0/t_0 is less than 20. Hence, service load truss deflections can be estimated by 1.15 times that predicted by a pin-jointed analysis.

The crane boom shown in Fig. 4.9 incorporates end-flattened circular HSS web members framing to both circular and square HSS chords.

FIGURE 4.9
Example of triangular Pratt truss with end-flattened circular HSS web members framing to both circular and square HSS chords

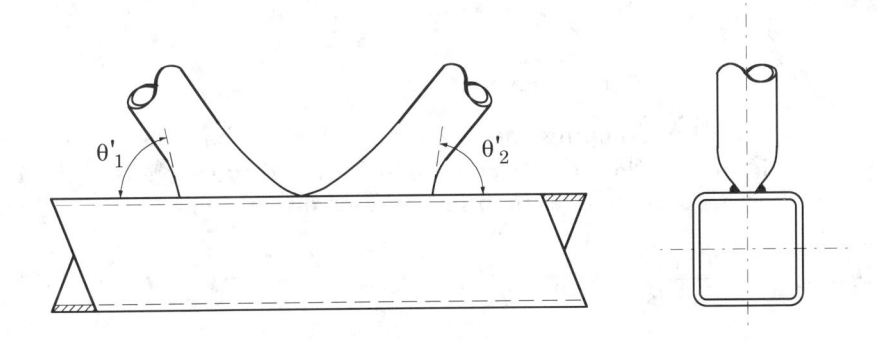

FIGURE 4.10
Zero gap K connection with cropped-end web members

Warren Connections

Figure 4.10 illustrates a Warren connection formed from identical cropped circular HSS web members and a square HSS chord. Toes of the web members touch so that there is no gap or overlap. Lau *et al.* (1985) conducted an experimental program on 45 such specimens to determine yield and ultimate strengths. Morris and Packer (1988) have published an analysis based on yield line theory, augmented by membrane considerations, to arrive at an ultimate strength expression for the same connections. Agreement is good since the expression was calibrated to produce a mean ratio of measured to predicted strengths of 1.00, with a coefficient of variation of the ratios of 13.0%.

Dimensions tested	Parameters tested
$102 \leq b_0 \leq 152$	$12 \leq b_0/t_0 \leq 32$
$42 \leq d_1 \leq 102$	$0.32 \leq d_1/b_0 \leq 0.88$
$4 \leq t_0 \leq 13$	$d_1/d_2 = 1.0$ and $t_1/t_2 = 1.0$
$3 \leq t_1 \leq 6$	$\theta_1 = \theta_2 = 60°$
$F_y \leq 400$ MPa	overlap = gap = 0.

Connection resistance is given by

$$N_2^* = 0.4 N_{y1} \left(1 + 0.021 \frac{b_0}{t_0}\right)\left(1 + 1.71 \frac{d_1}{b_0}\right) \quad [4.5]$$

where

$$N_{y1} = \frac{t_0^2 F_{y0}}{\sin \theta_1} \left(\frac{\pi}{2} + \frac{b_1' + 2h_1'}{b_0' - b_1'} + \frac{1.32}{t_0} \sqrt{\frac{F_{y1}}{F_{y0}} \tan \theta_1' b_0' t_1} \right) f(n) \quad [4.6]$$

$b_0' = b_0 - t_0$

$b_1' = $ width of flattened web member

With full cropping and flattening, this can be assumed to be $2t_1$. If fillet welding is used, this effective contact width can be increased to include the fillet weld leg dimensions.

$$h_1' = \frac{\pi(d_1 - t_1) + t_1}{2 \sin \theta_1}$$

$\theta_1' = $ slope of the web member face at the cropped end, relative to the chord (see Fig. 4.10)

Since θ_1' varies with the cropping geometry, it may not be accurately known at the design stage. Bearing in mind that $\tan \theta_1'$ is sensitive to the value of θ_1', one can always conservatively use θ_1 in place of θ_1'.

The term f(n) is the same reduction factor (to allow for axial loads in compression chords) as is usually used for rectangular HSS chords (see Table 3.2). Section 1.9 contains the definition.

Equations 4.5 and 4.6 apply to symmetrical connections where $\theta_1 = \theta_2$, $d_1 = d_2$, $t_1 = t_2$, $d_1/b_0 \geq 0.3$ and $b_0/t_0 \leq 32$.

The 0.8 factor at the beginning of [4.4] is the appropriate value of ϕ. Equation 4.5 has the value of ϕ built in.

4.3 Trusses with Web Members Framing onto Chord Corners

It is possible to have truss web members frame onto the corner of square HSS chord members, as shown in Fig. 4.11. This necessitates very careful profiling of the web member ends, particularly where corner radii are large, into so-called "bird mouth" or "bill-shaped" connections.

Such a member arrangement has been used occasionally in North America, for example in the Minneapolis Convention Center roof, and in the Minneapolis / St. Paul Twin Cities Airport skyway. It has also been used in Japan, where a robot was developed to profile the ends of the web members. By framing onto the corners of the square HSS chord member a high connection strength and stiffness is achieved, regardless of the web to chord member width ratio.

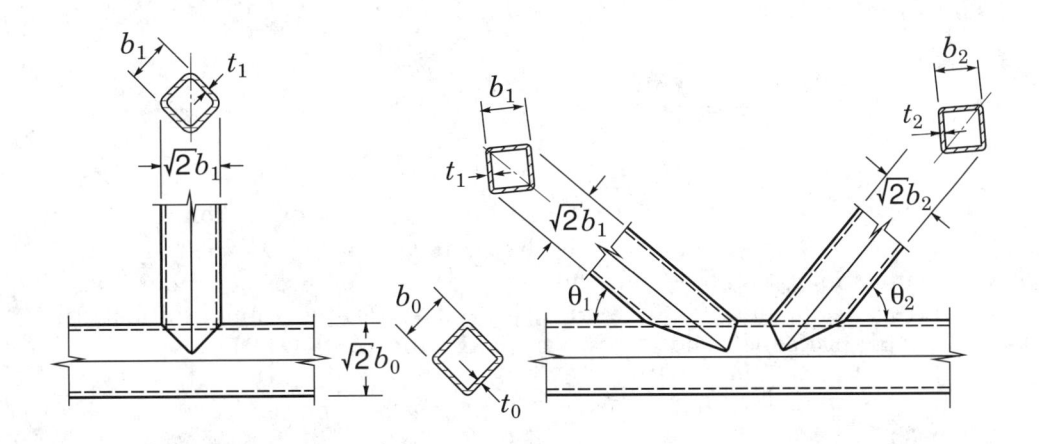

FIGURE 4.11

Square HSS "bird mouth" T and K connections

Ono *et al.* (1991) have undertaken an experimental study of such square HSS T and K connections. They found that these connections are not only much stronger than their conventional square HSS counterparts, but are also generally stronger than equivalent circular HSS connections with members having the same cross-sectional area as in the test specimens. The cases in which equivalent circular HSS connections are stronger are likely to occur for high web member to chord width ratios and low b_0/t_0 ratios.

A finite element study of bird mouth T connections by Owen *et al.* (1996) has also confirmed the superior performance of this connection type over equivalent circular HSS connections. They also found that bird mouth T connections between square HSS were stronger than their conventional square HSS counterparts, provided the length of the chord (between inflection points) was not greater than $18 b_0$.

All of the T connections (25) tested by Ono *et al.* (1991) had the web member loaded in compression, and the K connections (16) had all web members inclined at 45° to the chord. The orientations of the web and chord members are shown in Fig. 4.11, and one should note that the chord and web members are all rotated by 45° about their longitudinal axis. Ono *et al.* (1991) concluded that the connection ultimate strengths could be given as follows:

For T Connections

$$N_{1u} = t_0^2 F_{y0} \left(\frac{1}{0.211 - 0.147(b_1/b_0)} + \frac{b_0/t_0}{1.794 - 0.942(b_1/b_0)} \right) f(n') \qquad [4.7]$$

For K connections

$$N_{1u} = \frac{t_0^2 F_{y0}}{\sqrt{1 + 2\sin^2\theta_1}} 4\alpha \left(\frac{b_0}{t_0} \right) f(n') \qquad [4.8]$$

where the effective area coefficient α is given for 45° K connections in Fig. 4.12. The function $f(n')$ is the same as that used for circular HSS connections to allow for the influence of compression chord axial stresses not needed for equilibrium at the connection (in effect, the chord "preload"), and is given by Wardenier *et al.* (1991):

$$f(n') = 1 + 0.3 n' - 0.3 n'^2 \quad \text{for } n' < 0 \text{ (compression)} \qquad [4.9]$$

$$f(n') = 1.0 \quad \text{for } n' \geq 0 \text{ (tension)} \qquad [4.9a]$$

where $\quad n' = N_{0p}/(A_0 F_{y0}) \qquad [4.9b]$

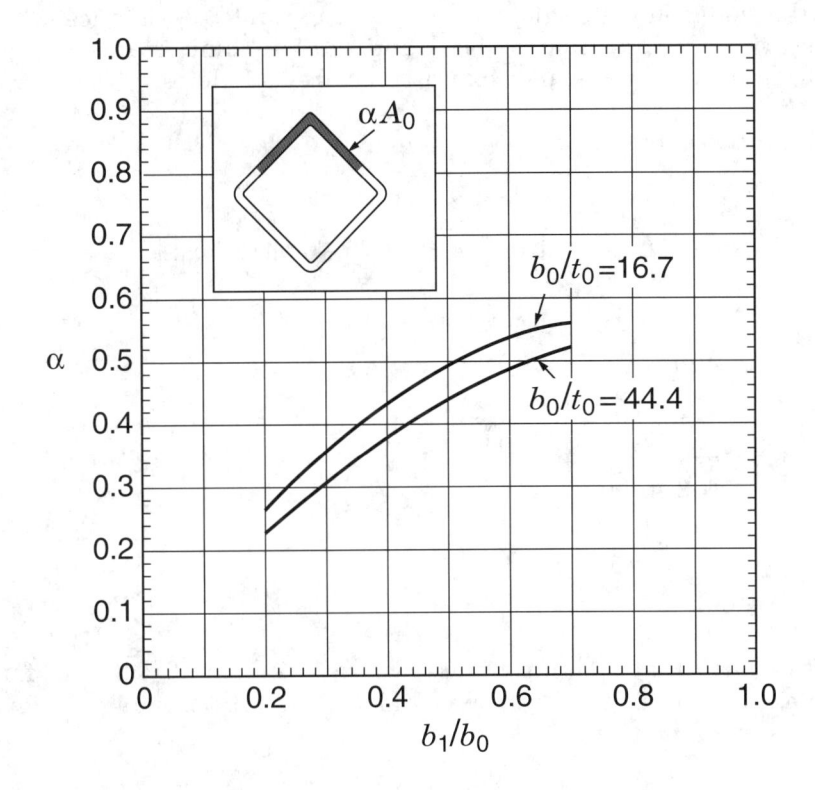

FIGURE 4.12

Effective area coefficient α for "bird mouth" 45° K connections

As these equations are based on a regression analysis of the test data, one should be careful to ensure that they are only applied within the approximate bounds of parameter ranges examined in the tests:

For T connections

$$16 \leq b_0/t_0 \leq 42 \qquad 0.3 \leq b_1/b_0 \leq 1.0$$

For K connections

$$16 \leq b_0/t_0 \leq 44 \qquad 0.2 \leq b_1/b_0 \leq 0.7 \qquad \theta \approx 45°$$

For example, the angle function in [4.8] is rather complex considering that the web member angle was not a test variable. Resistance factors are necessary for application of [4.7] and [4.8] to limit states design. The resulting factored resistance expressions are as follows:

For T connections

$$N_1^* = 0.9\, t_0^2\, F_{y0} \left(\frac{1}{0.21 - 0.15\,(b_1/b_0)} + \frac{b_0/t_0}{1.79 - 0.94\,(b_1/b_0)} \right) f(n') \quad [4.10]$$

For K connections

$$N_1^* = 0.9\, \frac{t_0^2\, F_{y0}}{\sqrt{1 + 2\sin^2\theta_1}}\, 4\alpha \left(\frac{b_0}{t_0} \right) f(n') \quad [4.11]$$

with α given by Fig. 4.12.

It is noted that a slightly more accurate version of [4.8], but far more complex, has been produced by Ishida *et al.* (1993) using a regression analysis fit to the test data.

4.4 Joists Fabricated from HSS Sections

A frequent configuration for floor and roof joists is the use of HSS chord sections with a pair of angles for each web member. The web angles are placed on the outer faces of the chords, one on each side, with routine fillet welds used to connect them to the chords. On a larger scale, the same concept applies readily to full size trusses. The pairs of web angles can be joined together with batten plates at suitable intervals if required. Figure 4.13 illustrates a typical example of such a truss. Conventional analysis and design procedures with $K = 1.0$ can be applied to member selection.

Connection design should incorporate a check for the chord shear failure mode (assuming that a "gapped connection" with positive eccentricity would typically be made). Equation 4.1 with $A_V = 2 h_0 t_0$ is applicable for the chord shear capacity. Also, the axial force in the "gap region" of the connection is limited by

$$N^*_{0(in\ gap)} \leq (A_0 - A_V)\, F_{y0} + A_V\, F_{y0} \left[1 - \left(\frac{V_f}{V_p} \right)^2 \right]^{0.5}$$

where $V_p = \dfrac{F_{y0}\, A_V}{\sqrt{3}}$

FIGURE 4.13

Trusses with HSS chords and angle web members

Downsized separated double chord HSS trusses are also used for joists. They are particularly suitable where long spans must be constructed with shallow joist depths. Connection considerations are similar to those for full size separated double chord HSS trusses discussed in Section 4.1.

Another common joist configuration is the use of HSS for web members between conventional double angle chords. Sometimes tension web members are double angles which are placed on the outside of the chord angles; this avoids nodal eccentricity since the compression web (HSS) can be "overlapped" with the tension web (double angles) on either side of the chord angles. Figure 4.14 shows examples.

(a)

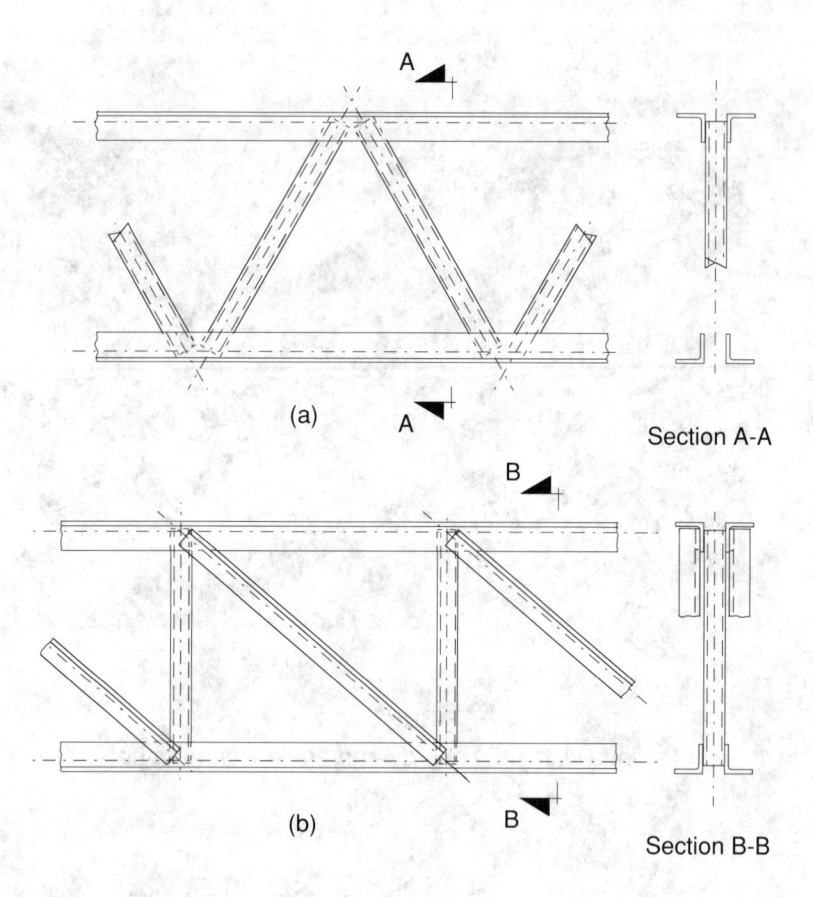

Section A-A

(b)

Section B-B

FIGURE 4.14
Joists consisting of: (a) Angle chords with HSS web members
(b) Angle chords and tension web members, with HSS compression web members

REFERENCES

Ishida, K., Ono, T., and Iwata, M. 1993. Ultimate strength formula for joints of new truss system using rectangular hollow sections. Proceedings, 5th. International Symposium on Tubular Structures, Nottingham, England, pp. 511–518.

Korol, R.M. 1983. The behaviour of HSS double chord Warren trusses and aspects of design. Canadian Symposium on Hollow Structural Sections, lecture tour to 10 Canadian cities, 26 pp.

Korol, R.M., and Chidiac, M.A. 1980. K-joints of double-chord square hollow sections. Canadian Journal of Civil Engineering, **7**: 523–539.

Korol, R.M., and Mitri, H.S. 1985. Strength analysis of RHS double chord joints with separated chords. Canadian Journal of Civil Engineering, **12**: 370–381.

Korol, R.M., Mirza, R.A., and Chiu, E.T.-C. 1983. An experimental investigation of double chord HSS trusses. Canadian Journal of Civil Engineering, **10**: 248–260.

Lau, W.C-P., and Dawe, J.L. 1982. Elasto-plastic finite element analysis of welded truss connections. Canadian Journal of Civil Engineering, **9**(3): 399–412.

Lau, B.L., Morris, G.A., and Pinkney, R.B. (1985). Testing of Warren-type cropped-web tubular truss joints. Proceedings, Canadian Society for Civil Engineering Annual Conference, Saskatoon, Saskatchewan.

Luft, R.T. 1991. DCTRUSS version 1.3 (a CIDECT computer program to aid in the analysis and design of separated HSS double chord Warren trusses).

Luft, R.T., Korol, R.M., and Huitema, H. 1991. An economic comparison of single chord and double chord RHS Warren trusses. Proceedings, 4th. International Symposium on Tubular Structures, Delft, The Netherlands, pp.11–20.

Morris, G.A. 1985. Tubular steel trusses with cropped webs. Journal of Structural Engineering, American Society of Civil Engineers, **111**(6): 1338–1357.

Morris, G.A., and Packer, J.A. 1988. Yield line analysis of cropped-web Warren truss joints. Journal of Structural Engineering, American Society of Civil Engineers, **114**(10): 2210–2224.

Ono, T., Iwata, M., and Ishida, K. 1991. An experimental study on joints of new truss system using rectangular hollow sections. Proceedings, 4th. International Symposium on Tubular Structures, Delft, The Netherlands, pp. 344–353.

Owen, J.S., Davies, G., and Kelly, R.B. 1996. A comparison of the behaviour of RHS bird beak T-joints with normal RHS and CHS systems. Proceedings, 7th. International Symposium on Tubular Structures, Miskolc, Hungary, pp. 173–180.

Wardenier, J., Kurobane, Y., Packer, J.A., Dutta, D., and Yeomans, N. 1991. Design guide for circular hollow section (CHS) joints under predominantly static loading. CIDECT (ed.) and Verlag TÜV Rheinland GmbH, Köln, Federal Republic of Germany.

5

MULTIPLANAR WELDED CONNECTIONS

5.1 Introduction

Even though structures built from HSS often incorporate multiplanar connection nodes (as shown in Fig. 5.1), there is much less design information available than for planar connections. This chapter relies on the summaries provided by the CIDECT Design Guides (Wardenier *et al.* 1991 and Packer *et al.* 1992), the American Welding Society (1996) and Kurobane (1995). The scope of the chapter is limited to predominantly static, axial loading on the braces.

Design criteria have not yet been established for multiplanar connections under in-plane or out-of-plane moment loading on the members, but most space frames are designed as pin-connected systems, which results in member axial forces only. Connection noding eccentricities induce primary, in-plane bending moments, but the effect of these on *connection* resistances is already incorporated in the connection resistance formulae, providing these eccentricities are within certain limits. These in-plane eccentricity moments should still be taken into account when designing the *members*. Hence struts will need to be treated as beam-columns. It is thus best to minimize connection noding eccentricities whenever possible.

5.2 Connections between Circular HSS

Basic multiplanar connections in which the braces are welded directly to main "through" members may be classified into four large groups as shown in Fig. 5.2. From the behaviour of these connections in the four groups, one can generalize about interactions between loads in different planes (Kurobane 1995). The resistances of multiplanar connections can

FIGURE 5.1
Canada Trust tower finial, Toronto, Ontario
with complex multiplanar circular HSS connections

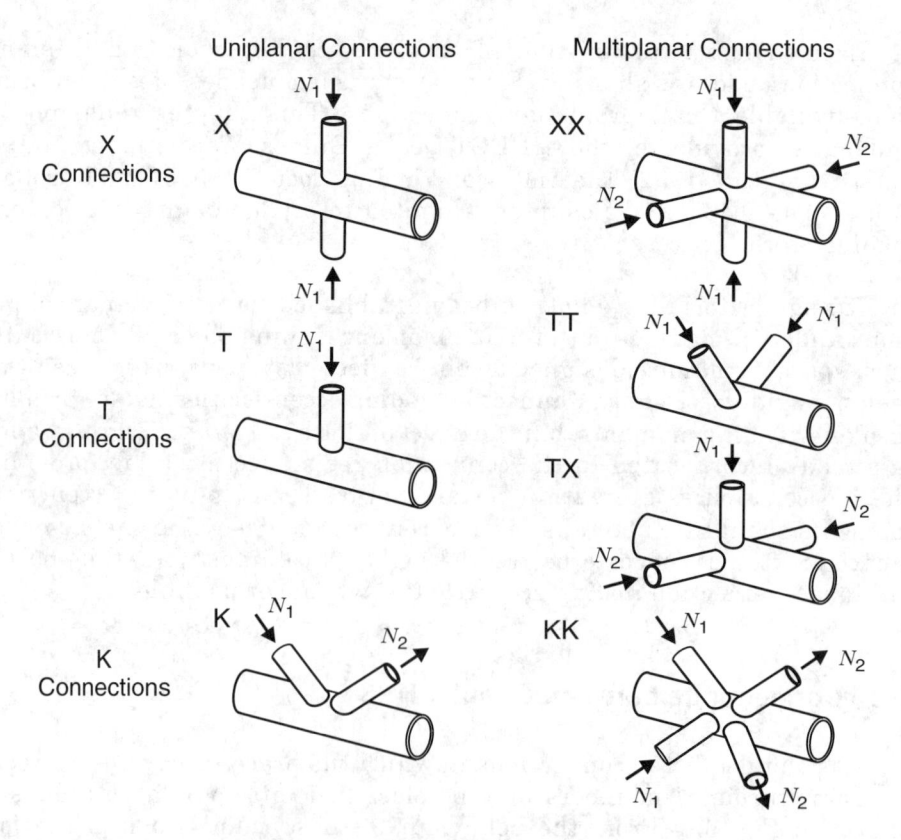

FIGURE 5.2
Basic types of multiplanar connections (Kurobane 1995)

Type of Connection		Correction Factors to Planar Connection Resistances from Table 3.1
TT		1.0
XX		$1 + 0.33 \dfrac{N_2}{N_1}$ Take account of the sign of N_1 and N_2 ($N_1 \geq N_2$)
KK		0.9

TABLE 5.1
Correction factors for circular HSS multiplanar connection resistances
$60° \leq \varphi \leq 120°$

best be predicted by resistances of corresponding uniplanar connections (for which reliable resistance prediction equations are available), multiplied by correction factors. The *simplified* correction terms for the TT, XX and KK ("symmetrically-loaded") connections are summarized in Table 5.1. Although the correction terms given in Table 5.1 are less accurate than more recent and complex versions (Kurobane 1995), they roughly correspond to the average values of those given by the more complex (and more accurate) prediction methods below.

5.2.1 XX Connections

If an XX connection has a pair of braces loaded in compression (due to load N_1 as shown in Table 5.1), then as the loads N_2 increase in compression, the connection resistance—expressed in terms of N_1^*—also increases, because the chord wall deflection (often called the chord ovalization) is supressed. On the other hand, tension N_2 out-of-plane loads have an adverse effect (for compression N_1), as they exacerbate the chord ovalization, and hence the connection resistance decreases. Similarly, if the in-plane braces are loaded in tension, then tensile loads in the out-of-plane braces have a beneficial effect on connection resistance, while compressive loads in the out-of-plane braces would decrease connection resistance.

The formula given in Table 5.1 roughly captures these multiplanar effects. Note that N_1 is the <u>numerically larger</u> pair of forces, and due regard is given to the sign of the forces (i.e., compression is negative and tension positive) (Kurobane 1995). More accurately (Kurobane 1995 and van der Vegte 1995), this multiplanar correction factor for XX connections can be expressed as

$$C_{XX} = \left[1 - \left(1.6\beta - 1.2\beta^2\right)\frac{N_2}{N_1} + \left(1.5\beta - 2.5\beta^2\right)\left(\frac{N_2}{N_1}\right)^2 \right]^{-1}$$

within the range of validity, $-0.6 \leq \frac{N_2}{N_1} \leq 1.0$

Design Example

Assume the main (or "through") member is an HSS 324x13 and four HSS 168x8 members are welded orthogonally to it, all in Grade 350W. One pair of opposite branches carries a 360 kN compression load and the other pair a 200 kN tension load. Let the main (chord) member be in tension. From Table 3.1(a), the members are within the validity range, so for the *planar X connection* the connection resistance (Table 3.1) is given by

$$N_1^* = \frac{F_{y0} t_0^2}{\sin\theta_1} \left(\frac{5.2}{1 - 0.81\beta}\right) f(n')$$

$$= \left[0.350\left(12.7^2\right)/1.0\right] \left[5.2/(1 - 0.81(168/324))\right] 1.0$$

$$= 56.5(8.97) = 506 \text{ kN}$$

This connection resistance now needs to be reduced to allow for multiplanar effects.

$N_1 = -360$ kN; $N_2 = +200$ kN; $N_2/N_1 = -0.556$; $\beta = 0.519$

From Table 5.1 (simple approach), $C_{XX} = 1 + 0.33(-0.556) = 0.817$

Using the "accurate" approach, $C_{XX} = (1.0 + 0.282 + 0.032)^{-1} = 0.761$
(only 7% lower)

Connection resistance of the *multiplanar XX connection* is therefore

$0.761(506) = 385$ kN $\quad \geq 360 \quad \therefore$ OK

5.2.2 TX (XT) Connections

These connections (Fig. 5.2) can act predominantly as a T connection (with the connection resistance being affected by the X connection action in the other plane), or predominantly as an X connection (with the connection resistance being affected by the T connection action in the other plane) (Kurobane 1995 and van der Vegte 1995). The correction factor to be applied to the connection resistance of the *planar T connection* (Table 3.1) is

$$C_{TX} = \left[1 - 1.9\beta^2 \frac{N_2}{N_1} - \left(0.5 - 3.5\beta + 3.7\beta^2\right)\left(\frac{N_2}{N_1}\right)^2 \right]^{-1}$$

The corresponding correction factor to be applied to the connection resistance of the *planar X connection* (Table 3.1) is

$$C_{XT} = \left[1 - \left(0.23\beta + 0.4\beta^2\right)\frac{N_1}{N_2} \right]^{-1}$$

Both of these correction factors C_{TX} and C_{XT} are subject to the validity range

$$-0.6 \leq \frac{N_2}{N_1} \leq 1.0$$

Design Example

Assume a TX connection is loaded by forces as shown in Fig. 5.3, with the chord (or "through") member loaded in tension. All the member sizes will be the same as in the previous example, so all parameters are within the range of validity. The connection resistance as a *planar T connection* (Table 3.1) is given by

$$N_1^* = \frac{F_{y0}\, t_0^2}{\sin\theta_1} \left(2.8 + 14.2\beta^2\right) \gamma^{0.2}\, f(n')$$

$$= \frac{0.350\,(12.7^2)}{1.0} \left(2.8 + 14.2\left(\frac{168}{324}\right)^2\right)\left(\frac{324}{2\,(12.7)}\right)^{0.2} 1.0$$

$$= 56.5\,(6.62)\,1.66 = 622\text{ kN}$$

This must now be corrected for multiplanar effects by the factor C_{TX}.

$$C_{TX} = (1 + 0.285 + 0.099)^{-1} = 0.723$$

Connection resistance of the *multiplanar TX connection* is therefore

$$0.723\,(622) = 450\text{ kN} \quad \geq 360 \quad \therefore \text{ OK}$$

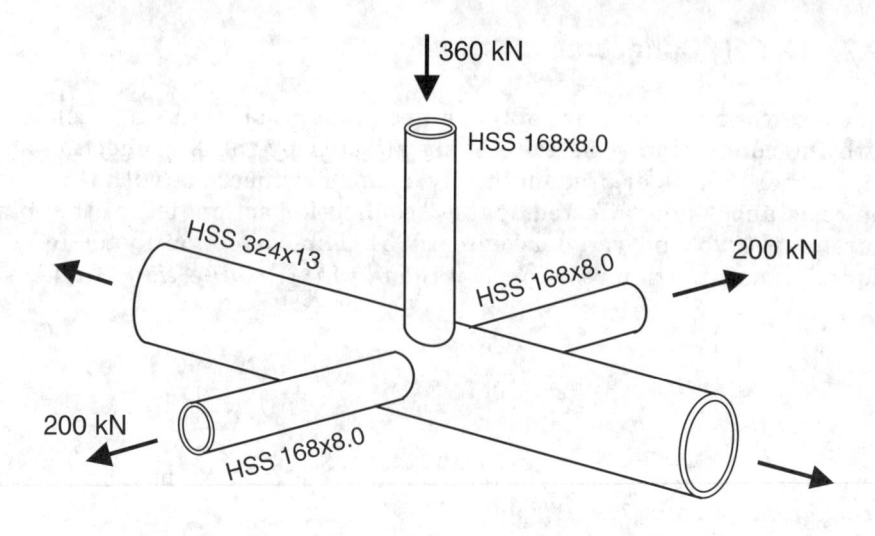

FIGURE 5.3
Loading for example TX connection (kN)

(a) Type 1 failure   (b) Type 2 failure

FIGURE 5.4
Failure modes for circular HSS KK connections under symmetrical axial loads
(Photographs courtesy of Professor Y. Kurobane, Kumamoto Institute of Technology, Japan)

The connection resistance as a *planar X connection* was calculated for the previous XX connection design example and is 506 kN. This must now be corrected for multiplanar effects by the factor C_{XT}.

$$C_{XT} = (1 + 0.409)^{-1} = 0.710$$

Connection resistance of the *multiplanar XT connection* is therefore

$0.710\ (506) = 359$ kN $\quad \geq 200 \quad \therefore$ OK

5.2.3 TT and Gap KK Connections under Symmetrical Loading

Both TT connections and gap KK connections can exhibit two possible failure modes, as shown in Fig. 5.4. The first mode, called Type 1 in Fig. 5.4(a), occurs when a pair of neighbouring braces, loaded in the same sense, collectively pushes the chord wall in or pulls it out. These neighbouring braces seem to act as one member to penetrate the chord wall together, and there is no distortion of the chord wall *between* these neighbouring braces.

The second mode, called Type 2 in Fig. 5.4(b), shows a radial deformation of the chord wall in the region *between* neighbouring braces, creating a pinching or fold between these braces. The Type 1 failure mode occurs when the transverse gap g_t or the angle φ is small (Fig. 5.5). As g_t or φ increases, Type 2 failure mode appears (Kurobane 1995).

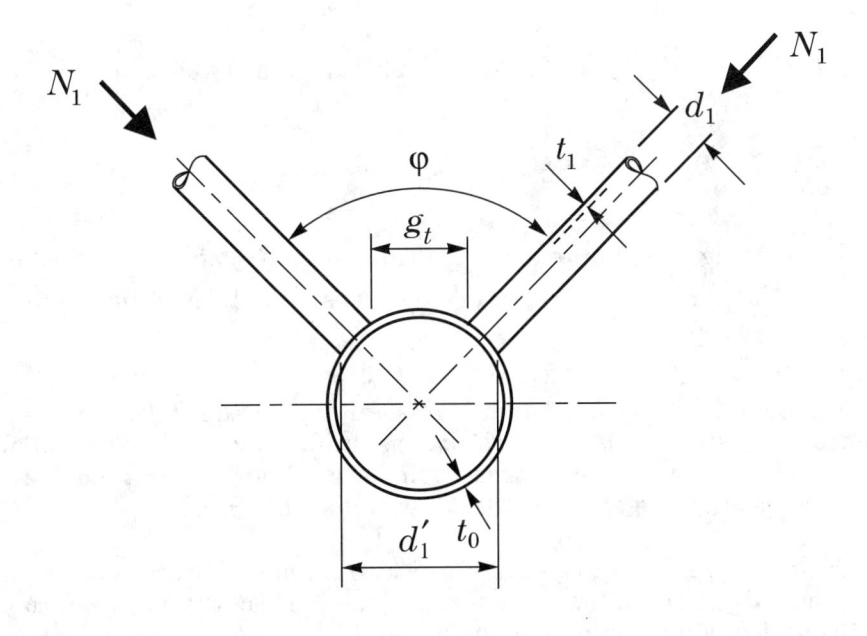

FIGURE 5.5
Illustration of parameters for gap KK connections

When a *Type 1* failure occurs, the ultimate resistance of the TT or KK connection is best represented by the resultant of axial loads in the two compression braces. The connection resistance is thus computed by the product of a multiplanar correction factor and the planar connection resistance, with the latter computed on the basis of the "combined footprint". The design equations are empirical ones reported by Paul *et al.* (1994).

Thus, for *TT connections* failing by the *Type 1* mode, the multiplanar connection resistance is

$$0.747\left(1 + 0.586 \frac{d_1}{d_0}\right) \cdot \text{(planar capacity of T connection [Table 3.1])}.$$

Note:

When using the <u>planar</u> formula (Table 3.1), β should be calculated as d_1'/d_0 where d_1' is the transverse distance between outsides of the braces, as shown on Fig. 5.5. Also, the angle θ_1 (Table 3.1) should be modified to θ_1', which is the angle between the chord axis and the plane in which two adjacent braces lie.

$$\theta_1' = \tan^{-1}\left(\tan\theta_1 \cdot \cos\frac{\varphi}{2}\right) \quad \text{(Kurobane 1995)}$$

For *KK connections* failing by the *Type 1* mode, the multiplanar connection resistance is

$$0.746\left(1 + 0.693\frac{d_1}{d_0}\right)\left(1 + 0.741\frac{g}{d_0}\right) \cdot \text{(planar capacity of K connection [Table 3.1])}.$$

Note:

When using the <u>planar</u> formula (Table 3.1), d_1 should be based on d_1' (Fig. 5.5). Also, the angle θ_1 (Table 3.1) should be modified to θ_1'.

For *both TT and KK connections* failing by the *Type 1* mode, the connection resistance is given as the resultant force in a pair of adjacent braces. Thus, this connection resistance should be greater than the combined resultant force acting in these two adjacent braces.

When a *Type 2 failure* occurs, the two planes of braces behave more independently. The ultimate connection resistance of the TT or KK connection is best predicted simply by using multiplanar correction terms to the planar T and K connection resistance. The design equations given are again the empirical ones by Paul *et al.* (1994).

Thus, for *TT connections* failing by the *Type 2* mode, the multiplanar connection resistance is

$$1.33\left(1 - 0.336\frac{g_t}{d_0}\right) \cdot \text{(planar capacity of T connection [Table 1])}.$$

For *KK connections* failing by the *Type 2* mode, the multiplanar connection resistance is

$$0.798(1 + 0.808\beta)\left(1 - 0.410\frac{g_t}{d_0}\right)\left(1 + 0.423\frac{g}{d_0}\right) \cdot \text{(planar capacity of K connection [Table 1])}.$$

For *both TT and KK connections* failing by the *Type 2* mode, the connection resistance is given as the limiting load in one brace; (one compression brace for KK connections). Thus, connection resistance of a single brace should be greater than the factored load in that brace.

5.2.4 Gap KK Connections under Non-Symmetrical Loading

"Triangular trusses", such as that shown in Fig. 5.6, are popular for relatively long spans, offering excellent stability, torsional stiffness and pleasing aesthetics. They can provide uncluttered clearances where lateral bracing systems are not desired or are not feasible. An array of these trusses can create the appearance of a space frame roof structure without the full expense of one. The triangular (or delta) truss typically has two compression chord members. The previous treatment of KK connections assumed

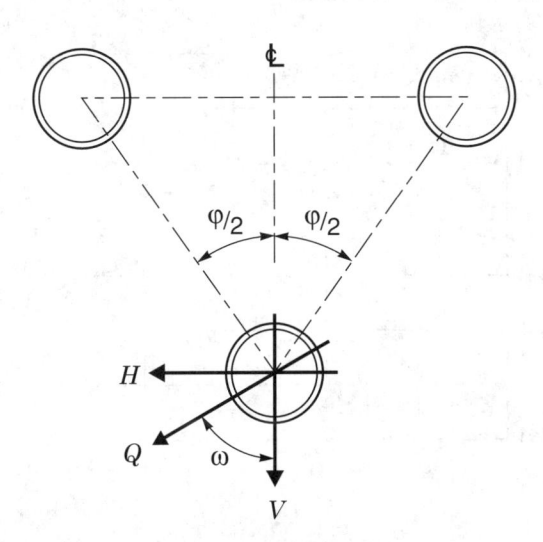

FIGURE 5.6
Triangular circular HSS truss under general loading direction

"symmetrical loading" on the two planes of web members; i.e., the loading was in the direction of V in Fig. 5.6.

Under horizontal loading (H in Fig. 5.6), the KK connection sustains axial brace loads that are anti-symmetrical about the vertical system plane shown in Fig. 5.6. In the KK connection under anti-symmetrical brace loads, diagonally opposite braces are loaded in the same force sense. Each pair of K connections behaves rather independently, and no Type 1 failure mode appears. The connection resistance of *KK connections under anti-symmetrical loads* can simply be predicted by the connection resistance equation for planar K connections (Table 3.1) with the following correction multipliers (Kurobane 1995):

$$C_{KK} = 0.858 \qquad \text{when } \frac{g_t}{d_0} \geq 0.16, \text{ and}$$

$$C_{KK} = \left(1.36 - 3.17\frac{g_t}{d_0}\right) \qquad \text{when } \frac{g_t}{d_0} < 0.16.$$

Under general loading in the direction Q (Fig. 5.6) at an angle to the vertical of ω, the connection resistance of KK connections can be determined using the following interpolation technique (Kurobane 1995). When $\omega = \varphi/2$, the braces in one plane carry all the load so the connection resistance is that of a planar K connection. The resistance of KK connec-

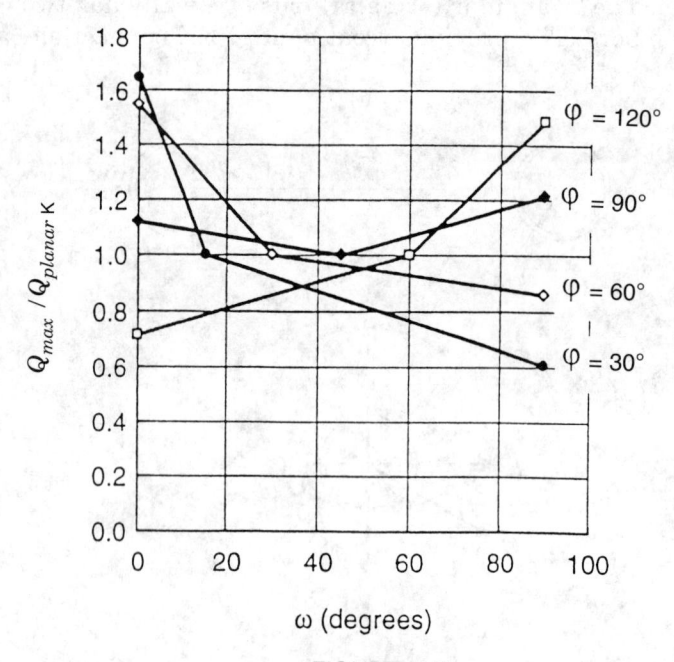

FIGURE 5.7

Connection resistance of KK connections under various load directions

tions under general load direction can be surmised by drawing straight lines linking the points at $\omega = 0$ (symmetrical loading), $\omega = \varphi/2$ (uniplanar K connection) and $\omega = 90°$ (anti-symmetrical loading), as shown in Fig. 5.7. In this figure the vertical axis shows the KK connection resistance non-dimensionalized by the planar K connection resistance, expressed in terms of the load Q.

The American Welding Society D1.1 (AWS 1996) is one prominent specification that also covers multiplanar circular HSS connections. It gives general design criteria for all non-overlapping, multiplanar tubular connections, without a need for connection classifications. However, the AWS approach should be treated with caution as it has been shown to be fairly inaccurate against a large database of tests. Although they exhibit a wide scatter, the AWS predictions have a tendency to become unsafe as d_0/t_0 decreases (i.e., the chord becomes stocky). This has been attributed by Kurobane (1995) to the "thickness squared" strength formulation in the AWS equation, whereas strength appears to vary with the 1.7 to 1.8 power of thickness.

Figure 5.8 shows the correlation with test results of the AWS method and of the method given in Table 3.1 (from Kurobane), for planar gap K connections. The AWS equation gives a lower bound and contains a built-in resistance factor of 0.74, so the AWS predicted strengths should be compared with the 1/0.74 line. It can be seen that the scatter is wide, and predictions err on the unsafe side for stocky chords. On the other hand, connection strengths predicted by the Kurobane approach (used throughout

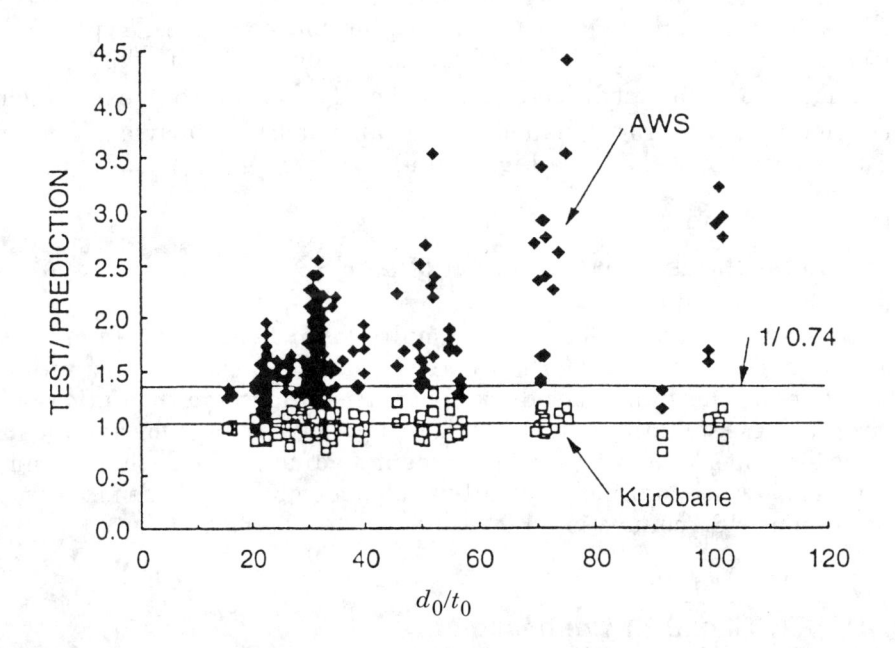

FIGURE 5.8
Ultimate strength predictions for planar K connections (Kurobane 1995)

this Design Guide) cluster closely about the 1.0 prediction line over the whole range of d_0/t_0, with a coefficient of variation of 9.4% (Kurobane 1995).

5.2.5 Fabrication Aspects

To avoid overlapping the branch members from one side plane onto those of the other, it may be necessary to introduce an offset perpendicular to the chord axis, as shown in Fig. 5.9. If the offset is less than 25% of the chord diameter it may be ignored when designing the connections, branch members and tension chord. However, moments due to the offset have to be distributed into a compression chord and considered in the chord design, (as for planar noding eccentricity moments).

Figure 5.10 (Wardenier et al. 1991) illustrates some geometric considerations. If the offset option is selected, the transverse gap g_t will be increased. If a single working point is chosen for the four branch members, the gap along the chord g will also increase with a larger g_t. The effects of g_t and g will be reflected in the connection resistance equations for the Type 1 and Type 2 failure modes. For the example in Fig. 5.10(b), the 50 mm offset perpendicular to the chord axis results in a 43 mm eccentricity in each of the planes of the web members.

If the two compression web members overlap each other and the two tension web members overlap each other, as shown in Fig. 5.10(a), but there is still a gap g along the chord between the tension and compression web members (under symmetrical loading), then the combined "footprint" can be treated as one unit and the KK connection checked for the Type 1 failure mode (see Section 5.2.3). As noted there, this entails using the transverse distance d'_1 and the modified angle θ'_1. The Type 2 failure mode (pinching between a pair of braces loaded in the same sense) clearly can not occur with an overlap as shown in Fig. 5.10(a).

5.3 Connections between Rectangular HSS

Less attention has been given to multiplanar rectangular (and square) HSS connections than to multiplanar circular HSS connections, although the former are now under study both experimentally and theoretically in a European research program. As little test evidence exists for axial loading on the branches, and even less for moment loading on them, the recommendations below are confined to axially-loaded members only (as for multiplanar circular HSS connections).

5.3.1 XX, TX and TT Connections

Davies and Morita (1991) initially showed by yield line modelling that little difference existed between the connection resistances of X and XX

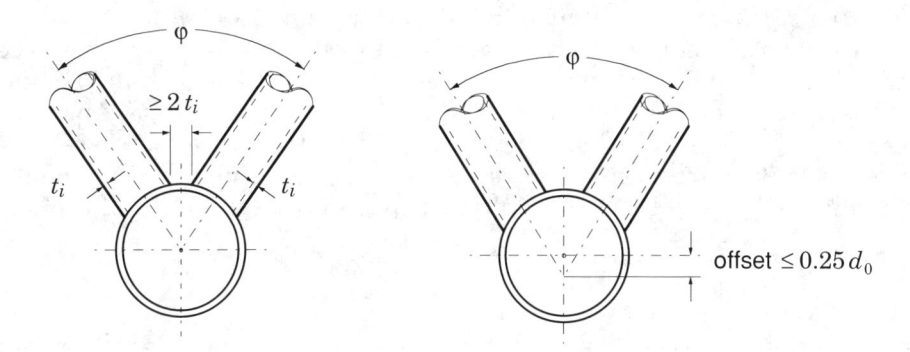

FIGURE 5.9
Transverse gap and offset for a KK connection

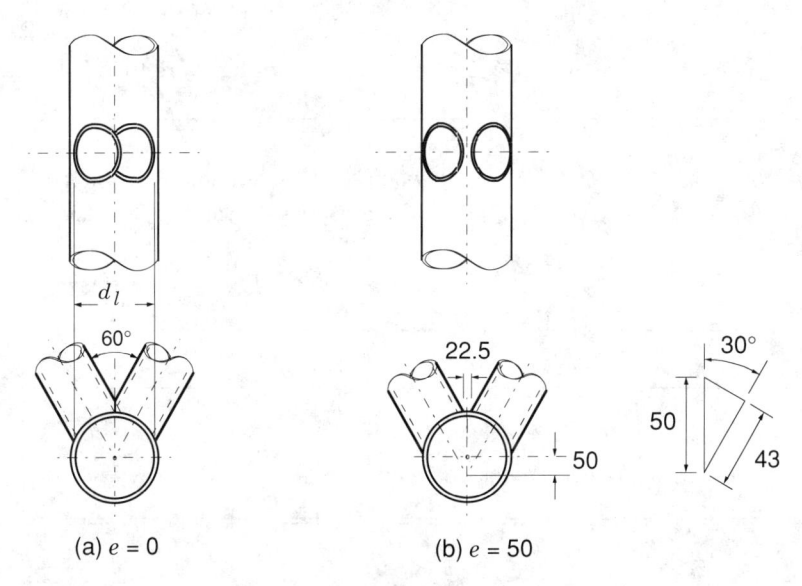

(a) $e = 0$

(b) $e = 50$

FIGURE 5.10
Geometry aspects of circular HSS KK connections

connections, and similarly between the connection resistances of 90° T and TT connections, *with loads in the same sense*. Strength enhancement was very modest, unlike XX connections of circular HSS. This difference from circular HSS connections can likely be attributed to the four flat rectangular HSS faces behaving somewhat more independently than the closed ring shape of the circular HSS.

Subsequent experiments on XX connections (Liu *et al.* 1993) and TX connections (Davies *et al.* 1993, Davies and Crockett 1996) have found that loads in the same sense do produce a small connection strength increase, while those in the opposite sense do cause a small connection strength decrease (relative to the planar case), but the range of connection parameters investigated is still limited. The act of welding brace members on in the out-of-plane direction serves to stiffen and strengthen the rectangular HSS side walls, merely by their physical presence regardlesss of the load

Type of Connection	Correction Factor to Planar Connection Resistances from Table 3.2 or Table 3.3
KK	0.9
	Also, for gap KK connections, check that: $$\left(\frac{N_{0\,(gap)}}{A_0 F_{y0}}\right)^2 + \left(\frac{V_f}{A_0 F_{y0}/\sqrt{3}}\right)^2 \leq 1.0.$$ (V_f is the total shear in the chord from both planes)
XX and TX	1.0 when $N_2/N_1 < 0$ and $\beta_{(in\ plane)} > 0.85$ 0.9 when $N_2/N_1 < 0$ and $\beta_{(in\ plane)} \leq 0.85$ 1.0 when $N_2/N_1 \geq 0$ (See Table 5.1 for N_1 and N_2)
Range of Validity	
$60° \leq \varphi \leq 90°$ and b_i/b_0, h_i/b_0 (Tables 3.2(a) and 3.3(a)) $\geq (0.1 + 0.015\, b_0/t_0)$	

TABLE 5.2

Correction factors for rectangular (and square) HSS multiplanar connection resistances

sense, particularly for larger β values (Liu *et al.* 1993, Davies *et al.* 1993, and Davies and Crockett 1993, 1996).

Evidence at present suggests that no adjustment be made to the planar connection resistance for multiplanar XX, TX and TT connections when all the brace loads act in the *same sense*. For brace loads acting in the *opposite sense* (i.e., N_2/N_1 in Table 5.1 being negative), a multiplanar reduction factor of at least 0.9 would be prudent for $\beta_{(in\ plane)} \leq 0.85$ (Davies and Crockett 1996). These factors are shown in Table 5.2.

5.3.2 KK Connections

Triangular trusses are frequently arranged in the form of a V that combines one bottom chord with two top chords, as seen in Fig. 5.11. Square chords have been used with web members framing on to the chord corners by shaping the web member ends, as for the Minneapolis Convention Center roof and the Minneapolis / St. Paul Twin Cities Airport skyway, discussed in Section 4.3. However, it is less expensive to rotate the bottom chord and to bevel cut the web ends so they land on the chord faces (Figs. 5.12 and 5.13). Purlins may not be necessary, as the top chords can be spaced at a suitable distance for the roof deck, which can be fastened directly to the flat surface of the chords. Incorporating diagonal members

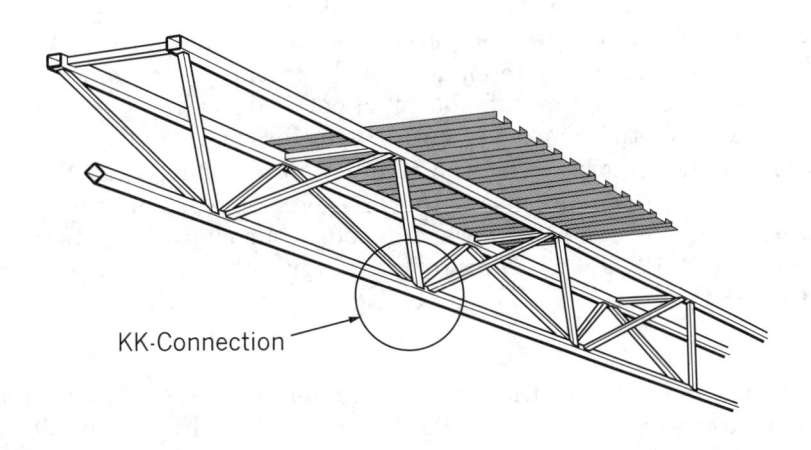

FIGURE 5.11
HSS triangular truss with two compression chords

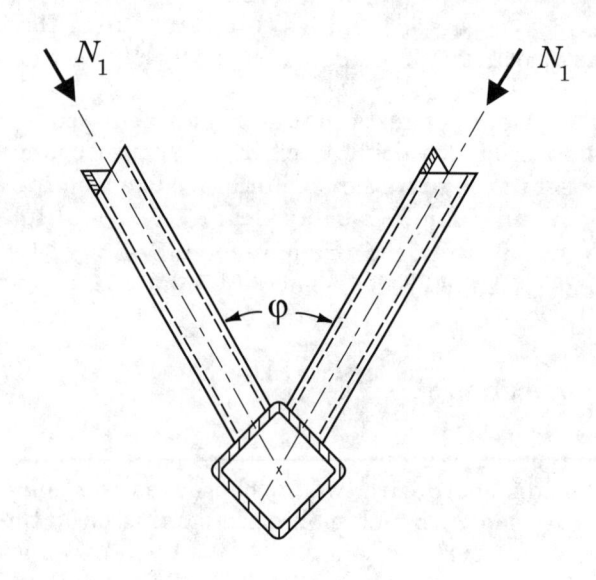

FIGURE 5.12
End view of KK connection to triangular truss tension chord

in the top plane of the triangular truss will increase torsional rigidity and lateral strength.

Initial tests by Coutie *et al.* (1983) found slight decreases in chord face strength for KK connections compared to planar K connections. Bauer and Redwood (1988) tested low to medium width ratio (β) KK connections like those in Fig. 5.12. They found that symmetrical loading (same sense) on adjacent walls of the chord produced little interactive effect. Indeed, in cases where the angle between web member planes (φ in Fig. 5.12) was less than 90° (leading to an increase in the effective value of β at the chord face), and when the web members were attached to the chord face off-centre (as drawn in Fig. 5.12), the strength of a triangular truss tension chord face was *greater* than that of a planar truss chord face with the same size of members.

Due to concern over the limited number of connection parameters tested, it was recently suggested (Packer *et al.* 1992) that a reduction factor of 0.9 be applied to the uniplanar K connection design formulae (Tables 3.2 and 3.3) to account for multiplanar action under symmetrical load conditions. Since then a further nine tests on rectangular HSS gap KK connections with $\varphi = 90°$ have been performed by British Steel (Yeomans 1993) and these have justified the reduction factor of 0.9, with the provision that small braces (low β) are not used with thin chord members.

To avoid the latter situation, it was recommended that the lower limit for the web to chord member width ratio be tightened from ($0.1 + 0.01\, b_0/t_0$)

FIGURE 5.13
Triangular truss KK connection with square HSS

to $(0.1 + 0.015\, b_0/t_0)$. In addition, Eurocode 3 (European Committee for Standardization 1992) suggests that one always perform a chord shear check for gap KK connections, even for square HSS members. These recommendations are incorporated in Table 5.2.

REFERENCES

AWS. 1996. Structural welding code—steel. ANSI/AWS D1.1-96, American Welding Society, Miami, Florida, U.S.A.

BAUER, D., and REDWOOD, R.G. 1988. Triangular truss joints using rectangular tubes. Journal of Structural Engineering, American Society of Civil Engineers, **114**(2): 408–424.

COUTIE, M.G., DAVIES, G., BETTISON, M., and PLATT, J. 1983. Development of recommendations for the design of welded joints between steel structural hollow sections or between steel structural hollow sections and H sections. Final report, Part 3—three dimensional joints. Report on ECSC Contract 7210.SA/814, University of Nottingham, England.

DAVIES, G., and CROCKETT, P. 1993. An interaction diagram for three-dimensional T-joints in rectangular hollow sections under both in-plane and out-of-plane axial loads. Proceedings, 5th. International Symposium on Tubular Structures, Nottingham, England, pp. 741–748.

DAVIES, G., and CROCKETT, P. 1996. The strength of welded T-DT joints in rectangular and circular hollow sections under variable axial loads. Journal of Constructional Steel Research, **37**(1): 1–31.

DAVIES, G., and MORITA, K. 1991. Three dimensional cross joints under combined axial branch loading. Proceedings, 4th. International Symposium on Tubular Structures, Delft, The Netherlands, pp. 324–333.

DAVIES, G., COUTIE, M.G., and BETTISON, M. 1993. The behaviour of three dimensional rectangular hollow section tee joints under axial branch loads. Proceedings, 5th. International Symposium on Tubular Structures, Nottingham, England, pp. 429–436.

EUROPEAN COMMITTEE for STANDARDIZATION. 1992. Eurocode 3: design of steel structures. Part 1.1—general rules and rules for buildings. ENV 1993-1-1:1992E, British Standards Institution, London, England.

KUROBANE, Y. 1995. Ultimate behaviour and design of multiplanar tubular joints. Proceedings, Workshop on Requalification of Tubular Steel Joints in Offshore Structures, Houston, Texas, U.S.A.

LIU, D.K., WARDENIER, J., de KONING, C.H.M., and PUTHLI, R.S. 1993. Static strength of multiplanar DX-joints in rectangular hollow sections. Proceedings, 5th. International Symposium on Tubular Structures, Nottingham, England, pp. 419–428.

PACKER, J.A., WARDENIER, J., KUROBANE, Y., DUTTA, D., and YEOMANS, N. 1992. Design guide for rectangular hollow section (RHS) joints under predominantly static loading. CIDECT (ed.) and Verlag TÜV Rheinland GmbH, Köln, Federal Republic of Germany.

PAUL, J.C., MAKINO, Y., and KUROBANE, Y. 1994. Ultimate resistance of unstiffened multiplanar tubular TT- and KK-joints. Journal of Structural Engineering, American Society of Civil Engineers, **120**(10): 2853–2870.

van der VEGTE, G.J. 1995. The static strength of uniplanar and multiplanar tubular T- and X-joints. Ph.D. thesis, Delft University of Technology, Delft, The Netherlands.

WARDENIER, J., KUROBANE, Y., PACKER, J.A., DUTTA, D., and YEOMANS, N. 1991. Design guide for circular hollow section (CHS) joints under predominantly static loading. CIDECT (ed.) and Verlag TÜV Rheinland GmbH, Köln, Federal Republic of Germany.

YEOMANS, N.F. 1993. Rectangular hollow section double K-joints - experimental tests and analysis. Proceedings, 5th. International Symposium on Tubular Structures, Nottingham, England, pp. 437–445.

6

HSS TO HSS MOMENT CONNECTIONS

6.1 Vierendeel Connections between Square or Rectangular HSS

In this section, the behaviour, strength and flexibility of Vierendeel connections are examined. In addition, an example truss is presented with the members selected first with plastic design, and then with elastic design.

6.1.1 Introduction to Vierendeel Trusses

Arthur Vierendeel first proposed Vierendeel trusses in 1896. They are comprised of chord members connected to web members which are nearly always at 90° to the chords, as shown in Fig. 6.1. The typical design premise with Vierendeel trusses has been to assume full connection rigidity, but this is very rarely the case with HSS to HSS Vierendeel connections. Unlike triangulated Warren or Pratt trusses, in which the connections approach a pinned condition at their ultimate limit state and cause the branch members to be loaded by predominantly axial forces, Vierendeel connections have branch (web) members subjected to substantial bending moments as well as axial and shear forces.

Until very recently, most of the testing performed on Vierendeel connections has been on isolated connection specimens with a lateral load applied to the vertical branch member while the connection is in an inverted T position, as shown in Fig. 6.2. Thus, the connection strength and moment-rotation behaviour have been assessed mainly by researchers under moment plus shear loading.

Square and rectangular HSS single chord connections loaded by in-plane bending moments have been studied by Duff (1963), Redwood (1965), Cute *et al.* (1968), Mehrotra and Redwood (1970), Lazar and Fang (1971),

FIGURE 6.1
Rectangular HSS Vierendeel trusses, Scotiabank building, Toronto, Ontario

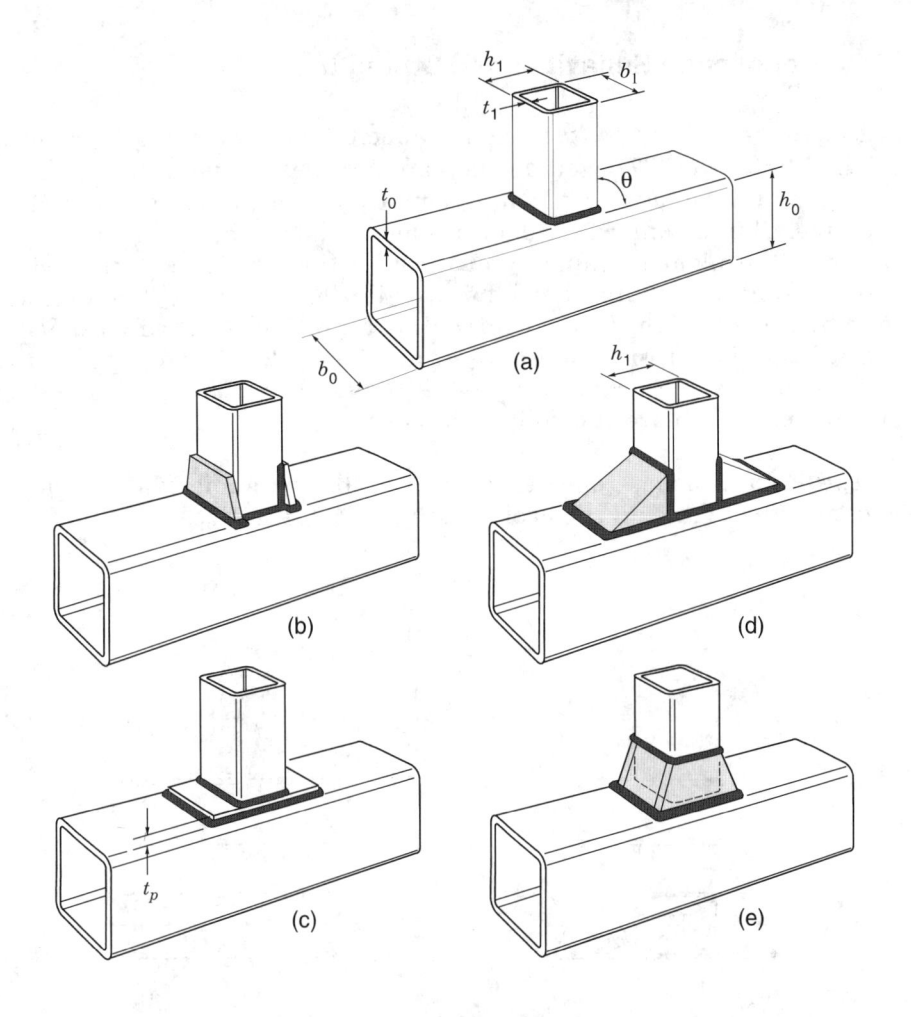

FIGURE 6.2

HSS Vierendeel connection types tested by Korol *et al.* (1977)
(Comments on reinforced types are in Section 6.4)

(a) Unreinforced
(b) With branch plate stiffeners (d) With haunch stiffeners
(c) With chord plate stiffener (e) With truncated pyramid stiffeners

Mehrotra and Govil (1972), Staples and Harrison (undated), Brockenbough (1972), Korol *et al.* (1977), Korol and Mansour (1979), Giddings (1980), Kanatani *et al.* (1980), Korol *et al.* (1982), Korol and Mirza (1982), Mang *et al.* (1983), Davies and Panjeh Shahi (1984), Szlendak and Brodka (1985, 1986a, 1986b), Szlendak (1986) and Kanatani *et al.* (1986).

Researchers concur that both the strength and flexural rigidity of an unstiffened connection decrease as the chord slenderness ratio b_0/t_0 increases, and as the branch to chord width ratio b_1/b_0 (or β) decreases. Connections with $\beta \approx 1.0$ and a low b_0/t_0 value approach full rigidity, but all other unstiffened connections can be classed as semi-rigid.

6.1.2 Connection Behaviour and Strength

The connection *ultimate* moment capacity in tests is typically recorded, and Korol et al. (1977) even develop an empirical formula for estimating the maximum connection moment, but this moment typically occurs at excessively large connection deformations. Thus, for all practical design purposes, the moment capacity of a connection can be determined in a manner similar to that used for axially-loaded HSS T connections, whereby the strength is characterized by an ultimate bearing capacity or by a deformation or rotation limit (Wardenier 1982). This design approach is more apparent if one considers the possible failure modes for such connections, which are shown in Fig. 6.3.

The failure modes represented in Fig. 6.3 presume that neither the welds nor the members themselves are critical (e.g., local buckling of the

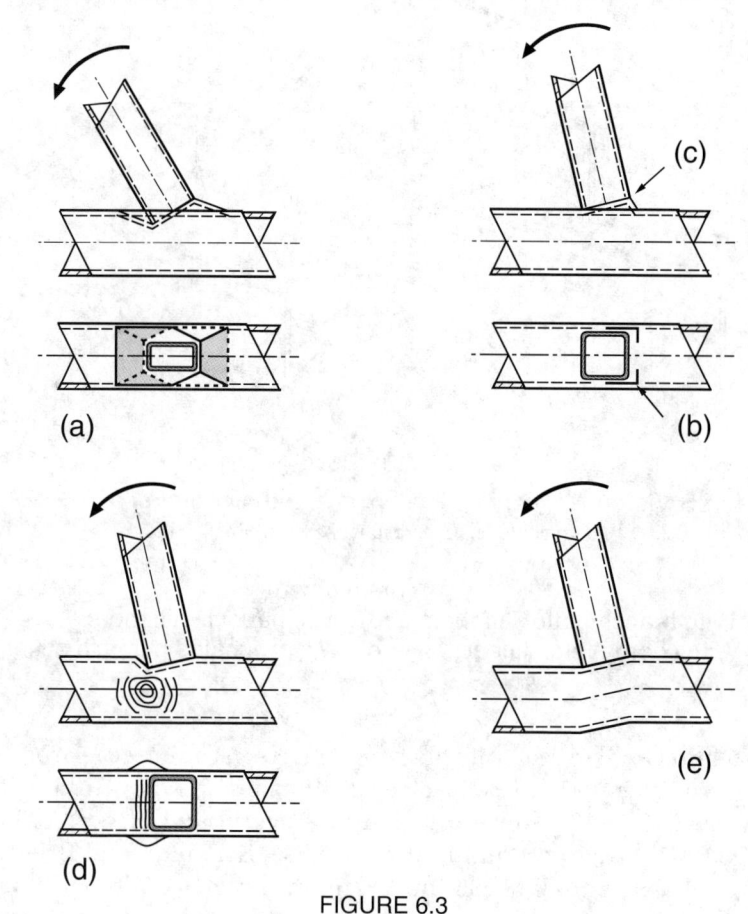

FIGURE 6.3
Possible failure modes for HSS connections loaded by
in-plane bending moments (Wardenier 1982)

(a) Chord face yielding
(b) Cracking in chord
(c) Cracking in branch member
(d) Crippling of the chord side walls
(e) Chord shear failure

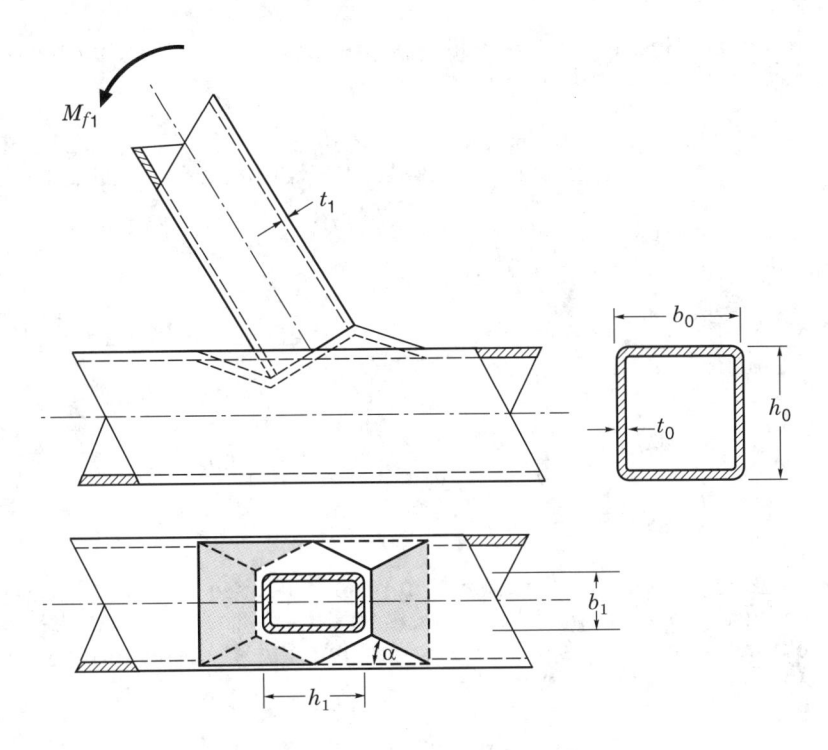

FIGURE 6.4
Yield line mechanism for chord face yielding under in-plane bending (failure Mode (a))

branch is precluded). Cracking in the chord (chord punching shear) has not actually been observed in any tests, and chord shear failure is strictly a member failure, so analytical solutions for failure Modes (b) and (e) are not considered herein.

For Mode (a), the moment capacity of connections with low to moderate values can be determined by the yield line model in Fig. 6.4. Neglecting the influence of membrane effects and strain hardening, the moment capacity for a Y connection where $\theta_1 \neq 90°$ is given by Wardenier (1982):

$$M_{r1}^* = 0.5 F_{y0} t_0^2 b_0 \left(1 + \frac{4h_1/b_0}{\sin\theta_1 \sqrt{1-\beta}} + \frac{2(h_1/b_0)^2}{\sin^2\theta_1 (1-\beta)} \right) f_2(n) \quad [6.1]$$

for $\beta \leq 0.85$.

The term $f_2(n)$ is a function to allow for the reduction in connection moment capacity in the presence of large *compression chord* forces. According to de Koning and Wardenier (1985):

$$f_2(n) = 1.2 + \left(\frac{0.5}{\beta}\right) n, \quad \text{but} \not> 1.0 \quad [6.1a]$$

where n is negative and is the axial compression load in the chord expressed as a fraction of the chord yield load, i.e.,

$n = N_0/(A_0 F_{y0})$. (The bending moment in the branch will be automatically distributed to the chord member, on either side of the connection, for equilibrium.)

For tension chords, $f(n) = 1.0$. Equation 6.1a is shown graphically in Fig. 6.5.

Nearly all Vierendeel connections have the branch to chord angle equal to 90°, which simplifies [6.1] to

$$M_{r1}^* = F_{y0} t_0^2 h_1 \left(\frac{1}{2h_1/b_0} + \frac{2}{\sqrt{1-\beta}} + \frac{h_1/b_0}{(1-\beta)} \right) f_2(n) \quad [6.2]$$

for $\beta \le 0.85$.

For Mode (c), an effective width approach is used to relate the reduced capacity of the branch member (considered to be the same on the tension and compression flanges of the branch member) to the applied branch moment as follows (Wardenier 1982):

$$M_{r1}^* = F_{y1} \left(Z_1 - \left(1 - \frac{b_e}{b_1} \right) b_1 t_1 (h_1 - t_1) \right) \quad [6.3]$$

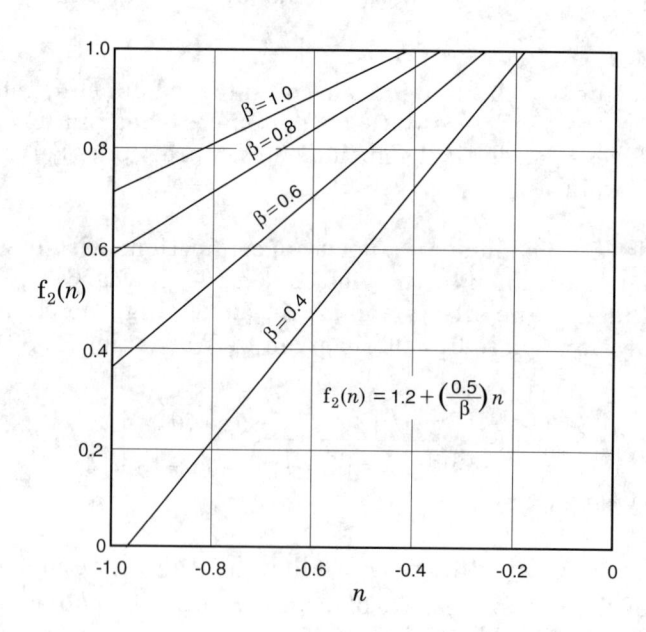

FIGURE 6.5

Connection strength reduction factor $f_2(n)$ as a function of the compressive load in the chord expressed as a fraction of the chord yield load ($n = N_0/A_0 F_{y0}$). For chord tension loading, $f_2(n) = 1.0$.

The term b_e in [6.3] is the effective width of the branch member flange, and is given by

$$b_e = \frac{10}{b_0/t_0} \frac{F_{y0} t_0}{F_{y1} t_1} b_1, \text{ but } \not> b_1 \qquad [6.4]$$

For Mode (d), a chord side wall bearing or buckling capacity can conservatively be given by [6.5] (Wardenier 1982), which is illustrated in Fig. 6.6:

$$M_{r1}^* = 0.5 F_k t_0 (h_1 + 5 t_0)^2 \qquad [6.5]$$

This moment is derived from stress blocks of twice (two walls) $F_k t_0 (h_1/2 + 2.5 t_0)$ acting as a couple at centres of $h_1/2 + 2.5 t_0$.

F_k is the buckling stress of the chord side walls, but since the compression is very localized, tests (de Koning and Wardenier 1985) have shown that buckling is less critical for moment-loaded T connections than for axially-loaded T connections. Hence, within the parameter range of validity given later, the chord yield stress can be used instead of the buckling stress for T connections. The fact that bearing rather than buckling will control is also supported by Davies and Packer (1987), since the ratio of bearing length ($h_1/2 + 2.5 t_0$) to chord height (h_0) is low. For X connections, F_{y0} is reduced by 20% to be consistent with Table 3.3.

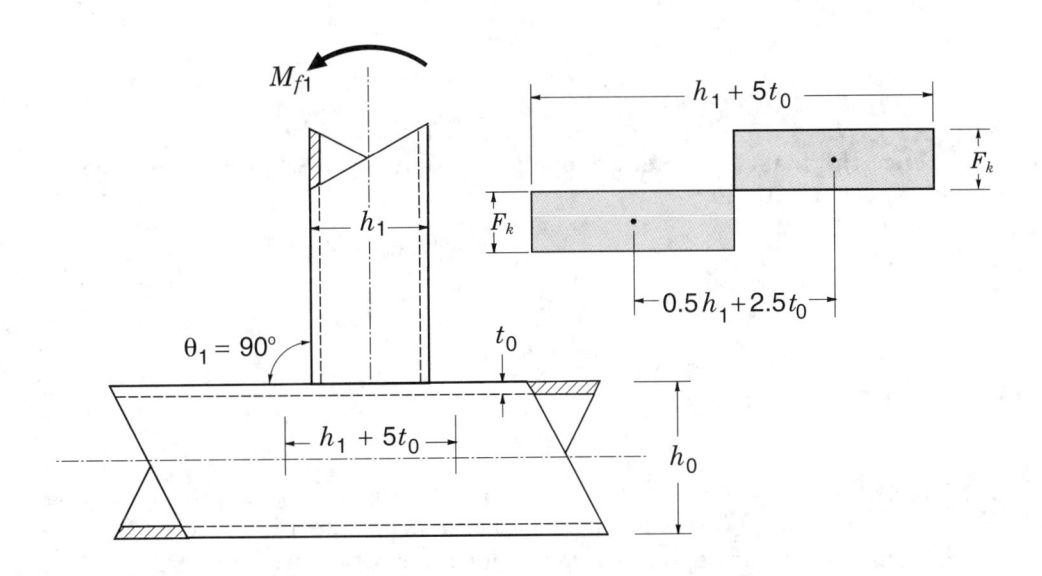

FIGURE 6.6
Chord side wall bearing or buckling failure model under in-plane bending (failure Mode (d))

Thus, for design purposes an estimate of the connection moment capacity can be obtained from the lower of the M_{r1}^* values obtained from [6.2], [6.3], and [6.5]. It can be seen that the moment capacity predicted by [6.2] tends towards infinity as β tends towards unity, and so this failure mode (which corresponds to a state of complete connection plastification and hence high joint deformations) is not critical in the high β range. This accounts for the β ≤ 0.85 limit attached to [6.2] and, for high β values, the web crippling failure criterion expressed by [6.5] will likely govern.

From the above expressions for M_{r1}^* it can be seen that full width (β ≈ 1.0) unstiffened HSS Vierendeel connections are capable of developing the full moment capacity of the branch member, providing b_0/t_0 is sufficiently low. For $h_0 = b_0 = h_1 = b_1$ and $h_0/t_0 \le 16$, the chord side wall web crippling capacity is given by Wardenier (1982) for hot and cold formed sections:

$$M_{r1}^* = 12 F_{y0} t_0^2 h_1. \qquad [6.6]$$

For Class C or Class H material and $h_0/t_0 = 12$, for example:

$$M_{r1}^* = 0.5 F_k t_0 (h_1 + 5t_0)^2 \qquad ([6.5])$$

$$= 0.5 F_{y0} t_0 \left(h_1 + \frac{5h_1}{12} \right)^2 \qquad \text{(since } t_0 = \frac{h_0}{12} = \frac{h_1}{12}\text{)}$$

$$= F_{y0} t_0 h_1^2$$

$$\therefore M_{r1}^* = 12 F_{y0} t_0^2 h_1$$

Since the plastic moment capacity of a square HSS branch member is given approximately by

$$M_{p1} \approx 1.5 b_1^2 t_1 F_{y1} \qquad [6.7]$$

then,

$$\frac{M_{r1}^*}{M_{p1}} \approx \frac{8}{b_0/t_0} \frac{F_{y0} t_0}{F_{y1} t_1} \qquad [6.8]$$

Therefore, for the same steel grades used throughout a truss and β ≈ 1.0, dimensional ratios of $b_0/t_0 = 16$ and $t_0/t_1 = 2$ will produce a connection with a moment capacity approximately equal to the plastic moment capacity of the branch (Wardenier 1982). In this case the branch member cross-section is fully effective ($b_e = b_1$ in [6.3] and [6.4]). The above is similar to the recommendation by Korol *et al.* (1977) for Class H sections that b_0/t_0 be less than 16 with β = 1 for full moment transfer to be assumed at the connection.

Where necessary, the resistance factor ϕ is already included in the above resistance expressions of M_{r1}^* for their use in a Canadian limit states design format. Even though a rigorous evaluation of the proposed equations against *all* available experimental data has not yet been performed, experience suggests that these proposed equations will prove to be practical lower bounds on the connection moment capacity.

The expressions for M_{r1}^* also have a limited range of validity which corresponds to the limits of the test data against which the equations have been checked. This validity range is

$$b_0/t_0 \leq 35$$
$$h_0/t_0 \leq 35$$
$$\theta_1 = 90°$$
$$F_{yi} \leq 355 \text{ MPa}$$

and the compression branch member is restricted to Class 1 sections.

The welds in HSS moment connections are loaded in a highly non-uniform manner, and should be capable of sustaining significant connection rotations. To enable adequate load redistribution to take place, the size of a fillet weld should be at least as large as that now specified for axially-loaded connections to develop the capacity of a branch member, as per CAN/CSA-S16.1-94, Clause 13.13.2.2. Under these provisions, with the branch member at 90° to the chord, the weld throat thickness t_w amounts to 1.10 times the branch wall thickness. (It is assumed that the HSS material has a specified ultimate stress of 450 MPa, as has CAN/CSA-G40.21, Grade 350W (CSA 1992).) Such fillet weld sizes are listed in Table 3-46 in the sixth edition of the CISC Handbook (CISC 1995). The reader is referred to Section 8.2.4.2.1 for an in-depth discussion of fillet weld use on HSS.

The previous expressions for moment capacity are based on moment loading only, whereas in Vierendeel trusses significant axial loads may also exist in the branch members. The effect of the axial load on the connection moment capacity depends on the critical failure mode, and so a complex set of interactions is developed. Consequently, Wardenier (1982) has conservatively proposed that a linear interaction relationship be used to reduce the in-plane moment capacity of a Vierendeel connection as follows:

$$\frac{N_1}{N_1^*} + \frac{M_{f1}}{M_{r1}^*} \leq 1.0 \qquad [6.9]$$

N_1 and M_{f1} are the applied axial load and bending moment respectively in the branch member, M_{r1}^* is the lower of the values obtained from [6.2], [6.3] and [6.5], and N_1^* is the connection resistance with only an axial load applied to the branch member.

The resistance of an HSS T connection under branch member axial load is given and discussed in Chapter 3, but is reproduced below for the most relevant case of $\beta \approx 1.0$. There are two failure modes to be checked. Web crippling of the chord member side walls is again the likely governing failure mode, and can be estimated by (IIW 1989):

$$N_1^* = 2F_k t_0 (h_1 + 5t_0) \qquad [6.10]$$

where F_k is determined as in Table 3.3. The value for F_k in [6.10] assumes that the branch member is in compression; if the branch is in axial tension, $F_k = F_{y0}$, which corresponds to chord wall tensile yielding. The other failure mode to check, for an HSS T connection with $\beta \approx 1.0$, is premature failure of the branch member. This is also termed an "effective width" check on the branch member, and is expressed by (IIW 1989):

$$N_1^* = F_{y1} t_1 (2h_1 - 4t_1 + 2b_e) \qquad [6.11]$$

where b_e is given by [6.4]. Thus the connection resistance as an axially-loaded HSS T connection, for $\beta \approx 1.0$, will be given by the lower of the N_1^* values from [6.10] and [6.11].

6.1.3 Connection Flexibility

In the foregoing, it was shown that unstiffened HSS connections with $\beta \approx 1.0$ and certain b_0/t_0 and t_0/t_1 values could achieve the full moment capacity of the branch member, but it should be remembered that any connection moment resistance calculated (M_{r1}^*) must be reduced to take account of the influence of axial load in the branch member (see [6.9]). Such *connections* which still develop a moment resistance which exceeds the moment capacity of the branch *member* can be considered as fully rigid for the purpose of analysis of the Vierendeel truss. All other connections, should be considered as semi-rigid.

To analyze a frame which is connected by semi-rigid connections, one needs the load-deformation characteristics of the connections being used. These can be obtained from finite element analysis, laboratory tests or published databases.

6.1.4 Design Example

The Vierendeel truss shown in Fig. 6.7 is to be designed for a factored panel point load P of 17 kN. All the connection points are laterally braced, perpendicular to the truss, by secondary members. The top and bottom chord members will be the same, and one section size will be used for all web members. All steel used has a yield stress F_y of 350 MPa. A statically admissible set of moments and shears follows in Fig. 6.8.

6.1.4.1 Plastic Design of Members

All members are loaded in double curvature, and plastic hinges will occur at the connections. Thus, the maximum unbraced length for chord members is 3 000 mm, and for web members 2 500 mm. By CAN/CSA-S16.1-94 the maximum unbraced length from a braced hinge location is given by Clause 13.7

$$L_{cr} = \frac{980 r_y}{\sqrt{F_y}} \quad \text{for } \kappa \geq -0.5 \quad [6.12]$$

where κ is the ratio of the smaller factored moment to the larger factored moment at opposite ends of the unbraced length, and is positive for double curvature. Hence, $\kappa = +1.0$ for all members, and L_{cr} is given by [6.12].

However, it should be noted that this restriction is <u>not</u> intended for application to square and circular sections, nor for rectangular sections bent about their minor axis, as lateral-torsional buckling of these flexural members need not be considered. Some steelwork design specifications (for example, AISC (1993) LRFD Clause F1) specifically point out this exclusion for these particular cross-sections.

Try HSS 152x152x9.5 for the chords

(Note that $b_0/t_0 = 152/9.53 = 15.9 \leq 16$.)

Check that this section is Class 1 (suitable for plastic design) at the worst axial load condition of $C_f = 5.1P = 86.7$ kN, using CAN/CSA-S16.1-94, Clause 11.3.

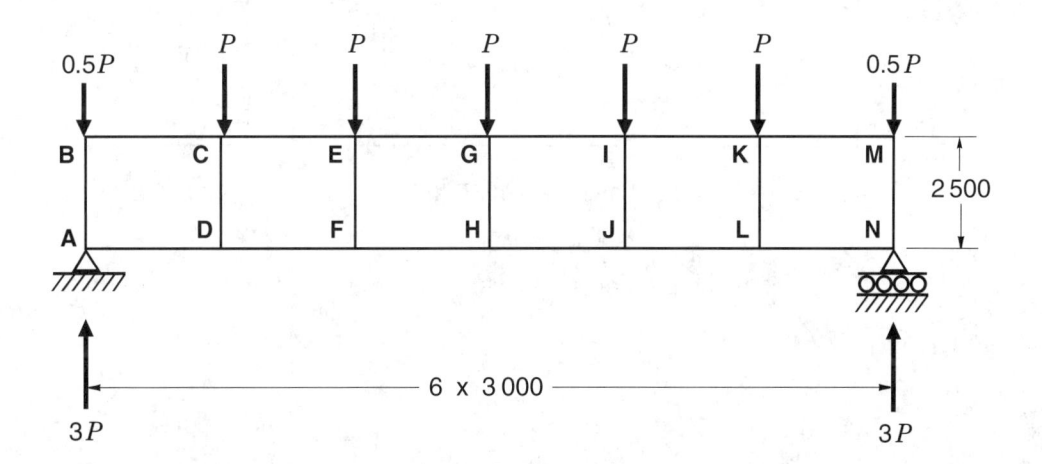

FIGURE 6.7
Example Vierendeel truss

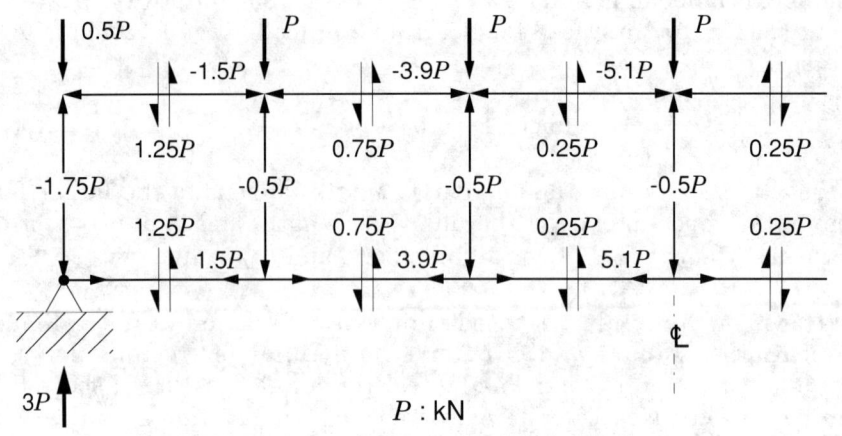

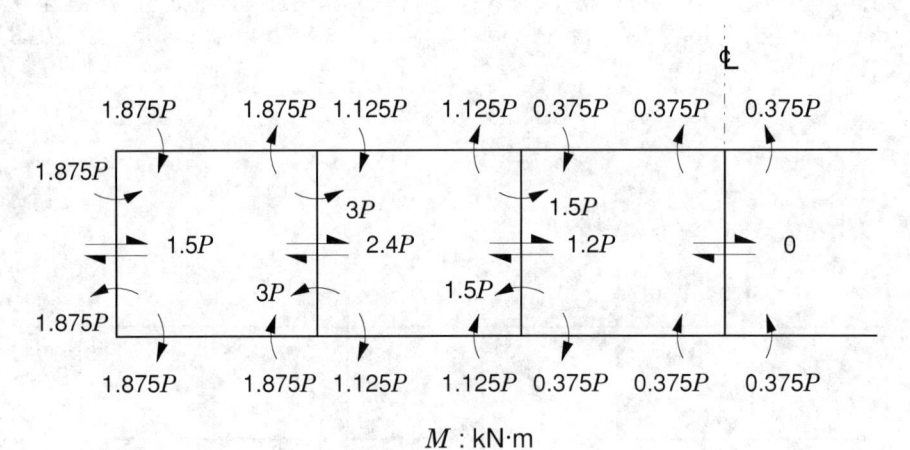

FIGURE 6.8
Forces and moments within Vierendeel truss
(shown applied to nodes)

For Class 1:
$$b/t \le \frac{420}{\sqrt{F_y}} = \frac{420}{\sqrt{350}} = 22.4$$

where b = flat width of the tube flange.

Assume an outside corner radius of $2t$.

$$b/t = (152 - 4(9.53))/9.53 = 12.0 \quad \le 22.4 \quad \therefore \text{OK}$$

Also for Class 1:
$$h/w \le \frac{1100}{\sqrt{F_y}}\left(1 - 0.39\left(\frac{C_f}{C_y}\right)\right)$$

$$= \frac{1100}{\sqrt{350}}\left(1 - 0.39\left(\frac{86.7}{5210(0.350)}\right)\right) = 57.7$$

where

h = clear height of the web between flanges of the member
w = web thickness ($= t$)

$$\therefore h/w = (152 - 2(9.53))/9.53 = 14.0 \quad \le 57.7 \quad \therefore \text{OK}$$

Further,

maximum $M_{f0} = 1.875P = 1.875(17) = 31.9$ kN·m

$M_{r0} = \phi Z_0 F_{y0} = 0.9(275)0.350 = 86.6$ kN·m $\quad \ge 31.9 \quad \therefore$ OK

Therefore, HSS 152x152x9.5 is Class 1, and is suitable for the chords.

Try HSS 152x152x6.4 for the web members

(Note that $\beta = 1.0$.)

$b/t = (152 - 4(6.35))/6.35 = 20.0 \quad \le 22.4 \quad \therefore$ OK

Maximum $C_{f1} = 1.75P = 29.8$ kN

$$h/w \le \frac{1100}{\sqrt{350}}\left(1 - 0.39\frac{29.8}{3610(0.350)}\right) = 58.3$$

$h/w = (152 - 2(6.35))/6.35 = 22.0 \quad \le 58.3 \quad \therefore$ OK

Maximum $M_{f1} = 3P = 3(17) = 51.0$ kN·m

$M_{r1} = \phi Z_1 F_{y1} = 0.9(195)0.350 = 61.4$ kN·m $\quad \ge 51.0 \quad \therefore$OK

Therefore, HSS 152x152x6.4 is Class 1, and is suitable for the webs.

Reductions in plastic moment capacity due to axial force or shear force can be shown to be negligible (Horne and Morris 1985).

Plastic collapse mechanism

Figure 6.9 illustrates the collapse mechanism. Let λ' be the additional multiplication factor by which the already factored loads have to be increased to cause plastic collapse. By the principle of Virtual Work,

$$17\lambda' (3\theta + 6\theta + 6\theta + 6\theta + 3\theta) = 86.6 (4\theta) + 61.4 (8\theta)$$

and $\lambda' = 2.05$.

Therefore, adequate reserve capacity exists for ultimate strength, as $\lambda' \geq 1.0$.

Connection capacity check

As β is 1.0, the moment resistance of the connection could be limited by Mode (c), cracking in the branch member, or Mode (d), chord side wall buckling.

Mode (c)

$$M_{r1}^* = F_{y1}\left[Z_1 - \left(1 - \frac{b_e}{b_1}\right) b_1 t_1(h_1 - t_1)\right] \quad ([6.3])$$

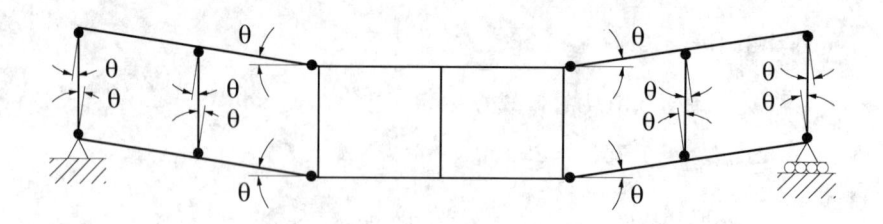

FIGURE 6.9
Plastic collapse mechanism

$$b_e = \frac{10}{b_0/t_0} \frac{t_0}{t_1} b_1 \quad ([6.4])$$

$$= \frac{10}{152/9.53} \left(\frac{9.53}{6.35}\right) 152.4 = 143 \text{ mm}$$

$$\therefore M_{r1}^* = 0.350 \left[195\,(10^3) - \left(1 - \frac{143}{152}\right) 152\,(6.35)\,(152 - 6.35) \right]/10^3$$

$$= (0.350 \text{ kN/mm}^2)\,(186 \text{ mm}^2\cdot\text{m})$$

$$= 65.1 \text{ kN·m} \quad \geq 51.0 \quad \therefore \text{ OK}$$

Mode (d)

$$M_{r1}^* = 0.5 F_k t_0 (h_1 + 5t_0)^2 \quad ([6.5])$$

For Class C or Class H,
$$M_{r1}^* = 0.5\,(0.350)\,9.53\,(152 + 5\,(9.53))^2/10^3$$

$$= 66.7 \text{ kN·m} \quad \geq 51.0 \quad \therefore \text{ OK}$$

Therefore, the limiting connection moment resistance would be determined by failure Mode (c) and is 65.1 kN·m. However, this connection resistance is greater than the resistance of the member itself (61.4 kN·m), so the connection moment resistance would be limited to $M_{r1}^* = 61.4$ kN·m.

Now check that the moment and axial force interaction is satisfied according to

$$\frac{N_1}{N_1^*} + \frac{M_{f1}}{M_{r1}^*} \leq 1.0 \quad ([6.9])$$

It will be assumed that Class C sections are chosen.

F_k, the buckling stress of the chord side walls is required in order to evaluate N_1^*, and this can be determined by considering each chord side wall to be a pin-ended strut of length $(h_0 - 2t_0)$ and having a radius of gyration of $t_0/3.46$ (Packer 1984). Thus, assuming that a Grade 350W Class 1, 2 or 3 HSS section is being used, F_k can be determined from CAN/CSA-S16.1-94, Clause 13.3.1 (with coefficient $n = 1.34$ for HSS manufactured to Class C requirements) as follows:

$$\lambda = \left(\frac{F_y}{\pi^2 E}\right)^{0.5} \frac{KL}{r}$$

$$= \left(\frac{350}{3.14^2 (200) 10^3}\right)^{0.5} \frac{3.46(h_0 - 2t_0)}{t_0}$$

$$= 0.0461 \frac{152 - 2(9.53)}{9.53} = 0.643$$

$$F_k = C_r/\phi A = F_{y0}\left(1 + \lambda^{2n}\right)^{-1/n}$$

$$= 350\left(1 + 0.643^{2(1.34)}\right)^{-1/1.34}$$

$$= 287 \text{ MPa}$$

$$N_1^* = 2F_k t_0 (h_1 + 5t_0) \quad ([6.10])$$
$$= 2(0.287)\, 9.53\, (152 + 5(9.53)) = 1090 \text{ kN}$$

or

$$N_1^* = F_{y1} t_1 (2h_1 - 4t_1 + 2b_e) \quad ([6.11])$$

$$b_e = \frac{10}{b_0/t_0} \frac{t_0}{t_1} \quad b_1 = 143 \text{ mm, as before}$$

$$\therefore N_1^* = 0.350\, (6.35)\, (2(152) - 4(6.35) + 2(143))$$
$$= 1260 \text{ kN}$$

Therefore, governing value of $N_1^* = 1090$ kN.

Hence, one should check the connections to the outside posts (maximum axial compression force of $1.75P = 29.8$ kN), and the connections to the most critical interior vertical (having a maximum moment of $3P = 51$ kN·m).

For outside posts:

$$(29.8/1090) + (31.9/61.4) = 0.03 + 0.52$$
$$= 0.55 \quad \leq 1.0 \quad \therefore \text{ OK}$$

For interior verticals:

$$(8.5/1090) + (51.0/61.4) = 0.01 + 0.83$$
$$= 0.84 \quad \leq 1.0 \quad \therefore \text{ OK}$$

Therefore, connection resistance is adequate; the truss is satisfactory.

6.1.4.2 Elastic Design of Members

Use CAN/CSA-S16.1-94, Clause 13.8.1 to check the top chord as a beam-column.

Axial force plus bending combination is

$$\frac{C_{f0}}{C_{r0}} + \frac{U_1 M_{f0}}{M_{r0}} \leq 1.0$$

Check overall member strength of length in end panel

$C_{f0} = 1.5P = 1.5\,(17) = 25.5\text{ kN}$

$C_{r0} = 1\,290\text{ kN}$ $\quad$ ($K = 1.0$ for S16.1, Clause 13.8.1(b), and $L = 3\,000$)

$$U_1 = \frac{\omega_1}{1 - \dfrac{C_{f0}}{C_e}} \quad \text{(S16.1, Clause 13.8.3)}$$

Since there are no transverse loads between panel points,

$\omega_1 = 0.6 - 0.4\kappa \geq 0.4$ $\quad$ (S16.1, Clause 13.8.4(a))

$ = 0.6 - 0.4\,(+1.0) = 0.2 \quad \therefore \omega = 0.4$

$$C_e = \frac{\pi^2 EI}{L^2}$$

$ = 3.14^2\,(200)\,17.3\,(10^6)/3\,000^2 = 3\,790\text{ kN}$

$\therefore U_1 = 0.4/(1 - (25.5/3\,790)) = 0.403$

$M_{f0} = 1.875P = 1.875\,(17) = 31.9\text{ kN·m}$

$M_{r0} = \phi Z_0 F_{y0} = 86.6\text{ kN·m}$, as earlier

Therefore, combination $\dfrac{C_{f0}}{C_{r0}} + \dfrac{U_1 M_{f0}}{M_{r0}}$

$ = (25.5/1\,290) + (0.403\,(31.9)/86.6)$

$ = 0.020 + 0.148 = 0.17 \quad \leq 1.0 \quad \therefore \text{OK}$

Check cross section strength of length in end panel

$C_{f0} = 25.5\text{ kN}$

$C_{r0} = \phi A_0 F_{y0} = 0.9\,(5\,210)\,0.350 = 1\,640\text{kN}$

$U_1 = 1.0$ (S16.1, Clause 13.8.1(a))

$M_{f0} = 31.9$ kN·m   and $M_{r0} = 86.6$ kN·m

∴ combination = $(25.5/1\,640) + (1.0\,(31.9)/86.6)$

$= 0.016 + 0.368 = 0.38$ ≤ 1.0 ∴ OK

Therefore, bottom chord will also be adequate.

Use S16.1, Clause 13.8.1 to check vertical web members as beam-columns.

Check overall member strength of first vertical from end

$C_{f0} = 0.5P = 0.5\,(17) = 8.5$ kN

$C_{r0} = 984$ kN $(L = 2\,500)$

$$U_1 = \frac{\omega_1}{1 - \dfrac{C_{f0}}{C_e}}$$

$$C_e = \frac{\pi^2 EI}{L^2}$$

$= 3.14^2\,(200)\,12.6\,(10^6)/2\,500^2 = 3\,980$ kN

∴ $U_1 = 0.4/(1 - (8.5/3\,980)) = 0.401$

$M_{f0} = 3P = 3\,(17) = 51.0$ kN·m

$M_{r0} = \phi Z_0 F_{y0} = 61.4$ kN·m, as earlier

Therefore, combination $\dfrac{C_{f0}}{C_{r0}} + \dfrac{U_1 M_{f0}}{M_{r0}}$

$= (8.5/984) + (0.401\,(51.0)/61.4)$

$= 0.009 + 0.333 = 0.34$ ≤ 1.0 ∴ OK

Check cross section strength of first vertical from end

$C_{f0} = 8.5$ kN

$C_{r0} = \phi A_0 F_{y0} = 0.9\,(3\,610)\,0.350 = 1\,140$ kN

$U_1 = 1.0$

$M_{f0} = 51.0$ kN·m   and $M_{r0} = 61.4$ kN·m

∴ combination = $(8.5/1\,140) + (1.0\,(51.0)/61.4)$

$$= 0.007 + 0.831 = 0.84 \quad \leq 1.0 \quad \therefore \text{OK}$$

Hence, the members would also be suitable by elastic design procedures. By either design method, the chord thickness is still enhanced to provide adequate connection strength. The end connections (at A, B, M and N) can be made by welding the vertical posts to the chords to form T connections, and then adding capping plates to the ends of the chord sections.

6.2 Knee Connections

Research on mitred square and rectangular HSS knee connections (such as those in Fig. 6.10) has been performed by Mang et al. (1980) at the University of Karlsruhe and subsequently incorporated into the German standard DIN 18 808 (1984). Their recommendations have also been reported by Wardenier (1982), CIDECT (1984), and Dutta and Würker (1988). They cover both stiffened and unstiffened knee connections, and are intended for use in corner connections of rigid frames.

The original test results and moment vs. rotation diagrams are not widely available, but DIN 18 808 makes its design recommendations applicable for "flexurally rigid frame corners". However, it could be expected that the rotation capacity of some *unstiffened* connections might be low, and in structures in which reasonable rotational capacity is required, a stiffened knee connection should be used (Wardenier 1982). In the Karlsruhe tests, simple unstiffened knee connections tended to fail by excessive deformation of the lateral HSS cross-wall in compression. On the other hand, for connections with a stiffening plate, excessive deformations appeared only for very thin walled members.

For thicker hollow sections, complete plastification was reached in the course of the tests (CIDECT 1984). Table 6.1 (adapted from DIN 18 808) gives the limiting HSS member dimensions which are permitted. In view of the uncertain moment vs. rotation properties, it would seem prudent to use, for *unstiffened* connections, only compact HSS members which satisfy plastic design requirements for rigid frames (i.e., Class 1 sections).

Analysis of the test results showed that, for design purposes, it was possible to estimate the total flexural and axial load capacity of the connection by applying a reduction factor to the material yield stress. Thus, adequate connection strength will be obtained for both stiffened and unstiffened 90° mitre connections providing [6.13] and [6.14] are satisfied (DIN 18 808 1984; Eurocode Committee for Standardization (1992a).

$$\frac{N_i}{N_{ri}} + \frac{M_{fi}}{M_{ri}} \leq \alpha \quad \text{for } i = 1 \text{ and } 2 \quad (\text{see Fig. 6.10}) \quad [6.13]$$

(a) Unstiffened

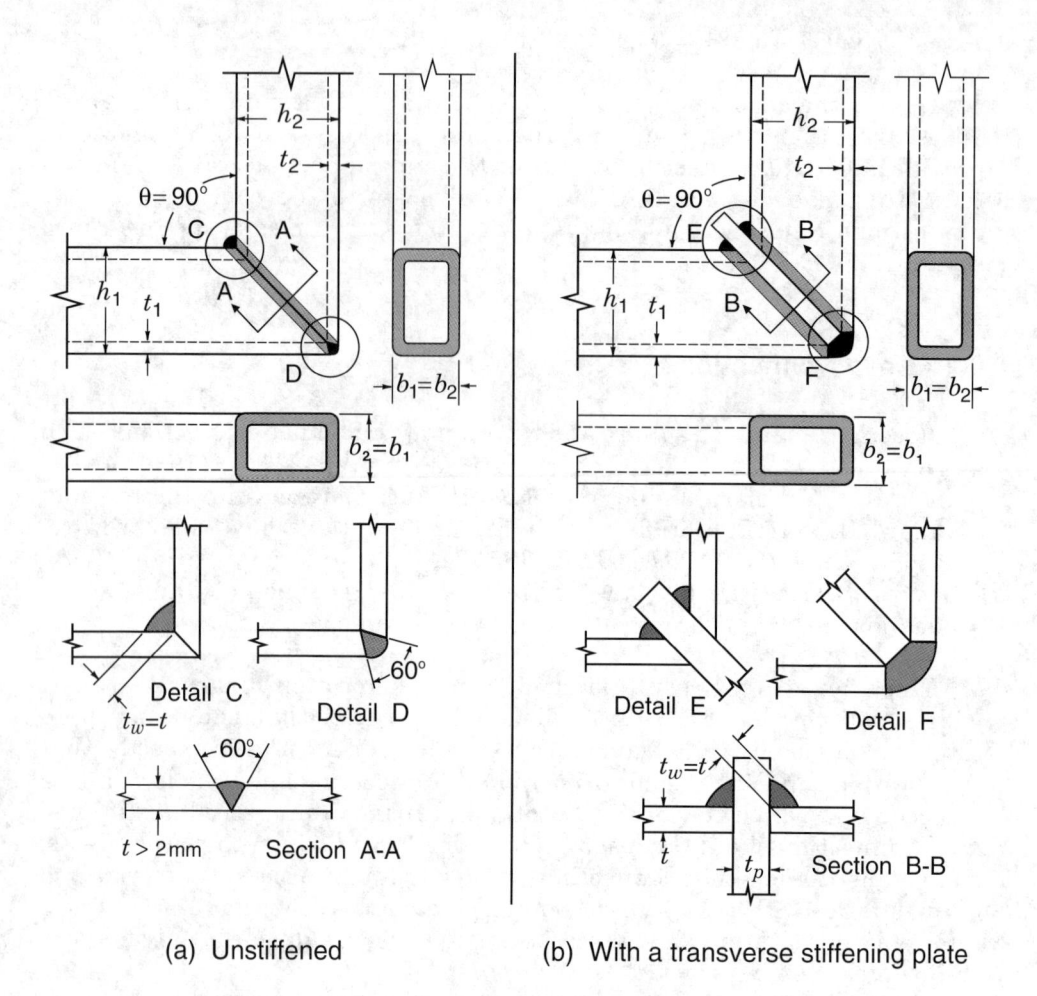

(b) With a transverse stiffening plate

FIGURE 6.10
Details of square and rectangular HSS knee connections
(from DIN 18 808, 1984)

b_i, h_i ≤ 400 mm, if stiffening plate is used (Fig. 6.10(b))
b_i, h_i ≤ 300 mm, if stiffening plate is not used (Fig. 6.10(a))
0.33 ≤ h_i/b_i ≤ 3.5
2.5 mm ≤ t_i ≤ 25 mm, for Grade 350 HSS
b_i/t_i, h_i/t_i ≤ 36, for Grade 350 HSS, if stiffening plate is used
b_i/t_i, h_i/t_i to meet Class 1 requirements if stiffening plate is not used

TABLE 6.1
Ranges of validity for HSS member dimensions in 90° mitred knee connections
(adapted from DIN 18 808, 1984)

N_{ri} is the axial yield resistance $\phi A_i F_{yi}$ of member i, and M_{ri} is the factored moment resistance of member i. The term α is a stress reduction factor which can be taken as 1.0 for mitre connections *with stiffening plates*. For mitre connections *without stiffening plates*, α is a function of the cross-sectional dimensions and is given in Figs. 6.11 and 6.12. Also for connections *without stiffening plates*, N_i should not exceed $0.2 N_{ri}$ (European Committee for Standardization (1992)). The shear force acting at the connection V_{fi} should also meet the requirement (European Committee for Standardization (1992)):

$$\frac{V_{fi}}{V_p} \leq 0.5 \qquad [6.14]$$

where V_p is the shear yield resistance in the member under consideration. This can be taken as $F_y/\sqrt{3}$ multiplied by the cross-sectional area of the HSS webs ($2 h_i t_i$). If [6.14] is not satisfied, the connection strength could still be deemed adequate, providing the combined stress does not produce failure according to the Von Mises failure criterion; in doing this check, the normal stresses (axial and bending) should be increased by a factor of $1/\alpha$.

For *stiffened* knee connections, the plate size should comply with the European Committee for Standardization (1992):

$$t_p \geq 1.5 t_i \quad (i = 1 \text{ or } 2), \text{ and } t_p \geq 10 \text{ mm}. \qquad [6.15]$$

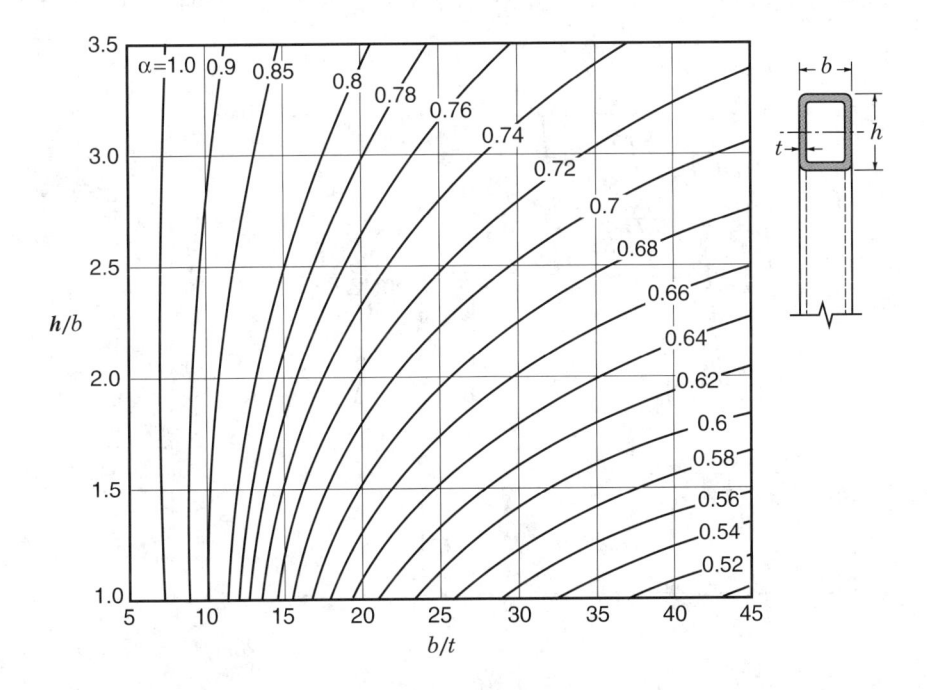

FIGURE 6.11

Stress reduction factors α for rectangular HSS subjected to bending about the Major Axis in 90° unstiffened mitred knee connections (DIN 18 808 1984)

The fabrication details shown in Fig. 6.10 should be followed, with t_w being the weld throat thickness. According to DIN 18 808, the welds can be considered a "pre-approved" size when the throat thickness is equal to the connected wall thickness, plus the factor α (for unstiffened knee connections) is ≤ 0.71 for Grade 350 HSS. (According to CAN/CSA-S16.1-94, the pre-approved fillet weld throat size would be 1.10 times the HSS wall thickness, for Grade 350 HSS, as discussed in Section 6.1.2.)

Since obtuse-angle (i.e., θ > 90° in Fig. 6.10) knee connections behave more favourably than right-angle connections, the same design checks can be undertaken for them as for right-angle knees (CIDECT 1984). For *unstiffened* knee connections with 90° < θ < 180° this strength enhancement can be used to advantage in [6.13] by increasing the value of α as follows:

$$\alpha = 1 - \left(\sqrt{2}\cos\frac{\theta}{2}\right)\left(1 - \alpha_{\theta=90}\right) \qquad [6.16]$$

$\alpha_{\theta=90}$ is the value obtained from Figs. 6.11 and 6.12.

An alternative form of connection reinforcement (other than a transverse stiffening plate) is a haunch on the inside of the knee. This haunch piece needs to be the same width as the two main members, and can easily be provided by taking a cutting from one of the HSS sections. Provided the haunch length is sufficient to ensure that the bending moment does not exceed the section yield moment ($S_i F_{yi}$) in either main member, the connection resistance will be adequate and does not require checking (CIDECT 1984).

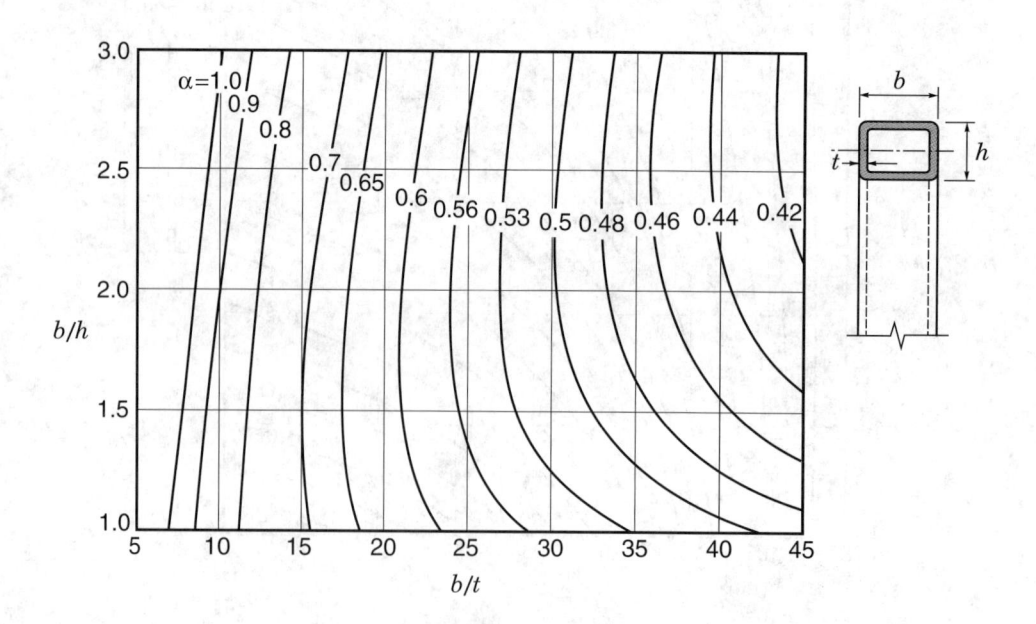

FIGURE 6.12

Stress reduction factors α for rectangular HSS subjected to bending about the Minor Axis in 90° unstiffened mitred knee connections (DIN 18 808 1984)

6.3 In-Plane and Out-of-Plane Moments for T and X Connections

Bending moments in the plane of the structure, and out of the plane of the structure are considered in the following sections.

6.3.1 In-Plane Bending Moments for Rectangular HSS

T Connections

The design criteria for square and rectangular HSS *T connections* with the branch member subjected to an in-plane bending moment M_{fi} are described in Section 6.1.2 under the topic of Vierendeel connections. To summarize, the moment resistances can be calculated as follows:

(a) For $\beta \leq 0.85$, design is governed by chord face yielding, with the connection moment of resistance given by:

$$M_{r1}^* = F_{y0} t_0^2 h_1 \left(\frac{1}{2h_1/b_0} + \frac{2}{\sqrt{1-\beta}} + \frac{h_1/b_0}{(1-\beta)} \right) f_2(n) \quad ([6.2])$$

where $f_2(n)$ is given by [6.1a].

(b) For $0.85 < \beta \leq 1.0$, design is governed by the more critical failure mode between 1) the reduced branch member capacity (an "effective width" failure mode) and 2) the chord side wall bearing or buckling capacity.

For "effective width" failure:

$$M_{r1}^* = F_{y1} \left[Z_1 - \left(1 - \frac{b_e}{b_1}\right) b_1 t_1 (h_1 - t_1) \right] \quad ([6.3])$$

In [6.3], Z_1 is the plastic section modulus about the correct axis of bending. The term b_e is defined by [6.4].

For chord side wall failure:

$$M_{r1}^* = 0.5 F_k t_0 (h_1 + 5t_0)^2 \quad ([6.5])$$

where $F_k = F_{y0}$

X Connections

The design criteria for square and rectangular HSS *X connections* subjected to equal and opposite (self-equilibrating) in-plane bending moments (M_{fi}) applied to the branch members are the same as those given above for T connections, with one exception. The difference is that the buckling stress for the chord side wall failure mode F_k should be reduced to $0.8\,F_{y0}$.

In-plane bending plus axial loading

For rectangular HSS connections loaded by a combination of in-plane bending moments plus axial loads, one should check the adequacy of the connections using [6.9].

6.3.2 In-Plane Bending Moments for Circular HSS

T, Y and X Connections

The design criterion for circular HSS *T, Y and X connections* with the branch member subjected to an in-plane bending moment M_{fi} is chord plastification according to [6.17] below, but also subject to a general punching shear check according to [6.18] (Wardenier *et al.* 1991):

$$M_{r1}^* = 4.85 F_{y0} t_0^2 \gamma^{0.5} \beta\, d_1 \frac{f(n')}{\sin\theta_1} \qquad [6.17]$$

where $f(n')$ is given by Table 3.1.

Punching shear check, for $d_1 \le d_0 - 2t_0$:

$$M_{r1}^* = \frac{F_{y0}}{\sqrt{3}} t_0\, d_1^2 \left(\frac{1 + 3\sin\theta_1}{4\sin^2\theta_1}\right) \qquad [6.18]$$

In a manner similar to axially loaded connections (Chapter 3), the above connection moment resistance expressions can be presented as the efficiency design chart given in Fig. 6.13 (Wardenier *et al.* 1991).

Figure 6.13 gives the uncorrected connection moment efficiency C_{ipb} which must be adjusted as shown on the figure to obtain M_{r1}^*/M_{p1} the connection moment resistance divided by the plastic moment capacity *of the branch member.*

If rigid connections are required, such as for Vierendeel trusses and portal frames, it is recommended that β be approximately 1.0 or that low d_0/t_0 ratios be used in combination with high t_0/t_1 ratios. Connections which do not meet these criteria would be classified as semi-rigid and the connection rotational stiffness may have a considerable influence on the moment distribution in a statically indeterminate structural system (Wardenier *et al.* 1991).

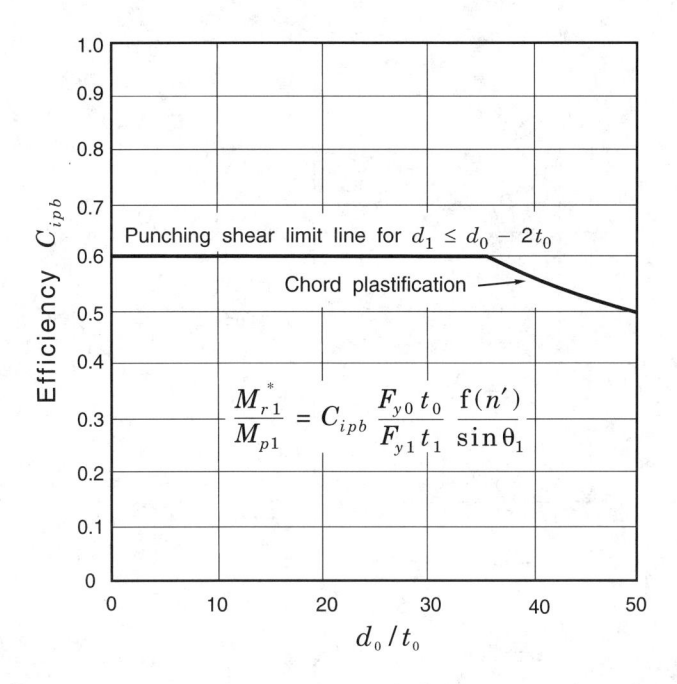

FIGURE 6.13
Efficiency of circular HSS T, Y and X connections under in-plane bending

In the figure:
$$\frac{M^*_{r1}}{M_{p1}} = C_{ipb} \frac{F_{y0} t_0}{F_{y1} t_1} \frac{f(n')}{\sin \theta_1}$$

6.3.3 Out-of-Plane Bending Moments for Rectangular HSS

T Connections

For square and rectangular HSS *T connections* with the branch member subjected to an out-of-plane bending moment (M_{opi}), such as shown in Fig. 6.14, there is very little test evidence available to support any design models. However, one can postulate analogous failure modes to those described before for in-plane moment loading, which has been done by the European Committee for Standardization (1992) and AWS (1996).

(a) <u>For β ≤ 0.85</u>, design would likely be governed by chord face yielding as shown in Fig. 6.14. For this yield line mechanism

$$M^*_{opr1} = F_{y0} t_0^2 \left[\frac{h_1(1+\beta)}{2(1-\beta)} + \sqrt{\frac{2b_0 b_1(1+\beta)}{1-\beta}} \right] f_2(n) \qquad [6.19]$$

where $f_2(n)$ is given by [6.1a].

(b) <u>For 0.85 < β ≤ 1.0</u>, design would likely be governed by the more critical failure mode between 1) the reduced branch member capacity (an "effective width" failure mode) and 2) the chord side wall bearing or buckling capacity (see Fig. 6.15), providing rhomboidal

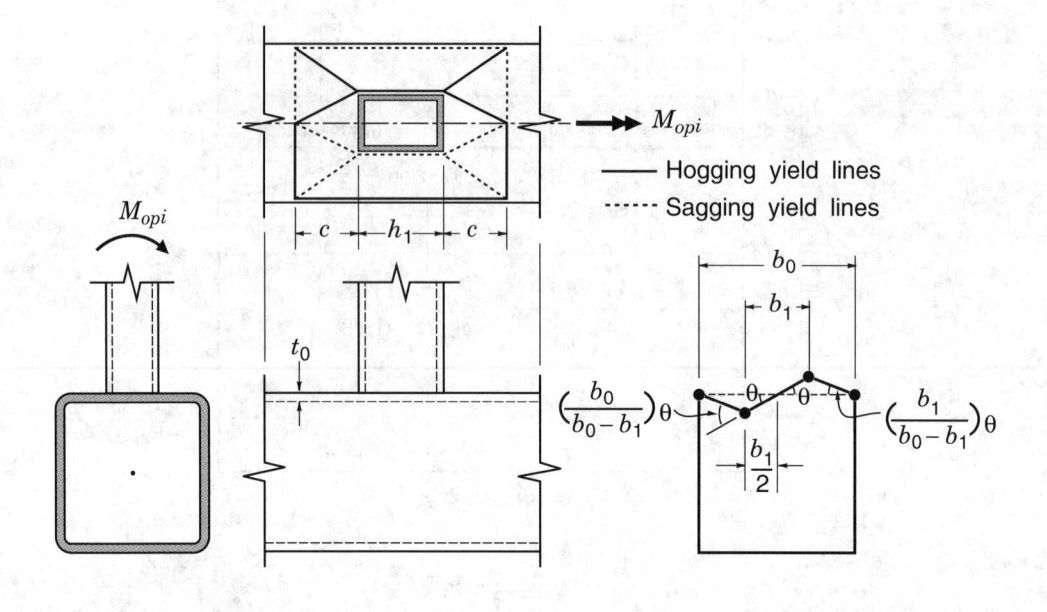

FIGURE 6.14
T Connection subject to an out-of-plane bending moment showing chord face yielding failure mode for $\beta \leq 0.85$

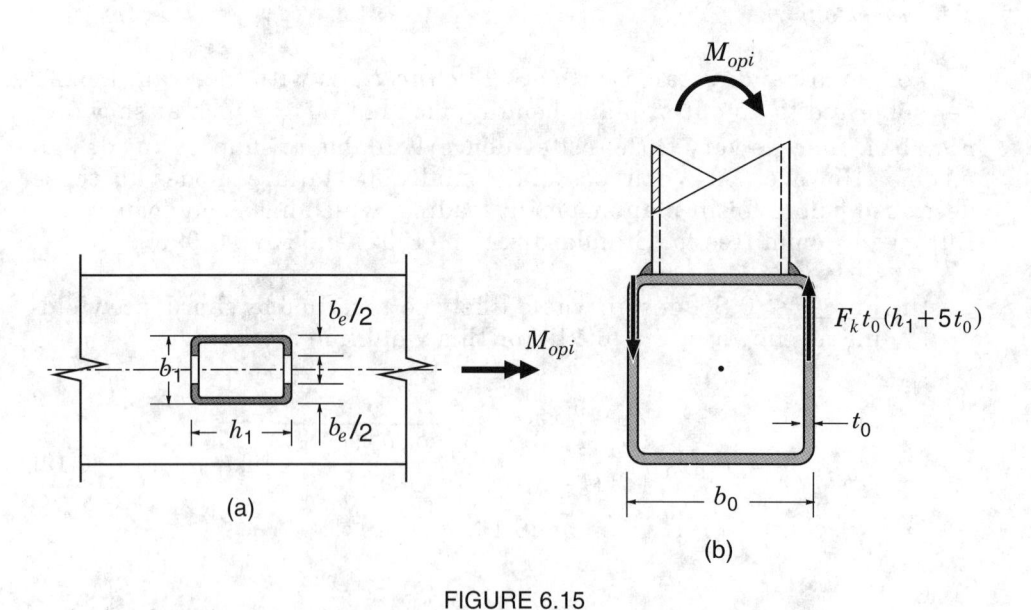

FIGURE 6.15
T Connection subject to an out-of-plane bending moment showing the basis of design models for: (a) effective width failure mode (b) chord side wall failure mode

distortion of the chord section is prevented (possibly by the use of diaphragm stiffening).

For "effective width" failure:

$$M^*_{opr1} = F_{y1}(Z_1 - 0.5t_1(b_1 - b_e)^2) \quad [6.20]$$

In [6.20], Z_1 is the plastic section modulus about the correct axis of bending and Class 1 sections should be selected for the branch member. The term b_e is defined by [6.4].

For chord side wall failure:

$$M^*_{opr1} = F_k t_0 (h_1 + 5t_0)(b_0 - t_0) \quad [6.21]$$

The term F_k is the compression buckling stress of the chord side walls for T connections, as determined from the Functions portion of Table 3.3.

X Connections:

One could speculate that the design criteria for square and rectangular HSS *X connections* subjected to equal and opposite (self-equilibrating) out-of-plane bending moments (M_{opi}) applied to the branch members are again the same as those given above for T connections, with one exception. The difference is that the buckling stress for the chord side wall failure mode F_k should be the reduced value for X connections specified in the Functions portion of Table 3.3.

6.3.4 Out-of-Plane Bending Moments for Circular HSS

T, Y and X Connections

The design criterion for circular HSS *T, Y and X connections* with the branch member subjected to an out-of-plane bending moment M_{opi} is chord plastification according to [6.22] below, but also subject to a general punching shear check according to [6.23] (Wardenier *et al.* 1991):

$$M^*_{opr1} = F_{y0} t_0^2 \left(\frac{2.7}{1 - 0.81\beta}\right) \frac{f(n')}{\sin\theta_1} d_1 \quad [6.22]$$

where $f(n')$ is given by Table 3.1.

Punching shear check, for $d_1 \leq d_0 - 2t_0$:

$$M^*_{opr1} = \frac{F_{y0}}{\sqrt{3}} t_0 d_1^2 \left(\frac{3 + \sin\theta_1}{4\sin\theta_1}\right) \quad [6.23]$$

In a manner similar to axially loaded connections (Chapter 3), the above connection moment resistance expressions can be presented as the efficiency design chart given in Fig. 6.16 (Wardenier et al. 1991).

Figure 6.16 gives the uncorrected connection moment efficiency C_{opb} which must be adjusted as shown on the figure to obtain M^*_{opr1}/M_{p1} the connection moment resistance divided by the plastic moment capacity of the branch member.

A comparison of Fig. 6.16 with Fig. 6.13 shows that in most cases the in-plane bending moment resistance is considerably better than the out-of-plane bending moment resistance for circular HSS connections.

Combinations of bending moment plus axial loading

In some structures, connections may be loaded by combinations of axial loads and bending moments. Several investigations have been carried out to study this problem and as a result many interaction formulae exist. All investigations have shown that in-plane bending is less severe than out-of-plane bending and a reasonable simplified lower bound interaction function for circular HSS is given by (Sedlacek et al. 1989, Wardenier et al. 1991):

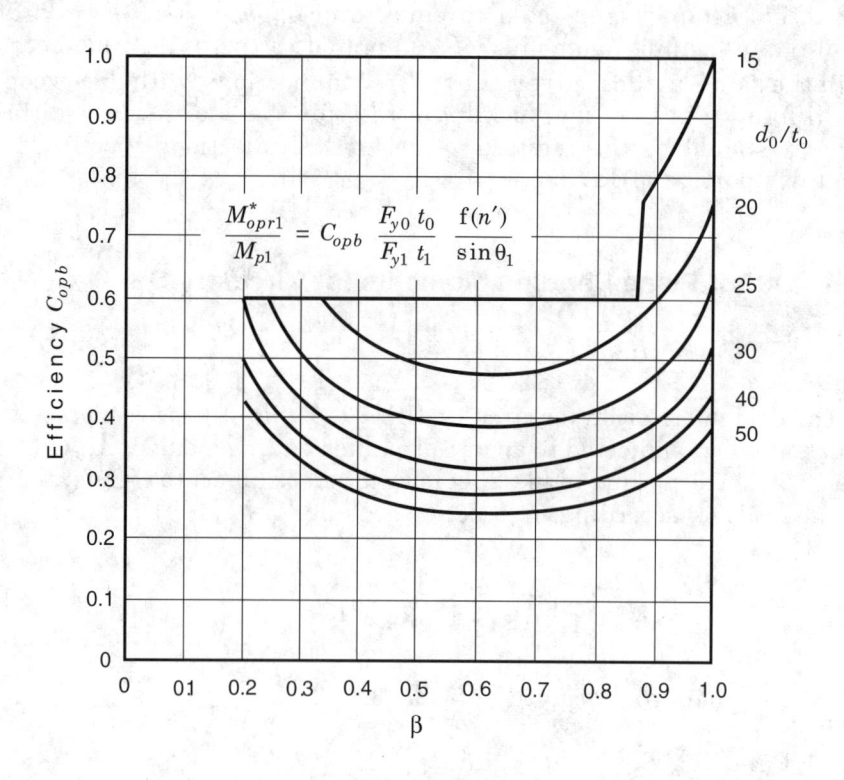

FIGURE 6.16

Efficiency for circular HSS T, Y and X connections under out-of-plane bending

$$\frac{N_1}{N_1^*} + \left(\frac{M_{f1}}{M_{r1}^*}\right)^2 + \left(\frac{M_{op1}}{M_{opr1}^*}\right) \leq 1.0 \qquad [6.24]$$

6.4 Reinforced Connections between Rectangular/Square HSS

In Section 6.1.1 it is noted that Vierendeel-type connections subjected to in-plane moment loading approach full connection ridigity when $\beta \approx 1.0$ and b_0/t_0 has a low value. Furthermore, in Section 6.1.2 it is shown that connections with $\beta \approx 1.0$, $b_0/t_0 = 16$ and $t_0/t_1 = 2$ will develop a moment resistance approximately equal to the plastic moment capacity of the branch. Unstiffened connections which do not meet the criteria of $\beta \approx 1.0$ and $b_0/t_0 \leq 16$ can be classified as semi-rigid. For such semi-rigid connections, Figs. 6.2(b) to (e) illustrate various stiffening methods which have been used to achieve full rigidity. From these alternatives, Figs. 6.2(c) and (d) are recommended since the resistance of Fig. 6.2(b) is limited by the "effective width" criterion and Fig. 6.2(e) is rather expensive to fabricate. Design recommendations for connection types illustrated by Figs. 6.2(c) and (d) are given in Chapter 9, Section 9.2.3.

REFERENCES

AISC. 1993. Load and resistance factor design specification for structural steel buildings. American Institute of Steel Construction, Chicago, Illinois, U.S.A.

AWS. 1996. Structural welding code—steel. ANSI/AWS D1.1-96, American Welding Society, Miami, Florida, U.S.A.

BROCKENBOUGH, R.L. 1972. Strength of square tube connections under combined loads. Journal of the Structural Division, American Society of Civil Engineers, **98**(ST12): 2753–2768.

CIDECT. 1984. Construction with hollow steel sections. British Steel plc., Corby, Northants., England.

CISC. 1995. Handbook of Steel Construction, 6th. ed. Canadian Institute of Steel Construction, Willowdale, Ontario.

CSA. 1992. Structural quality steels, CAN/CSA-G40.21-M92. Canadian Standards Association, Rexdale, Ontario.

CSA. 1994. Limit states design of steel structures, CAN/CSA-S16.1-94. Canadian Standards Association, Rexdale, Ontario.

CUTE, D., CAMO, S., and RUMPF, J.L. 1968. Welded connections for square and rectangular structural steel tubing. Research Report No. 292-10, Drexel Institute of Technology, Philadelphia, Pennsylvania, U.S.A.

DAVIES, G., and PACKER, J.A. 1987. Analysis of web crippling in a rectangular hollow section. Proceedings of the Institution of Civil Engineers. Part 2, **83**: 785–798.

DAVIES, G., and PANJEH SHAHI, E. 1984. Tee joints in rectangular hollow sections (RHS) under combined axial loading and bending. Proceedings, 7th. International Symposium on Steel Structures, Gdansk, Poland.

DIN 18 808. 1984. Stahlbauten: tragwerke aus hohlprofilen unter vorwiegend ruhender beanspruchung. (Steel structures: structures made from hollow sections subjected to predominantly static loading.) Deutsches Institut für Normung, Berlin, Federal Republic of Germany.

DUFF, G. 1963. Joint behaviour of a welded beam-column connection in rectangular hollow sections. Ph.D. thesis, The College of Aeronautics, Cranfield, England.

DUTTA, D., and WÜRKER, K. 1988. Handbuch hohlprofile in stahlkonstruktionen. Verlag TÜV Rheinland GmbH, Köln, Federal Republic of Germany.

EUROPEAN COMMITTEE for STANDARDIZATION. 1992. Eurocode 3: design of steel structures. Part 1.1—general rules and rules for buildings. ENV 1993-1-1:1992E, British Standards Institution, London, England.

GIDDINGS, T.W. 1980. The design of RHS vierendeel girders. British Steel Corporation Tubes Division Draft Report, Corby, England.

HORNE, M.R., and MORRIS, L.J. 1985. Plastic design of low-rise frames. Collins, London, England, Chapter 3.

IIW. 1989. Design recommendations for hollow section joints—predominantly statically loaded, 2nd. ed. International Institute of Welding Subcommission XV-E, IIW Doc. XV-701-89, IIW Annual Assembly, Helsinki, Finland.

KANATANI, H., KAMBA, T., and TABUCHI, M. 1986. Effect of the local deformation of the joints on RHS vierendeel trusses. Proceedings, International Meeting on Safety Criteria in Design of Tubular Structures, Tokyo, Japan, pp. 127–137.

KANATANI, H., FUJIWARA, K., TABUCHI, M., and KAMBA, T. 1980. Bending tests on T-joints of RHS chord and RHS or H-shape branch. CIDECT Report 5AF-80/15, Croydon, England.

de KONING, C.H.M., and WARDENIER, J. 1985. The static strength of welded joints between structural hollow sections or between structural hollow sections and H-sections. Part 2: joints between rectangular hollow sections. Stevin Report 6-84-19, Delft University of Technology, Delft, The Netherlands.

KOROL, R.M. and MANSOUR, M.H. 1979. Theoretical analysis of haunch-reinforced T-joints in square hollow sections. Canadian Journal of Civil Engineering, **6**(4): 601–609.

KOROL, R.M., and MIRZA, F.A. 1982. Finite element analysis of RHS T-joints. Journal of the Structural Division, American Society of Civil Engineers, **108**(ST9): 2081–2098.

KOROL, R.M., EL-ZANATY, M., and BRADY, F.J. 1977. Unequal width connections of square hollow sections in vierendeel trusses. Canadian Journal of Civil Engineering, **4**(2): 190–201.

KOROL, R.M., MITRI, H., and MIRZA, F.A. 1982. Plate reinforced square hollow section T-joints of unequal width. Canadian Journal of Civil Engineering, **9**(2): 143–148.

LAZAR, B.E., and FANG, P.J. 1971. T-type moment connections between rectangular tubular sections. Research Report, Sir George Williams University, Montreal, Quebec.

MANG, F., BUCAK, Ö., and WOLFMULLER, F. 1983. The development of recommendations for the design of welded joints between steel structural hollow sections (T- and X-type joints). Report on ECSC Agreement 7210 SA/109 and CIDECT Program 5AD Final Report, University of Karlsruhe, Federal Republic of Germany.

MANG, F., STEIDL, G., and BUCAK, Ö. 1980. Design of welded lattice joints and moment resisting knee joints made of hollow sections. International Institute of Welding Document XV-463-80, University of Karlsruhe, Federal Republic of Germany.

MEHROTRA, B.L., and GOVIL, A.K. 1972. Shear lag analysis of rectangular full-width tube connections. Journal of the Structural Division, American Society of Civil Engineers, **98**(ST1): 287–305.

MEHROTRA, B.L., and REDWOOD, R.G. 1970. Load transfer through connections between box sections. Canadian Engineering Institute, C-70-BR and Str. 10.

PACKER, J.A. 1984. Web crippling of rectangular hollow sections Journal of Structural Engineering, American Society of Civil Engineers, **110**(10): 2357–2373.

REDWOOD, R.G. 1965. The behaviour of joints between rectangular hollow structural members. Civil Engineering and Public Works Review, pp. 1463–1469.

SEDLACEK, G., WARDENIER, J., DUTTA, D., and GROTMANN, D. 1989. Evaluation of test results on hollow section lattice girder connections in order to obtain strength functions and suitable model factors. Background report to Eurocode 3: Common unified rules for steel structures. Document 5.07, Eurocode 3 Editorial Group.

STAPLES, C.J.L., and HARRISON, C.C. (Undated). Test results of 24 right-angled branches fabricated from rectangular hollow sections. University of Manchester Institute of Science and Technology (UMIST) Report, England.

SZLENDAK, J. 1986. Interaction curves for M-N loaded T RHS joints. Proceedings, International Meeting on Safety Criteria in Design of Tubular Structures, Tokyo, Japan, pp. 169–174.

SZLENDAK, J., and BRODKA, J. 1985. Strengthening of T moment of RHS joints. Proceedings of the Institution of Civil Engineers, Part 2, **79**: 717–727.

SZLENDAK, J., and BRODKA, J. 1986a. Design of strengthen frame RHS joints. Proceedings, International Meeting on Safety Criteria in Design of Tubular Structures, Tokyo, Japan, pp. 159–168.

SZLENDAK, J., and BRODKA, J. 1986b. Reply to discussion by J.A. Packer on the paper, "Strengthening of T-moment of RHS joints". Proceedings of the Institution of Civil Engineers, Part 2, **81**: 721–725.

WARDENIER, J. 1982. Hollow section joints. Delft University Press, Delft, The Netherlands.

WARDENIER, J., KUROBANE, Y., PACKER, J.A., DUTTA, D., and YEOMANS, N. 1991. Design guide for circular hollow section (CHS) joints under predominantly static loading. CIDECT (ed.) and Verlag TÜV Rheinland GmbH, Köln, Federal Republic of Germany.

7
BOLTED HSS CONNECTIONS

7.1 Circular HSS Flange-Plate Connections

Tubular flange-plate connections under tensile loading as shown in Fig. 7.1 have been studied by a number of HSS researchers since Rockey and Griffiths (1970), and many of these investigations are cited elsewhere (Stelco 1981, Dutta and Würker 1988, and Packer *et al.* 1989). Early structural design methods attempted to reduce prying forces at the flange extremity to zero by providing the connection with thick rigid flange plates, as was the procedure for mechanical pipe flanges. More economical structural connections can be obtained if prying action is permitted at the ultimate limit state with the connection proportioned on the basis of a failure mechanism involving yielding of the flange plates.

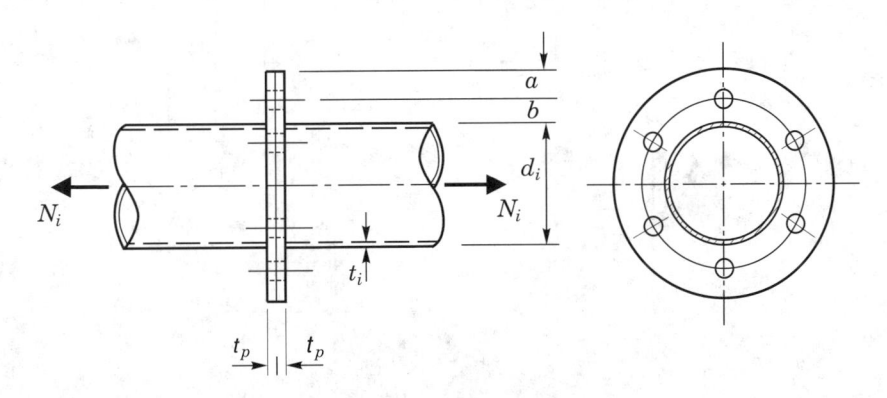

FIGURE 7.1
Bolted circular HSS flange-plate connection

Such an analysis has been performed by Igarashi *et al.* (1985) in Japan, on both reinforced and unreinforced flange-plate connections in conjunction with experimental verification. However, the addition of welded rib plates in an attempt to decrease the required flange plate thickness is not recommended, as such reinforcing adversely induces local bending in the tube walls at the top of the ribs. This movement of the rib plates can be prevented by a further stiffener ring around the tube at the end of the ribs, but such a highly fabricated connection will be quite expensive.

Flange plate thickness for unreinforced flanges is given by Igarashi *et al.* as

$$t_p \geq \sqrt{\frac{2N_i}{\phi F_{yp} \pi f_3}} \qquad [7.1]$$

where $\phi = 0.9$, and the connection geometry parameter f_3 is given in Fig. 7.2. More exactly, f_3 is defined by

$$f_3 = \frac{1}{2k_1}\left(k_3 + \sqrt{k_3^2 - 4k_1}\right) \qquad [7.2]$$

where $k_1 = \ln\left(\frac{r_2}{r_3}\right)$ (ln = natural logarithm)

$$r_2 = \frac{d_i}{2} + b \qquad r_3 = \frac{d_i - t_i}{2} \quad \text{and} \quad k_3 = k_1 + 2.$$

The dimension b (Fig. 7.1) should always be kept as low as possible (around $1.5d$ to $2.0d$), but the clearance between the nut face and the weld should not be less than 5 mm.

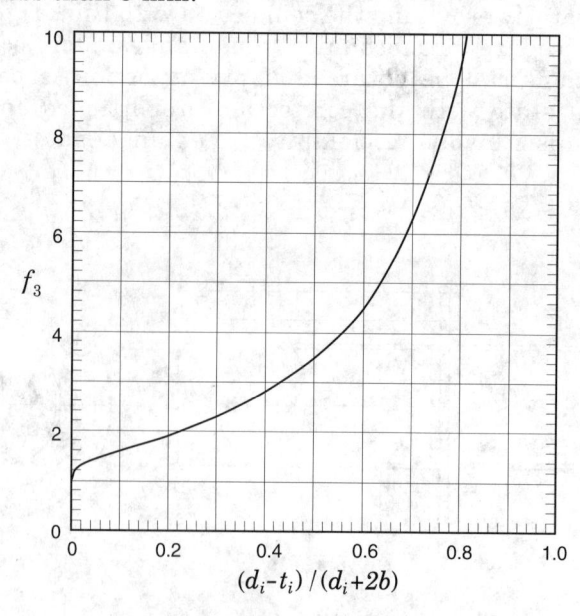

FIGURE 7.2

Parameter f_3 for use in equations 7.1 and 7.3

Equation 7.1 will result in flange plate thicknesses considerably less than those recommended earlier by Timoshenko and Woinowsky-Krieger's (1959) elastic analysis.

Igarashi *et al.* provide a method compatible with [7.1] for determining the required number of fasteners:

$$n \geq \frac{N_i}{\phi T_r}\left(1 - \frac{1}{f_3} + \frac{1}{f_3 \ln (r_1/r_2)}\right) \qquad [7.3]$$

where

ln = natural logarithm

$$r_1 = \frac{d_i}{2} + 2b \qquad r_2 = \frac{d_i}{2} + b$$

$a = b$ (see Fig. 7.1) and $\phi = 0.90$.

This method presumes that the flange is continuous, the bolts are arranged symmetrically and the connection is predominantly statically loaded.

Equations 7.1 and 7.3 have been used to design a series of optimized flange connections, with 300 MPa plates, which will develop the tensile resistance of circular HSS sections. Results are shown in Tables 7.1(a) and 7.1(b) for A325M and A490M bolts respectively. CAN/CSA-S16.1-94 defines in Clause 22.5 that minimum bolt spacing be 2.7 times the bolt diameter. Implicit in these connection details is an allowance for prying forces amounting to 1/3 the total bolt force at the ultimate limit state, and the assumption that the tube yield strength is fully developed by the welds.

A worked example is presented in Section 13.3.8.

HSS SECTION	PLATE THICK	BOLT DIAMETER	No. of BOLTS	EDGE a = b	HSS SECTION	PLATE THICK	BOLT DIAMETER	No. of BOLTS	EDGE a = b
27 x 2.5	8	16	4	25	168 x 4.8	18	20	10	30
x 3.2	10	16	4	25	x 6.4	22	20	13	40
					x 8.0	25	24	12	40
33 x 2.5	10	16	4	25	x 9.5	28	27	11	45
x 3.2	10	16	4	25					
					219 x 4.8	18	20	13	30
42 x 2.5	10	16	4	25	x 6.4	25	22	15	40
x 3.2	10	16	4	25	x 8.0	28	24	15	45
					x 9.5	30	30	12	45
48 x 2.8	10	16	4	25	x 11	35	36*	9	55
x 3.2	12	16	4	25	x 13	40	36*	10	70
x 3.8	12	16	4	25					
x 4.8	14	16	4	30	273 x 6.4	22	22	18	35
					x 8.0	28	27	15	45
60 x 3.2	12	16	4	25	x 9.5	35	30	14	60
x 3.8	12	16	5	25	x 11	35	36*	12	55
x 4.8	14	16	6	25	x 13	40	36*	13	65
x 6.4	16	16	7	25					
					324 x 6.4	25	24	18	40
73 x 3.2	12	16	5	25	x 8.0	28	27	18	45
x 3.8	14	16	6	25	x 9.5	35	30	17	55
x 4.8	14	16	7	25	x 11	35	36*	14	55
x 6.4	18	16	9	25	x 13	45	36*	16	70
89 x 3.8	14	16	7	25	356 x 6.4	25	24	20	40
x 4.8	16	16	8	25	x 8.0	30	27	20	50
x 6.4	18	20	7	30	x 9.5	35	36*	13	55
x 8.0	22	20	8	40	x 11	38	36*	15	60
					x 13	45	36*	17	70
102 x 3.8	14	16	8	25					
x 4.8	16	16	9	30	406 x 6.4	25	24	23	40
x 6.4	20	20	8	30	x 8.0	28	30	18	45
x 8.0	22	22	8	35	x 9.5	35	36*	15	55
					x 11	40	36*	18	70
114 x 4.8	16	16	11	25	x 13	45	36*	20	70
x 6.4	20	20	9	30					
x 8.0	22	22	9	35					
141 x 4.8	16	16	13	25					
x 6.4	20	20	11	30	* Confirm availability of bolts and feasibility of tightening by the erector.
x 8.0	25	24	10	40
x 9.5	28	27	9	45

TABLE 7.1(a): A325M Bolts

Flange and bolt details for circular HSS flange connections to develop section capacities for HSS F_y = 350 MPa and plate F_y = 300 MPa

HSS SECTION	PLATE THICK	BOLT DIAMETER	No. of BOLTS	EDGE a = b	HSS SECTION	PLATE THICK	BOLT DIAMETER	No. of BOLTS	EDGE a = b
27 x 2.5	8	16	4	25	168 x 4.8	16	16	13	25
x 3.2	10	16	4	25	x 6.4	20	20	11	30
					x 8.0	25	20	13	35
33 x 2.5	10	16	4	25	x 9.5	28	22	13	40
x 3.2	10	16	4	25					
					219 x 4.8	18	16	17	25
42 x 2.5	10	16	4	25	x 6.4	20	20	14	30
x 3.2	10	16	4	25	x 8.0	25	22	15	35
					x 9.5	30	24	15	45
48 x 2.8	10	16	4	25	x 11	32	27	13	45
x 3.2	12	16	4	25	x 13	35	30*	12	50
x 3.8	12	16	4	25					
x 4.8	14	16	4	25	273 x 6.4	22	20	18	30
					x 8.0	25	22	18	35
60 x 3.2	12	16	4	25	x 9.5	32	27	14	50
x 3.8	12	16	4	25	x 11	35	30*	14	50
x 4.8	14	16	5	25	x 13	38	30*	15	60
x 6.4	16	16	6	25					
					324 x 6.4	22	22	17	35
73 x 3.2	12	16	4	25	x 8.0	28	24	18	40
x 3.8	14	16	5	25	x 9.5	30	27	17	45
x 4.8	14	16	6	25	x 11	35	30*	16	50
x 6.4	18	16	7	25	x 13	38	36*	13	55
89 x 3.8	14	16	6	25	356 x 6.4	25	22	19	35
x 4.8	16	16	7	25	x 8.0	28	24	20	40
x 6.4	18	16	9	25	x 9.5	32	27	19	45
x 8.0	20	16	11	30	x 11	38	30*	18	60
					x 13	38	36*	14	55
102 x 3.8	14	16	6	25					
x 4.8	16	16	8	25	406 x 6.4	25	22	22	35
x 6.4	18	16	10	25	x 8.0	28	24	23	45
x 8.0	22	20	8	30	x 9.5	35	30*	17	60
					x 11	38	36*	14	55
114 x 4.8	16	16	9	25	x 13	40	36*	16	55
x 6.4	18	16	11	25					
x 8.0	22	20	9	30					
141 x 4.8	16	16	11	25					
x 6.4	18	16	14	25	* Confirm availability of bolts and feasibility of tightening by the erector.				
x 8.0	22	20	11	30
x 9.5	25	22	11	35

TABLE 7.1(b): A490M Bolts

Flange and bolt details for circular HSS flange connections to develop section capacities for HSS F_y = 350 MPa and plate F_y = 300 MPa

7.2 Rectangular HSS Flange-Plate Connections

Rectangular (including square) flange-plate connections have generally been bolted along all four sides of the HSS; however, the option of bolting along only two sides as seen in Fig. 7.3 has been investigated during the 1980's, and shown to be effective.

7.2.1 Connections Bolted along Two Sides of the HSS

Preliminary tests on flange-plate connections bolted along two sides of the HSS as drawn in Fig. 7.4 were performed by Mang (1980) and Kato and Mukai (1985), followed by a more extensive study by Packer *et al.* (1989). The latter showed that, by selecting specific connection parameters, one could fully develop the tensile resistance of the member by bolting along only two sides of the tube. This form of connection lends itself to analysis as a 2-dimensional prying problem, but the application of traditional prying models developed for T-stubs was found to correlate poorly with the test results. One reason for this was the location of the hogging plastic hinge lines, which tended to form within the width of the tube as shown on Fig. 7.4.

FIGURE 7.3
Panel point with flange-plate HSS splices
— some bolted along only two sides of the HSS members

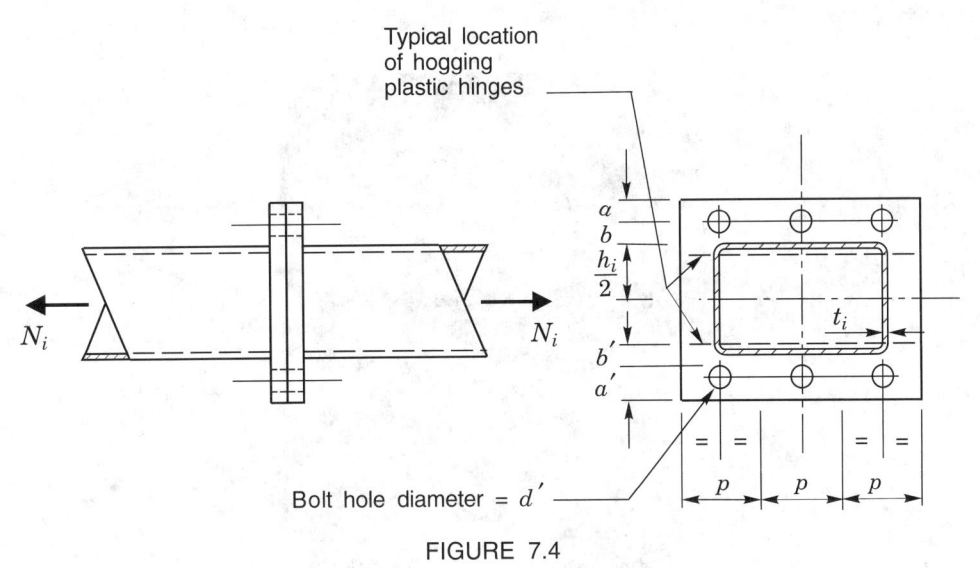

FIGURE 7.4
Flange-plate connection with bolts along two sides of rectangular HSS

A modified T-stub design procedure was consequently proposed (Birkemoe and Packer 1986) and verified against a set of possible failure mechanisms based on the observed failure modes. The design procedure involves a redefinition of various parameters in the T-stub design method of Struik and de Back (1969) currently adopted by many structural steelwork codes. To reflect the observed location of the inner (hogging) plastic hinge line, and also represent the connection behaviour illustrated by the more complex analytical models, the distance b was adjusted to b' as shown on Fig. 7.4, where

$$b' = b - \frac{d}{2} + t_i \qquad [7.4]$$

The term α has been used in Struik and de Back's T-stub prying model to represent the ratio of the bending moment per unit plate width at the bolt line to the bending moment per unit plate width at the inner (hogging) plastic hinge. For the limiting case of a rigid plate $\alpha = 0$, and for the limiting case of a flexible plate in double curvature with plastic hinges occurring both at the bolt line and the edge of the T-stub web $\alpha = 1.0$. Hence, the term α in Struik and de Back's model was restricted to the range $0 \leq \alpha \leq 1.0$.

For bolted HSS flange-plate connections, this range of validity for α was changed to simply $\alpha \geq 0$. This implies that the sagging moment per unit width at the bolt line is allowed to exceed the hogging moment per unit width, which is proposed because the HSS member tends to yield adjacent to the hogging plastic hinge and participate in the general failure mechanism. This behaviour is confirmed by the inward movement of the hogging plastic hinge line (Figs. 7.4 and 7.5).

FIGURE 7.5

Location of plastic hinges in flange plates (University of Toronto)

Thus, a suitable design method for this connection is as follows:

1. Estimate the number, grade and size of bolts required, knowing the applied tensile force N_i and allowing for some amount of prying. In general, the applied external load per bolt should be only 60% to 80% of the bolt tensile resistance in anticipation of bolt load amplification due to prying. Hence, determine a suitable connection arrangement. The bolt pitch p should generally be about 4 to 5 bolt diameters (although closer pitches are physically possible if required), and the edge distance a about 1.25 b, which is the maximum allowed in calculations. Prying decreases as a is increased up to 1.25 b, beyond which there is no advantage. Then determine

$$\delta = 1 - \frac{d'}{p} \qquad [7.5]$$

where

d' = bolt hole diameter, and

p = length of flange plate tributary to each bolt, or bolt pitch.

Also determine a trial flange plate thickness t_p from:

$$\sqrt{\frac{KP_f}{1+\delta}} \leq t_p \leq \sqrt{KP_f} \qquad [7.6]$$

where $P_f = \dfrac{N_i}{n}$, the external factored tensile load on one bolt, and

$$K = \dfrac{4b' \, 10^3}{\phi_p \, F_{yp} \, p} \qquad (F_{yp} \text{ in MPa}) \qquad [7.7]$$

where ϕ_p = flange plate resistance factor = 0.9.

2. With the number, size and grade of bolts preselected, plus a trial flange plate thickness, calculate the ratio α necessary for equilibrium by

$$\alpha = \left(\dfrac{K T_r}{t_p^2} - 1 \right) \left(\dfrac{a + (d/2)}{\delta \, (a + b + t_i)} \right) \qquad [7.8]$$

T_r = factored tensile resistance of one bolt

3. Calculate the connection factored resistance N_i^* by using α from [7.8], except set $\alpha = 0$ if $\alpha < 0$.

$$N_i^* = \dfrac{t_p^2 \, (1 + \delta\alpha) \, n}{K} \qquad [7.9]$$

where n is the number of bolts. N_i^* must be $\geq N_i$.

The actual total bolt tension, including prying, can usefully be examined by

$$T_f \approx P_f \left(1 + \dfrac{b'}{a'} \left(\dfrac{\delta\alpha}{1 + \delta\alpha} \right) \right) \qquad [7.10]$$

where

T_f = the total bolt tension,

a' = "effective a" plus $d/2$, and

$$\alpha = \left(\dfrac{K P_f}{t_p^2} - 1 \right) \dfrac{1}{\delta} \qquad \text{(becomes [7.6] for } \alpha = 0 \text{ or } \alpha = 1.0\text{)}.$$

"Effective a" is simply the "a" dimension (see Fig. 7.4) up to a maximum of $1.25b$.

Note that this value of α is not necessarily the same as that from [7.8] which was premised on the bolts being loaded to their full tensile resistance.

This design method should be restricted to the range of flange plate thicknesses over which it has been validated experimentally and analytically (Packer *et al.* 1989; Birkemoe and Packer 1986), namely 12 to 26 mm. It should be borne in mind that when a connection with bolts in tension is subject to repeated loads, the flange must be made thick enough and stiff enough so that deformation of the flange is virtually eliminated ($\alpha \leq 0$).

A worked example is presented in Section 13.2.7.

Canadian standard CAN/CSA-S16.1-94 requires that bolts with tensile loads be pretensioned, an essential requirement for fatigue situations. Spacers placed between the plates in line with the HSS webs parallel to the bolt lines can preclude prying action and improve fatigue performance.

7.2.2 Connections Bolted along Four Sides of the HSS

CIDECT research programs on flange-plate connections bolted along all four sides of the HSS (as in Fig. 7.6) have been undertaken by Mang (1980) and Kato and Mukai (1982), but a reliable connection design procedure has not yet evolved. Kato and Mukai proposed a complex model based on yield line theory with an estimate of the prying force. Depending on the relative strengths of the flange plate to the bolts, the ultimate strength of the connection was determined by one of six failure modes.

Failure modes 1 to 3 involved failure of the flange plates, while modes 4 to 6 involved bolt failure. However, two recent connection tests of this type (Caravaggio 1988) have indicated that the model overestimated the strength of the connections by about 25%, so further investigation of this connection is still warranted.

For connections on 150 and 200 mm square HSS, Kato and Mukai summarized that the flange plate thickness should not be less than the bolt diameter for connections with four bolts, nor less than the bolt diameter plus 3 mm for connections having eight bolts. They used 16, 20 and 24 mm diameter bolts equivalent in strength to ASTM A490 bolts, and suggested that the "yield load" of the connection is 0.8 times the sum of the original tensions in the bolts.

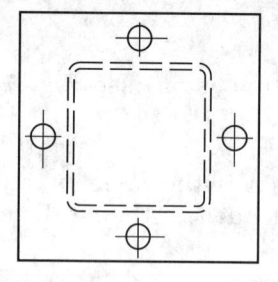

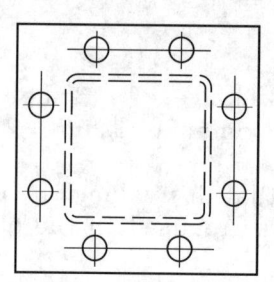

FIGURE 7.6
Four and eight bolt configurations for bolting on all sides of a rectangular HSS

When $\phi = 0.9$ is applied to this yield load in order to obtain the factored resistance of the connection, and since initial tension in a bolt is 0.70 times its ultimate strength T_u (CAN/CSA-S16.1-94, Clause 23.4.1), the *connection* resistance (per bolt) is $0.9(0.8)0.7T_u = 0.50T_u$. The tensile resistance T_r of a bolt is $0.67T_u$ (since $\phi_b = 0.67$). Therefore, Kato and Mukai recommended that the external load per bolt be kept to no more than $0.50/0.67 = 75\%$ of the bolt's factored resistance T_r. (Clause 13.11.3 of CAN/CSA-S16.1-94 ($T_r = 0.75\,\phi_b A_b F_u$) reduces to $T_r = 0.67T_u$, because the factor 0.75 defines the stress area of the bolt (ASTM A325, Table 4)).

Kato and Mukai's method for proportioning flange plate thickness does not consider the plate yield strength. This fact and the evidence that some connection strengths are overestimated suggest that a more conservative approach be taken pending additional experimental work and comprehensive recommendations.

7.3 Other Bolted Splice Connections

Various alternative splices may be used when flange-plate connections are not desired.

Figure 7.7 shows a separated double chord truss with chords that are spliced with the use of interior splice plates for the top and bottom surfaces of each chord section. The side wall hand hole for bolting had not been closed when the photo was taken.

FIGURE 7.7

Separated double chord truss with two interior splice plates in each section of the chord

Interior splice plates with bolts in double shear possible if sealing (cap plates) is not necessary

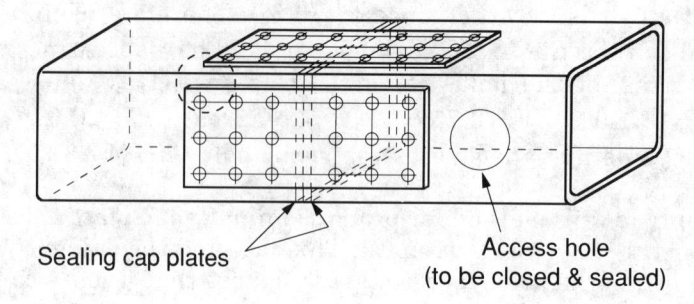

Sealing cap plates

Access hole (to be closed & sealed)

(a) Splice plates on large rectangular HSS

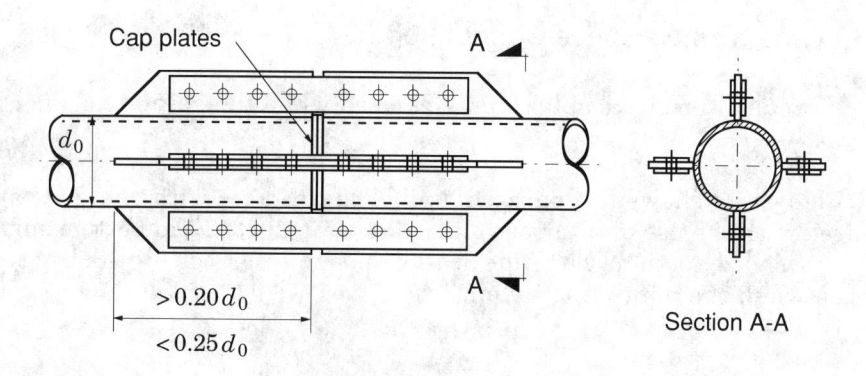

Cap plates

d_0

$> 0.20 d_0$
$< 0.25 d_0$

Section A-A

(b) Splice plates on circular HSS

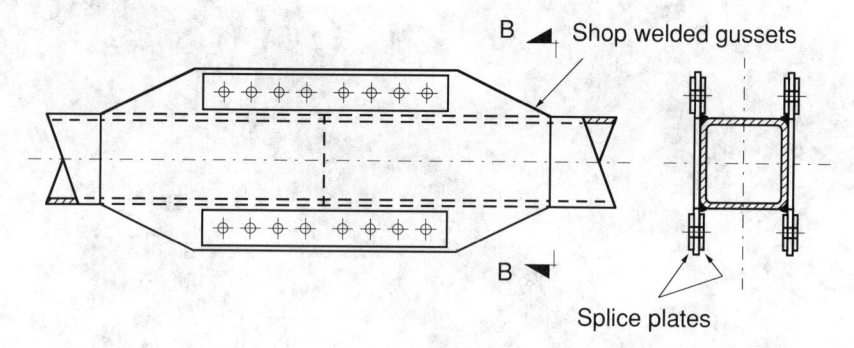

Shop welded gussets

Splice plates

Section B-B

(c) Splice plates on rectangular HSS

FIGURE 7.8
Examples of bolted HSS splices

Large rectangular sections can be spliced with conventional splice plates as indicated in Fig. 7.8(a), but sealing considerations may complicate details. Cap plates on the HSS sections, exterior splice plates only, and sealing of the hand holes will be necessary if the sections must be closed. However, both interior and exterior plates may be used with the bolts in double shear when sealing is not required.

Configurations using gussets with external splice bars have been suggested as shown in Fig. 7.8(b) for circular HSS, and Fig. 7.8(c) for rectangular HSS (Stelco 1981). Both permit sealing with cap plates on the sections.

7.4 Blind Bolting into HSS Sections

The lack of access to the interior of HSS (other than near their ends) has limited the scope for bolting connecting plates or other sections to HSS members. Creating (and plugging when necessary) access holes in order to install bolts can be economically and aesthetically unattractive.

Henderson (1996) described some recent initiatives that may provide solutions for this problem. Two of them follow:

7.4.1 Huck Ultra-Twist Blind Bolts

Huck International, Inc., headquartered in Ogden, Utah, has developed and is now marketing blind bolts (Sadri, 1994) that have tensile strengths and installed tensions meeting those specified for A325 bolts. Known as Ultra-Twist fasteners, they are available in sizes equivalent to ¾, ⅞ and 1

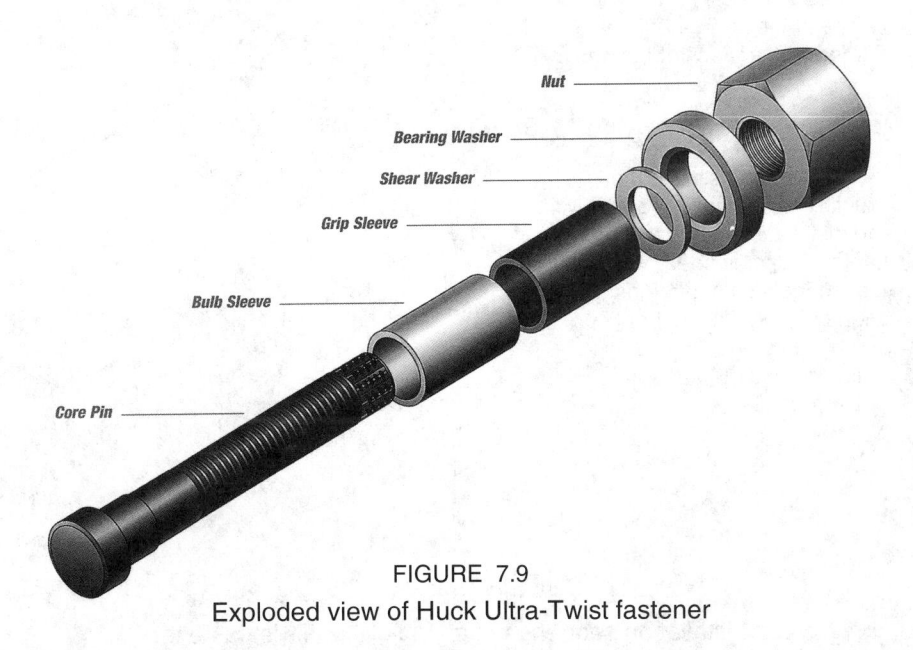

FIGURE 7.9
Exploded view of Huck Ultra-Twist fastener

inch diameter A325 bolts. Figure 7.9 shows an exploded view of an Ultra-Twist fastener, and Fig. 7.10 illustrates the installation sequence.

Installation is by the use of an electric bolting wrench (as for twist-off type bolts), rather than by the use of the hydraulic wrench that was required for an earlier Huck high strength blind bolt (the HSBB). Also, the Ultra-Twist fasteners are used in holes $1/16$ inch larger than the outer diameter of the units, which provides conventional clearances for fit-up. It is expected that erectors will find them a more attractive product than the original Huck HSBB because of these features.

4
Continued torquing of the unit develops the required clamp and the torque pintail shears off, completing the installation. Using a standard S60EZ shear wrench, installation time for a $3/4$ in. fastener is approximately 30 seconds.

3
As the installation load increases, a special internal washer shears, allowing the backside bulb to come into contact with the work surface and for all clamp load to go into the work structure.

2
The backside bulb is fully formed in the air to a uniform diameter regardless of the grip

1
The Ultra-Twist blind bolt is installed from one side of the structure by a single operator. The installation tool is the standard electric shear wrench tooling used for installation of Twist-Off Control (T-C) type fasteners. The fastener is inserted and the tool engaged.

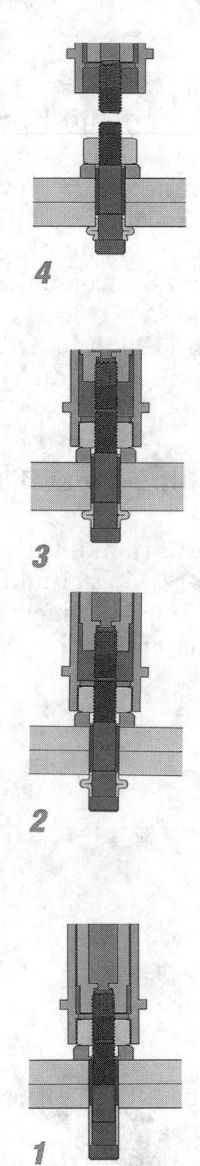

FIGURE 7.10
Installation sequence of a Huck Ultra-Twist fastener

7.4.2 Flowdrilling

The Flowdrill method of creating holes in steel involves the use of a tungsten carbide smooth-sided drilling bit that tapers from a point to a diameter the size of the intended hole. Contact of the high speed rotating bit against the HSS face generates heat to soften the metal so that it extrudes to form a protruding "sleeve" fused to the inside surface of the tube as the bit is forced through the wall. The hole in the wall and its sleeve are then threaded with a rolling Flowtap tapping tool, without removal of material, to accept a conventional high strength bolt as shown in Fig. 7.11. In effect, the hole and sleeve are a nut for the bolt.

The Flowdrill bit in cross section is actually not perfectly round, but somewhat flattened on four sides to produce four lobes as indicated in Fig. 7.12, a shape that aids the extrusion process as the metal of the hole is displaced. A slight upset or boss is created on the outside surface of the material, but that is removed as part of the drilling operation, while the metal is still soft, by the use of a bit incorporating a milling collar.

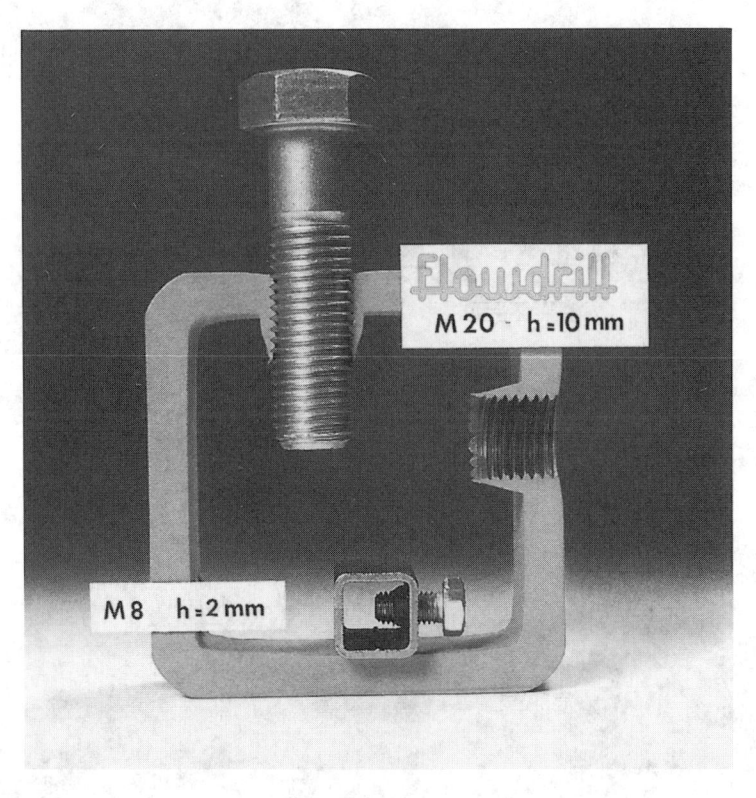

FIGURE 7.11
Samples of bolts in Flowdrilled holes

The Flowdrill system was originally developed in the Netherlands (Dekkers 1993) for light manufacturing, but has been examined for structural HSS applications in the USA (Sherman 1989, 1995), in England (Banks 1993, Yeomans 1996), and in Italy (Ballerini et al. 1995a, 1995b).

Sherman's work of 1989 (reported more publicly in 1995) concluded that Flowdrilling has potential for blind bolting to HSS columns. He pointed out that the fabricator would need drilling equipment with suitable rotational speed, torque and thrust, but the system permits field bolt installation with conventional tools.

Banks (1993) determined that threads produced by the Flowtap tool are metallurgically sound with good toughness. A substantial increase in the strength of material around Flowdrilled holes resulted from partial refining of the microstructure in the threaded area due to heat generated by the process (approaching 800°C). Thickness of the parent metal had little effect on the length of the extruded "sleeve", which was generally 11 to 13 mm long. Rather, the increased amount of displaced metal from thicker material produced sleeves with thicker walls.

Ballerini et al. (1995a, 1995b) established that the average length of effective thread in 6, 8 and 10 mm material was 12.4, 15.3 and 17.5 mm

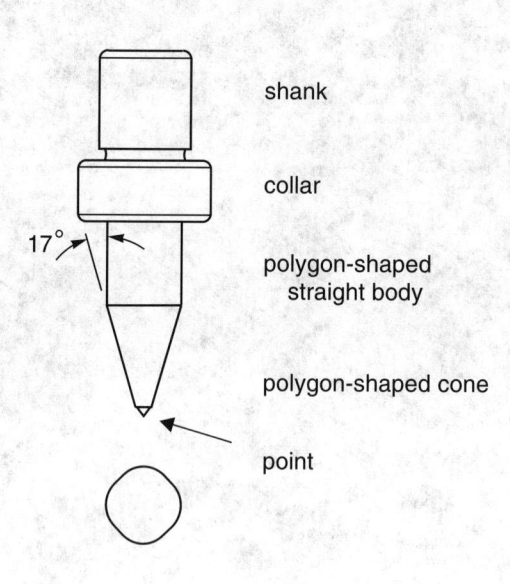

FIGURE 7.12
Flowdrill drilling bit

respectively and was only slightly sensitive to the diameter of the holes. Flatness of 6 mm walls in 140 mm square HSS was not affected, even when Flowdrilling for M20 bolts.

They determined that Flowdrilled holes in HSS with a wall thickness ≥ 0.4 times the hole diameter do not fail by thread stripping, but by bolt (A325 equivalent) rupture. However in those tests, the HSS wall was supported against distortion. In testing where a single bolt was located in the middle of the wall of a 140 mm square HSS and loaded in tension, the t/d ratio had to be ≥ 0.67 to avoid bolt pullout from the deformed hole in the distorted HSS face. In practical connections limited to a serviceability wall distortion of 1%, the bolts in Flowdrilled holes always developed the capacity of the HSS wall.

Yeomans (1996) has reported that Flowdrilling produces sound threaded holes in both hot and cold-formed HSS, on or off the welded seam, in thicknesses from 5 to 12.5 mm, for M16 to M24 bolts. He suggests that optimum drilling speeds are in the range of 500 to 1100 r.p.m. for M24 to M16 bolts, with a spindle feed rate of 0.1 to 0.15 mm/rev., resulting in rapid hole drilling. The minimum required drill rating is 2.5 and 4.5 kW for M16 and M24 bolts respectively. Shear connections were investigated and can be designed using normal practice.

Flowdrilling provides somewhat less tolerance for field assembly than conventional bolting since the bolt has clearance on only one ply of material, not two. However, the Flowdrill system is now regarded as suitable for the production of threaded holes in HSS for structural use.

7.4.3 Connection Failure Modes with Blind Bolts

Comparative tests (Korol et al. 1993) on bolted, extended end-plate moment connections between wide flange beams and square HSS columns have been performed using both regular ASTM A325 bolts and the Huck HSBB. The connection performance in terms of stiffness, moment capacity and ductility was found to be similar for the two types of bolts, and the authors of this Design Guide believe that the Huck Ultra-Twist blind bolt would produce similar results.

If Huck HSBB (or presumably Huck Ultra-Twist fasteners) are used in a rectangular or square HSS column face and are loaded in tension, a potential failure mode is punching shear of the fastener through the column face, in which case the column thickness becomes a critical parameter (Korol et al. 1993). To avoid this failure mode, the factored resistance of the blind bolt in tension should be less than the factored resistance of the column face in punching shear. Hence:

$$T_r < 0.6\, \phi_b\, \pi\, d'_f\, t\, F_u \qquad [7.11]$$

where

T_r = tension resistance of the fastener

ϕ_b = 0.67

d'_f = diameter of the HSBB primary sleeve after installation, or diameter of the Ultra-Twist fastener + 6 mm (estimated effective bulb diameter)

t = wall thickness of hollow section

F_u = specified minimum tensile strength of the HSS material.

Thus, $t > 0.79 \dfrac{T_r}{d'_f F_u}$

Another critical failure mode for an unstiffened HSS column face loaded by point tension loads at the fastener positions is yielding of the HSS connecting face. This failure mechanism can occur due to the flexibility of the column face at medium to large wall slenderness ratios. The resistance for this failure mode can be calculated by assuming that the column wall is loaded like a 90° T-connection with the web member in tension. (See "Chord face yielding" failure mode in Table 3.2, which is based on a yield line mechanism.) In this case, the "web member" can be assumed, for two bolts in tension, to be of width $(w + d'_f)$ and depth d'_f, where w = distance across the HSS column face between the bolt hole centres.

If either of these failure modes (punching shear or column face yielding) produces an inadequate resistance for the HSS column face in the connection tension region, the column face will need to be reinforced. This is best achieved by welding on a doubler plate to the column face. Unfortunately, column reinforcement will nearly always be necessary for practical moment connections, as discussed in Chapter 9, but may not be needed for simple shear connections.

The problem of requiring a very thick column wall to achieve an unreinforced moment connection is recognized in Japan where recent research has focused on developing a method for increasing the column thickness just in the connection region, by using an induction heating device and jack (Tanaka *et al.* 1996).

7.5 Rectangular HSS to Gusset-Plate Connections

Rectangular HSS web or bracing members can be field bolted to gusset plates which have been shop welded to rectangular HSS chord members, thus producing bolted shear connections as shown in Fig. 7.13. Such configurations are regularly used when shipping constraints compel field connections, and bolting has been selected.

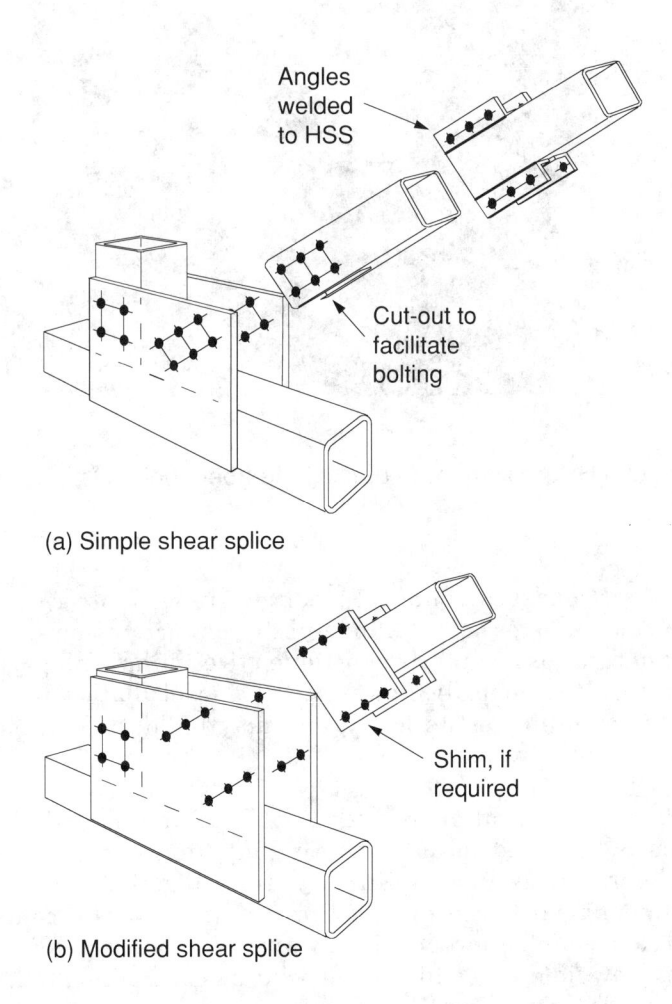

(a) Simple shear splice

(b) Modified shear splice

FIGURE 7.13
Bolted rectangular HSS gusset-plate connections

FIGURE 7.14
Compound HSS column with bracing members bolted to a gusset plate

In dynamic load environments, bolted gusset-plate connections may have an advantage over bolted flange-plate connections for member splices because flange plates must be proportioned to eliminate all prying if fatigue loads are present. Generally, however, the gusset-plate connection is considered less desirable aesthetically, and may be more expensive than the flange-plate alternative.

An important limitation to the use of rectangular HSS gusset-plate connections is the need to have closely matching member widths. Equal width members are connected directly as in Fig. 7.13(a), either with or without clip angles depending on the geometry of the connection. The gussets often need to be spread slightly by jacking after welding is complete in order to allow field assembly (since welding contraction tends to pull the gussets inwards). Small width differences can be adjusted by the use of filler plates welded to the sides of the branch member. Larger differences allow the option of using bolting plates (Fig. 7.13(b)) which can be more convenient in the field.

Single gusset plate connections for bracing members are shown in the photos used for Figs. 7.14 and 7.15.

FIGURE 7.15
Three examples of bolted bracing connections

7.5.1 Net Area, Effective Net Area and Reduced Effective Net Area

Standard CAN/CSA-S16.1-94, Clause 13.2(a) uses the concepts of gross area, net area, effective net area and reduced effective net area to describe various failure modes for a tension member with holes or openings. The three basic checks are

$$T_r = \phi A_g F_y \quad \text{(yielding on gross area)} \quad [7.12]$$

$$T_r = 0.85\phi A_{ne} F_u \quad \text{(rupture on effective net area)} \quad [7.13]$$

$$T_r = 0.85\phi A'_{ne} F_u \quad \text{(rupture on effective net area, reduced for shear lag)} \quad [7.14]$$

where $\phi = 0.9$.

The 0.85 factor represents a minimum margin between factored loads and the factored ultimate resistance for failure modes involving fracture of the tension members.

Effective Net Area

The effective net area A_{ne} (CAN/CSA-S16.1-94, Clause 12.3) is the sum of individual net areas A_n along a potential critical section of the member. Such a critical section may comprise net area segments loaded in tension, segments loaded in shear and segments with a combination of the two loads. This method provides a means of checking against "block shear" failures, whereby a chunk of material tears away from the piece by a combination of shear and tension ruptures. An illustrative example that includes all three types of net area segments is the gusset plate in Fig. 7.16.

The tension segment perpendicular to the load has

$$A_{n1} = w_{n1} t = (g_1 - d'/2) t$$

Shear segments parallel to the load have

$$A_{n2} = 0.6 L_n t = 0.6 (L - 2.5 d') t$$

The 0.6 factor relates shear strength to tensile strength.

Each inclined segment has

$$A_{n3} = \left(w_{n3} + \frac{s^2}{4g}\right) t = (g_2 - d') t + \frac{s^2}{4g_2} t$$

The effective net area A_{ne} of the gusset plate for the potential critical section being examined is the sum of the net area segments above,

$$A_{n1} + A_{n2} + 2A_{n3}$$

For the calculation of w_n and L_n, it should be noted that CAN/CSA-S16.1-94, Clause 12.3.2 requires that the width of bolt holes be taken as 2 mm larger than the specified hole diameter unless it is known that the holes will be drilled. This is an allowance for the rough edges of punched holes. As per CAN/CSA-S16.1-94, Clause 23.3.2, the diameter of bolt holes is generally 2 mm larger than the bolt diameter, so calculations are normally based on bolt holes either 2 mm or 4 mm (drilled or punched) larger than the bolt diameters.

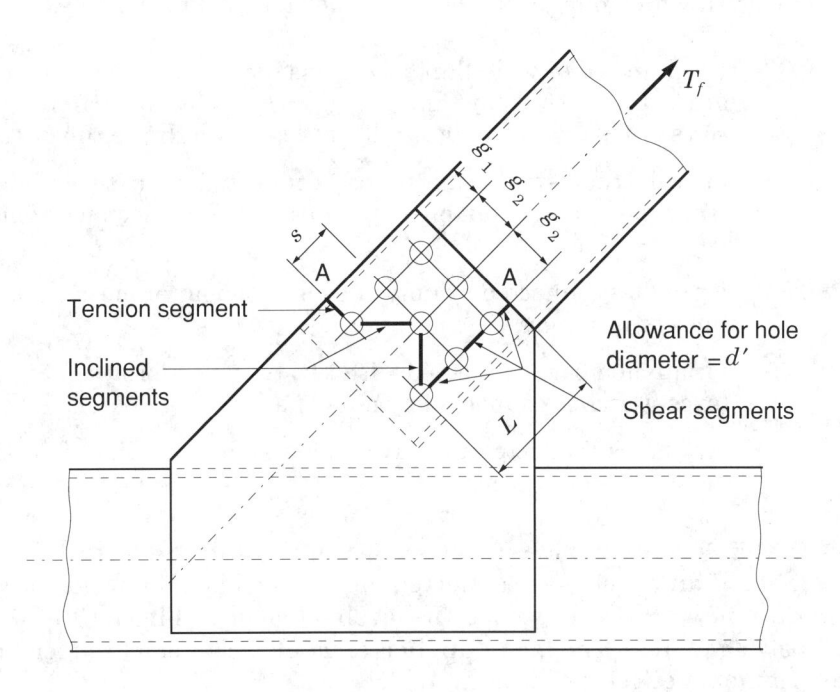

Effective net area A_{ne} for critical section A-A is the sum of A_n for individual segments:

A_{n1} for tension segment is $w_{n1} t = (g_1 - d'/2) t$

A_{n2} for shear segments is $0.6 L_n t = 0.6 (L - 2.5 d') t$

A_{n3} for <u>each</u> inclined segment is $\left(w_{n3} + \dfrac{s^2}{4g} \right) t = (g_2 - d') t + \left(\dfrac{s^2}{4 g_2} \right) t$

FIGURE 7.16
Calculation of effective net area A_{ne} for a gusset plate

Effective Net Area Reduced for Shear Lag

The effective net area reduced for shear lag A'_{ne} is the effective net area A_{ne} multiplied by a shear lag factor as per CAN/CSA-S16.1-94, Clause 12.3.3, ($A'_{ne} = A_{ne} \times$ shear lag factor). This factor comes into play when a member is connected by some but not all of its cross-sectional elements, if the critical net section includes elements which are not connected. The critical net section in such an instance may include net area segments A_n that are perpendicular to the load or inclined to it, but not those that are parallel to it. (This is not a check against block shear tear out.) The shear lag factor with which one multiplies A_{ne} in order to obtain A'_{ne} is

- 0.90   for shapes like wide flange sections (or tees cut from them) that are connected only by their flanges with at least three transverse rows of fasteners (flange width at least two thirds the depth)
- 0.85   for all other structural shapes connected with three or more transverse rows of fasteners (except angles connected by only one leg)
- 0.80   for angles connected by only one leg with four or more transverse lines of fasteners
- 0.75   for all members connected with two transverse rows of fasteners (except angles connected by only one leg)
- 0.70   for angles connected by only one leg with fewer than four transverse lines of fasteners.

Effective net area reduced for shear lag A'_{ne} also applies to welded connections if an element is connected along its edge(s) by welds *parallel to the direction of load* (e.g., the HSS web member in Fig. 7.13(b) welded to the bolting plates). The factor applied to wt of each element of a member to determine its effective net area A_{ne} is

» For an element that is connected by welds along two parallel edges,

- 1.00   when the average weld length is at least twice the distance between the welds
- 0.87   when the average weld length is from one and a half times to less than twice the distance between them
- 0.75   when the average weld length is from one to less than one and a half times the distance between them

» For an element that is connected by a single line of weld,

$1 - \dfrac{\bar{x}}{L}$ where $\bar{x}$ is the eccentricity of the weld with respect to the centroid of the element and L is the weld length. (The outstanding leg of an angle would be considered as being connected by the line of weld along its heel.)

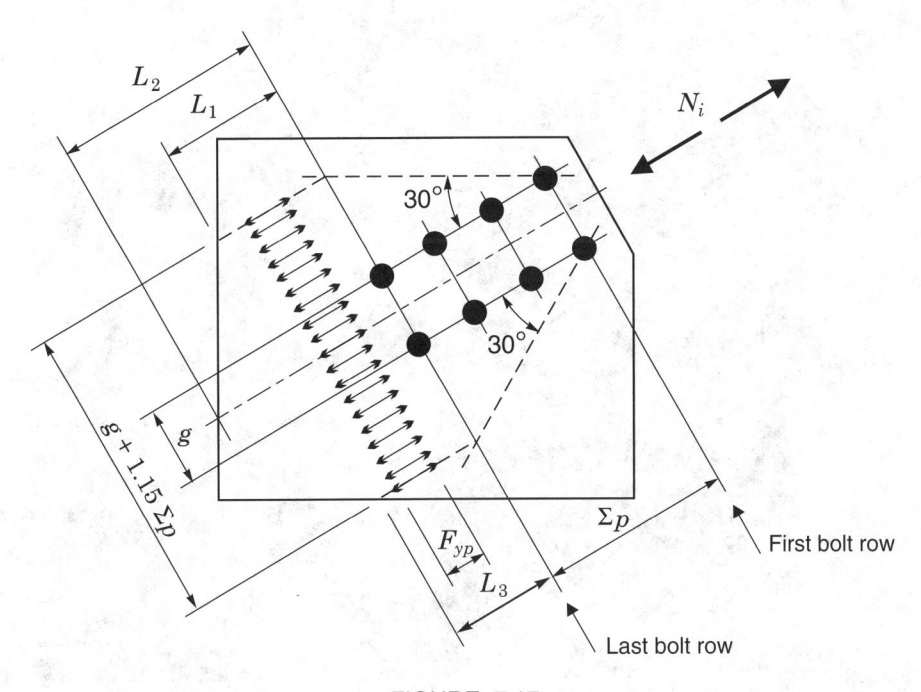

FIGURE 7.17
Whitmore criterion for gusset plate yielding or buckling

If *transverse welds* are used to transmit load to some elements of a member, the effective net area is taken as the area of the connected elements.

The effective net area reduced for shear lag A'_{ne} of the member is then taken as the sum of the effective net areas A_{ne} of all the elements.

Another failure mode which must be checked for gusset plates is yielding across an effective dispersion width of the plate, which can be calculated using the Whitmore (1952) effective width concept illustrated in Fig. 7.17. For this failure mode, (for two gusset plates),

$$N_i^* = 2 \phi F_{yp} t_p (g + 1.15 \Sigma p) \qquad [7.15]$$

This check applies to the tension load case. The term Σp represents the sum of the bolt pitches in a bolted connection or the length of the weld in a welded connection.

If the member is in compression, buckling of the gusset plate must also be prevented. A suitable method for checking this is given by Thornton (1984). The gusset plate compressive resistance is the column resistance given by CAN/CSA-S16.1-94 for a column having a width of $(g + 1.15 \Sigma p)$, a depth of t_p, a length equal to the minimum of L_1, L_2 and L_3, and an effective length factor of 0.65.

FIGURE 7.18
Complex node at a wheel bogie of the SkyDome roof, Toronto, Ontario

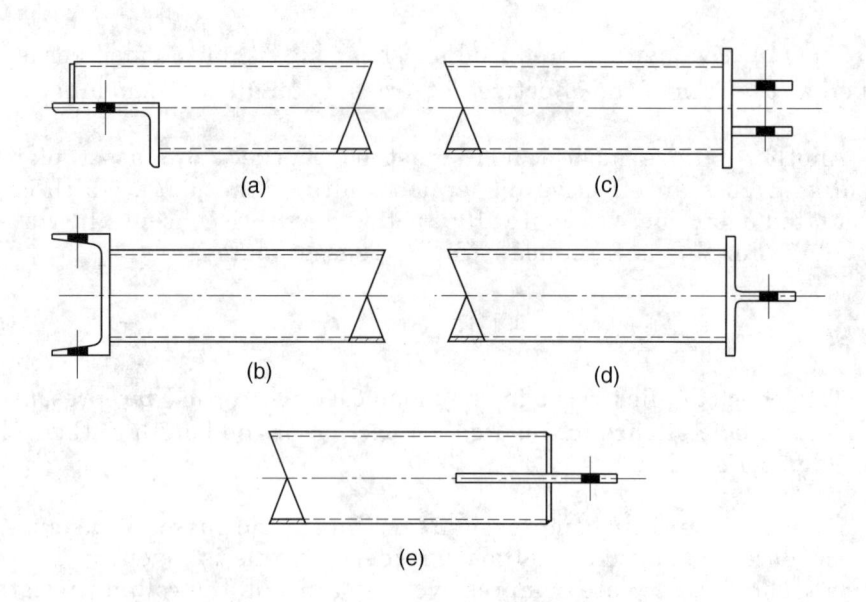

FIGURE 7.19
Examples of bolted end connections for HSS members

7.6 Other Bolted Connections

Many configurations for other bolted connections are feasible with HSS construction. Most use bolts in shear and the considerations already discussed are applicable. The following examples illustrate the range of possibilities. As a matter of interest, an extremely complex HSS node from the Toronto SkyDome roof is shown in Fig. 7.18.

End Connections with Welded Attachments

Five different end attachments welded to HSS sections are shown in Fig. 7.19. All function with either rectangular or circular sections.

One needs to keep in mind that the flange of the tee in Fig. 7.19(d) must be sufficiently thick to distribute the load effectively to all portions of the HSS cross section (if the load requires the involvement of all portions). A conservative distribution angle can be assumed as 2.5 to 1 from each face of the tee's web, as illustrated in Fig. 7.20 (Wardenier *et al.* 1991, Kitipornchai and Traves 1989). Resistance of the connection is thus:

$$N_i^* = 2\phi\,(5\,t_p + t_w)\,F_y\,t_{HSS} \qquad [7.16]$$

for both circular and square/rectangular HSS and where $\phi = 0.9$. In this expression, the size of the weld legs has been conservatively ignored. If the weld leg size is know, it is acceptable to assume load dispersion from the toes of the welds.

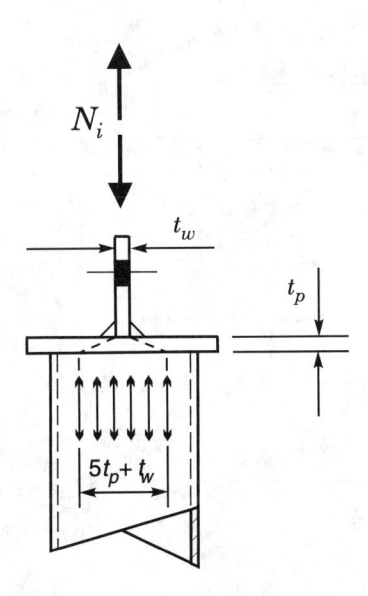

FIGURE 7.20

Load dispersion for tee connection on end of HSS member

If a heavily loaded plate is slotted into an HSS as in Fig. 7.19(e), there is a possibility that shear lag requirements may govern the capacity of the welded connection. (Effective net area reduced for shear lag was discussed in Section 7.5.) A worked example which shows shear lag in this situation is presented in Section 7.7.2.

Connections for Continuous Members (purlins)

Continuous members such as purlins may be connected directly by means of threaded studs that have been welded to the HSS member, or by connecting plates or clips. Figure 7.21 shows both approaches.

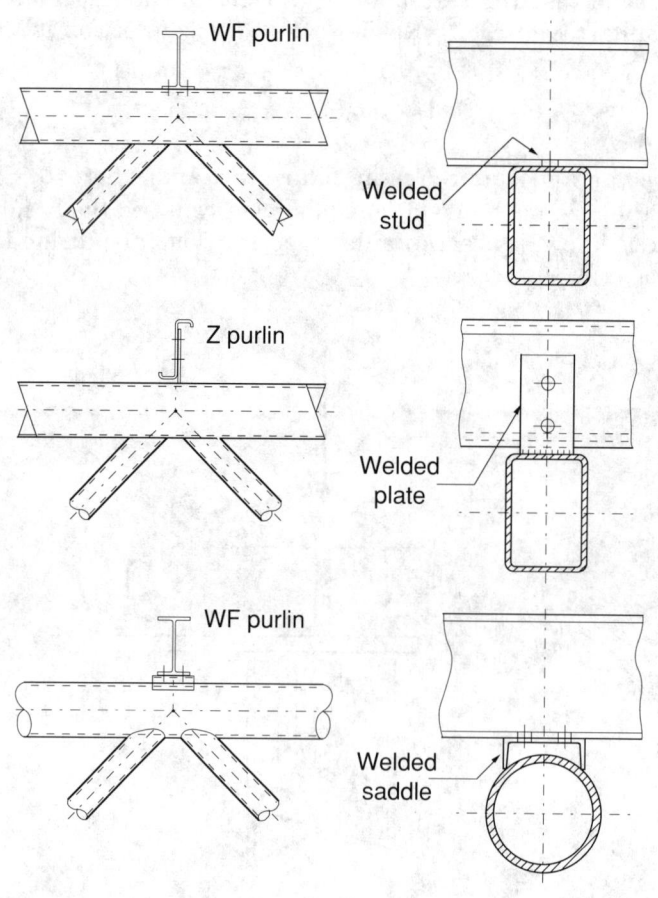

FIGURE 7.21
Examples of bolted purlin connections

End Flattened Connections

The flattening of HSS ends in order to make bolted or welded connections is discussed in Section 8.1.1. Some general arrangements for their use are depicted in Fig. 7.22.

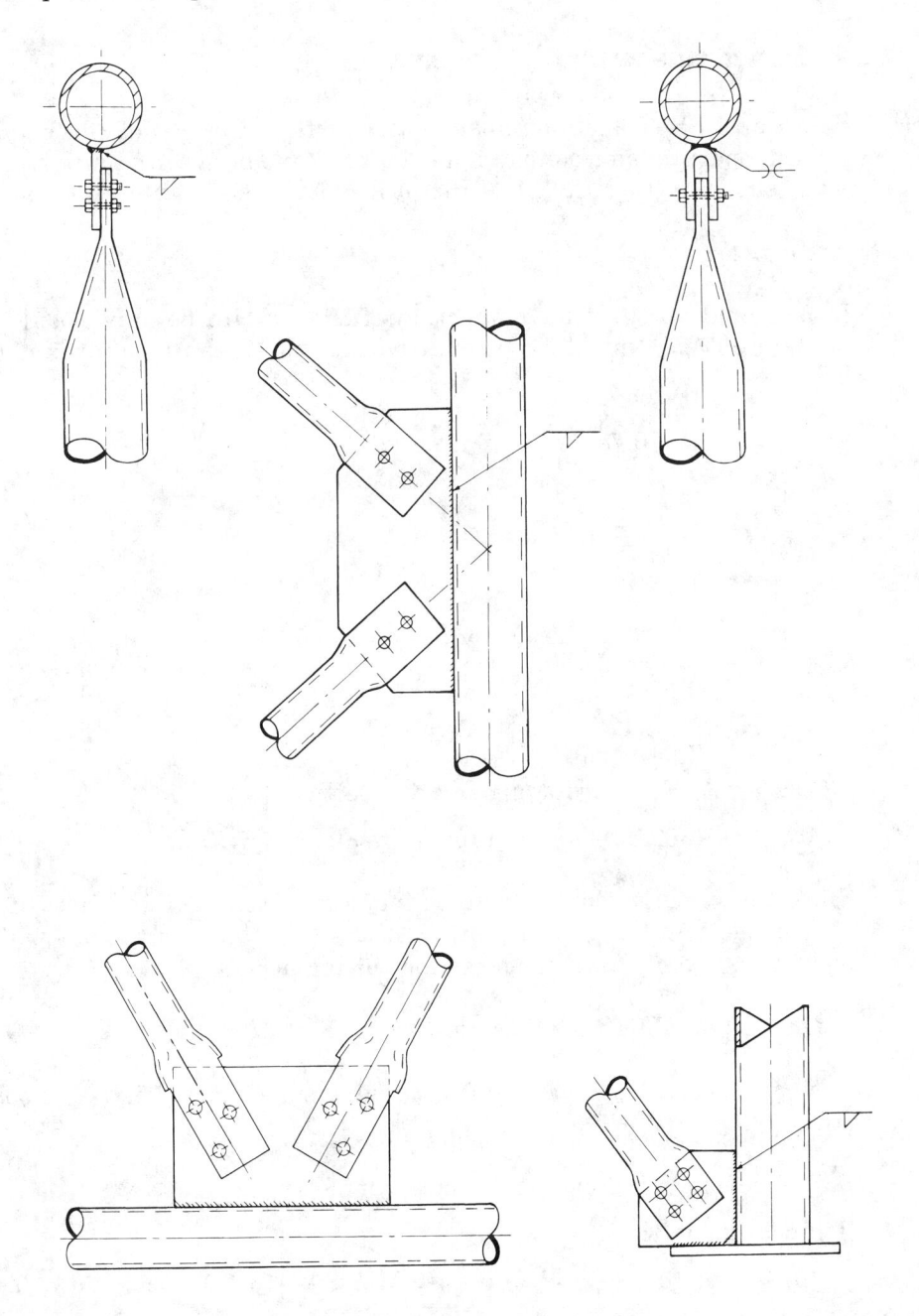

FIGURE 7.22
Examples of end-flattened HSS connections

243

7.7 Design Examples

Design examples of bolted flange-plate connections for square and circular HSS are provided in Sections 13.2.7 and 13.3.8 respectively.

7.7.1 Bolted Gusset-Plate Connection

Select an HSS tension member to carry a factored load of 800 kN, and design a compact tension connection to bolt one end between a pair of gusset plates as shown in Fig. 7.23. Use the provisions of CAN/CSA-S16.1-94.

Select Trial Section

In anticipation that the connection efficiency will be only about 70% (50% to 80% is typical), the required gross area can be approximated by

$$T_r = 0.85 \phi (0.7 A_g) F_u \quad \text{(from [7.13])}$$

$$\therefore A_g = \frac{T_r}{0.6 \phi F_u}$$

$$= \frac{800}{0.6(0.9)0.450} = 3\,290 \text{ mm}^2$$

Try an HSS 127x127x8.0. $A = 3\,620 \text{ mm}^2$

$$T_r = \phi A_g F_y \quad \text{(yielding on gross section, [7.12])}$$
$$= 0.9(3620)0.350 = 1\,140 \text{kN} \geq 800 \quad \therefore \text{ OK}$$

Since a physically compact connection is required, bolt two faces of the HSS directly to the gusset plates. Assume that an end cut out will be needed in <u>one</u> of the other faces for bolting access.

Consider M20 A325M High Strength Bolts

In single shear, with threads intercepted, resistance = 73.5 kN per bolt.

Number = 800/73.5 = 10.9 bolts.

Therefore, use 12 bolts for 2-face symmetry.

Bolt bearing on HSS:

$$B_r = 3 \phi_b t d n F_u \quad \text{(CAN/CSA-S16.1-94, Clause 13.10.(c))}$$
$$= 3 (0.67) 7.95 (20) 0.450 = 144 \text{ kN per bolt} \geq 73.5 \quad \therefore \text{ OK}$$

One may wonder whether bolts that are less than three bolt diameters from material edges can develop the full bearing value just calculated.

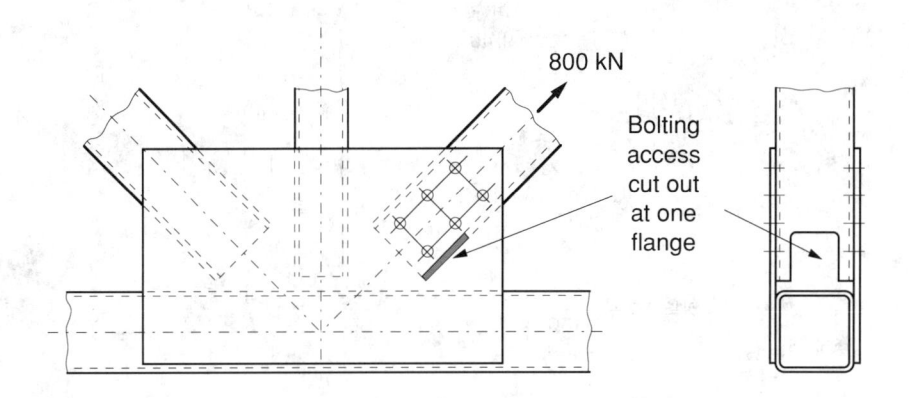

FIGURE 7.23
Design example for a bolted HSS connection

Actually, such bolts do have less than the calculated bearing resistance, but the failure mode becomes one of material tearing, which involves more bolts than just those near the edge. The design check is then for block tear-out, based on effective net area, not bearing.

See Fig. 7.23 for the trial arrangement.

Effective Net Area

One possible critical section is shown in Fig. 7.24(a), a tearing out of the bolt hole patterns on the two bolted faces of the HSS.

Hole allowance is 22 mm (drilled for M20 bolt).

For tension segment (one face):

$$A_{n1} = w_{n1} t$$
$$= (65 - 22)7.95 = 342 \text{ mm}^2$$

For shear segments (one face):

$$A_{n2} = 0.6 L_n t$$
$$= 0.6 \, (2 \, (70 + 70 + 40) - 5(22)) \, 7.95 = 1\,190 \text{ mm}^2$$

Therefore, this total net area is $2 \, (342 + 1\,190) = 3\,060 \text{ mm}^2$.

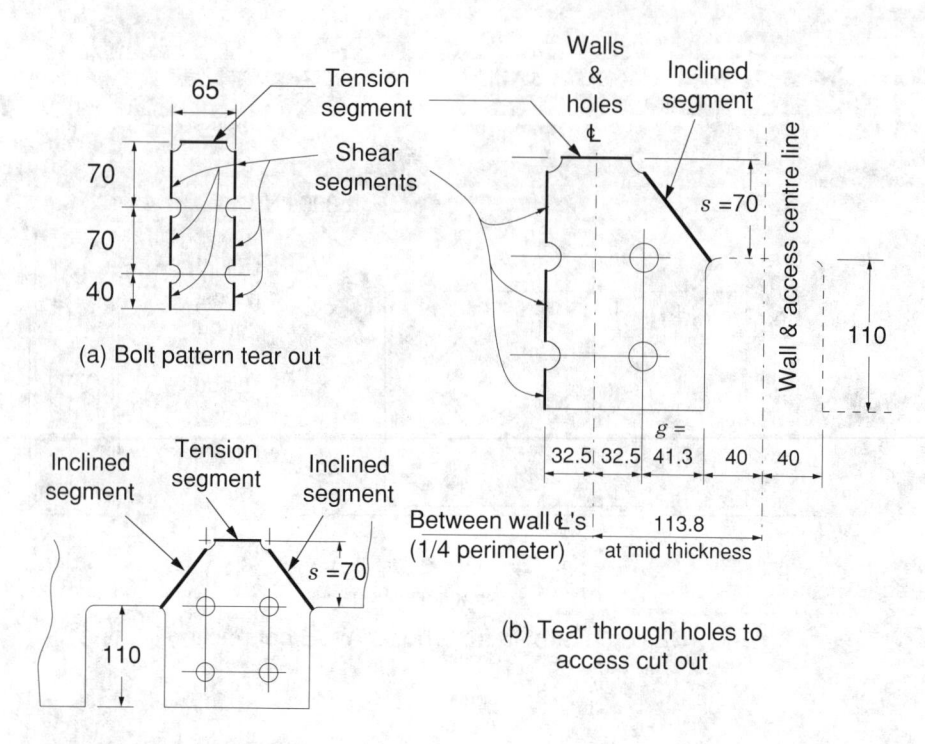

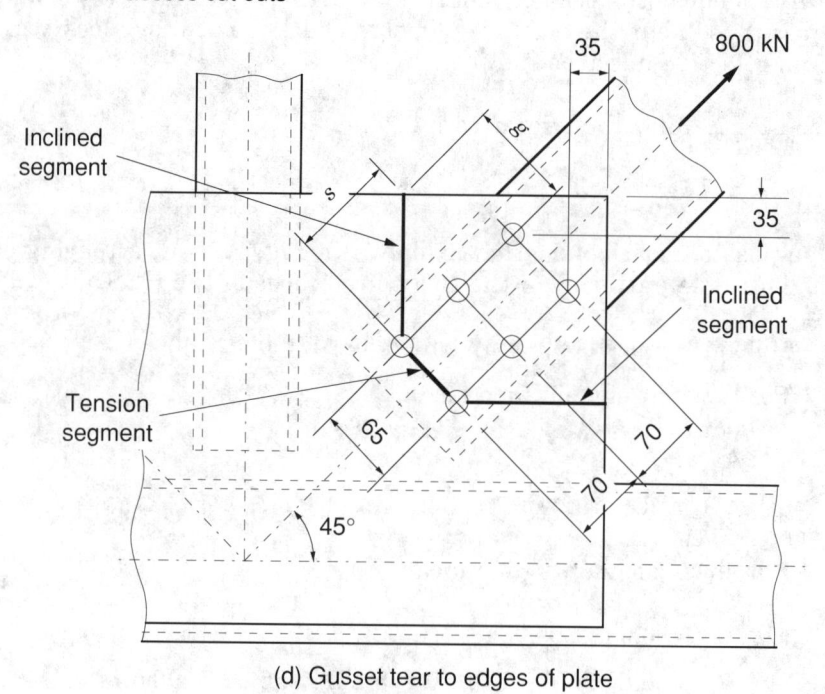

FIGURE 7.24
Effective net area examples

Another possible critical section, Fig. 7.24(b), is the tearing away of the hole patterns over to the bolting access cut out on the third face of the HSS.

For tension segment (one face):

$$A_{n1} = w_{n1} t = 342 \text{ mm}^2, \text{ as previously}$$

For shear segments (one face):

$$A_{n2} = 0.6 L_n t$$
$$= 0.6 (180 - 2.5 (22)) 7.95 = 596 \text{ mm}^2$$

For inclined segment (one face), the net width can be arrived at by considering one quarter the perimeter, minus half the access opening width, minus half the distance between the bolt lines, minus half a hole diameter:

$$w_{n3} = \frac{3\,620}{4\,(7.95)} - \frac{80}{2} - \frac{65}{2} - \frac{22}{2} = 30.3 \text{ mm}$$

$$A_{n3} = w_{n3} t + \frac{s^2}{4g} t$$

$$= 30.3 (7.95) + \frac{70^2}{4\,(41.3)} 7.95 = 477 \text{ mm}^2$$

Therefore, this total net area is $2\,(342 + 596 + 477) = 2\,830 \text{ mm}^2$.

If there had been <u>two</u> bolting access cut outs (one each, on opposite faces) another possible critical section including both cut outs (Fig. 7.24(c)) would have governed, with a net area:

$$2\,(477 + 342 + 477) = 2\,590 \text{ mm}^2$$

Therefore, consider that $A_{ne} = 2\,590 \text{ mm}^2$.

$$T_r = 0.85 \phi A_{ne} F_u \qquad ([7.13])$$
$$= 0.85 (0.90)\,2\,590\,(0.450) = 892 \text{ kN} \quad \geq 800 \quad \therefore \text{OK}$$

Examine Shear Lag

Consider the net section through the uppermost row of four holes.

$$A_{ne} = 3\,620 - 4\,(22)\,7.95 = 2\,920 \text{ mm}^2$$

$$A'_{ne} = 0.85 A_{ne} \quad \text{(HSS with 3 rows of bolts on 2 faces)}$$
$$= 0.85 (2\,920) = 2\,480 \text{ mm}^2$$

$$T_r = 0.85 \phi A'_{ne} F_u \quad ([7.14])$$
$$= 0.85 (0.90)\, 2\,480\, (0.450) = 854 \text{ kN} \geq 800 \quad \therefore \text{ OK}$$

Therefore, the HSS member is satisfactory.

Gusset Plates

Try 10 mm thickness. $\quad (F_y = 300 \text{ MPa})$

Two possible critical sections are apparent:

1. The first is similar to the bolt pattern tear out examined for the HSS member, and is not expected to be critical with gusset plates which are 25% thicker than the HSS wall (both materials have F_u = 450 MPa), and need not be evaluated here.

2. Figure 7.24(d) illustrates the second section, which consists of a tear from the top edge of the gusset through the two end bolts to the side of the gusset. If $\theta = 45°$ and the hole edge distances are 35 mm, the resistance of this section is as follows.

   Hole allowance is 24 mm. $\quad$ (22 mm punched for M20 bolts)

   For tension segment (one gusset):

   $$A_{n1} = w_{n1} t = (65 - 24)10 = 410 \text{ mm}^2$$

   For each inclined segment (one gusset), the net width can be arrived at by calculating g in Fig. 7.24(d) and deducting half a hole diameter.

   $$g = [((70 + 70)\, 0.707) + 35]\, 0.707 = 94.7 \text{ mm}$$

   Therefore, $w_{n3} = 94.7 - 24/2 = 82.7 \text{ mm}$.

   $$s = g \quad \text{(from 45° symmetry)}$$

   $$A_{n3} = w_{n3} t + \frac{s^2}{4g} t$$

   $$= 82.7\,(10.0) + \frac{94.7^2}{4\,(94.7)}\, 10.0$$

   $$= 1\,060 \text{ mm}^2 \quad \text{(one inclined segment on one gusset)}$$

   Therefore, this total net area is $2\,(2\,(1\,060) + 410) = 5\,060 \text{ mm}^2$.

$$T_r = 0.85\,\phi\,A_{ne}\,F_u \quad ([7.13])$$
$$= 0.85\,(0.9)\,5\,060\,(0.450) = 1\,740\text{ kN} \geq 800 \quad \therefore \text{OK}$$

Check for gusset plate yielding according to [7.15].

$$N_i^* = 2\,\phi\,F_{yp}\,t_p\,(g + 1.15\,\Sigma p)$$
$$= 0.9\,(0.300)\,2\,(10)\,(65 + 1.15\,(140))$$
$$= 1\,220\text{ kN} \geq 800 \quad \therefore \text{OK}$$

Therefore, the gusset plates also are satisfactory.

7.7.2 Connection with Plate Slotted into End of HSS

Figure 7.24 illustrates a plate of 300W steel for a bolted HSS connection. It is welded into longitudinal slots in the short faces at the end of an HSS 203x152x6.4 member (A = 4 250 mm^2). Confirm that the connection meets the requirements of CAN/CSA-S16.1-94 for the factored tension load of 560 kN.

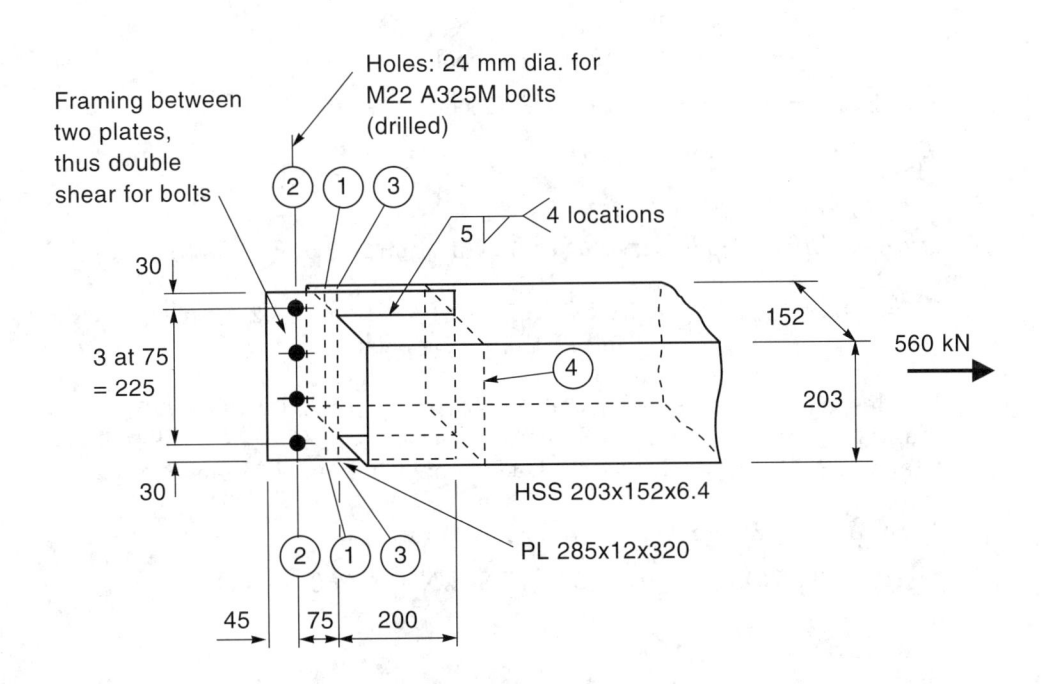

FIGURE 7.25

Connection with plate slotted into end of HSS

Bolts

CISC Handbook (Table 3–4) lists the factored shear resistance of an M22 A325M bolt (threads intercepted) as 88.9 kN.

Resistance of 4 bolts in double shear is

$\qquad$ 4 (2) 88.9 = 711 kN $\quad \geq 560 \quad \therefore$ OK

Bolt tear-out

Bolt tear-out (a form of block tear-out) becomes a consideration when the edge distance is less than $3d$. It can be conservatively treated as a pair of material shear failures in line with the sides of a bolt for a length equal to the clear distance from the bolt (centred in the hole) to the edge of the material, and is expressed thus:

$\qquad T_r = 0.85 \phi A_{ne} F_u$ per bolt $\qquad$ ([7.13])

$\qquad \quad = 0.85 (0.9) \, 2 \, (0.6) \, (45 - 11) \, 12 \, (0.450) = 169$ kN

Bearing load per bolt is $560/4 = 140$ kN $\quad \leq 169 \quad \therefore$ OK

Yield on Gross Area of Plate (section 1–1 of Fig. 7.25)

$\qquad T_r = \phi A_g F_y \qquad$ (CAN/CSA-S16.1-94, Clause 13.2(a) (i)
$\qquad \qquad \qquad \qquad$ and [7.12] in this book)

Plate is 12×285 mm $\quad \therefore A_g = 3\,420$ mm^2

$\qquad T_r = 0.9 \, (3\,420) \, 0.300 = 923$ kN $\quad \geq 560 \quad \therefore$ OK

Fracture on Effective Net Area of Plate (section 2–2 of Fig. 7.25)

$\qquad T_r = 0.85 \phi A_{ne} F_u \qquad$ (CAN/CSA-S16.1-94, Clause 13.2(a) (ii)
$\qquad \qquad \qquad \qquad \quad$ and [7.13] in this book)

If holes are known to be drilled rather than punched, the specified hole diameter can be used for the net section. $\quad$ (CAN/CSA-S16.1-94, Clause 12.3.2, plus Clause 23.3.2 for specified size of hole)

Then $A_{ne} = 12 \, (285 - (4 \times 24)) = 2\,270$ mm^2

$\qquad T_r = 0.85 \, (0.9) \, 2\,270 \, (0.450) = 781$ kN $\quad \geq 560 \quad \therefore$ OK

Welds

CISC Handbook (Table 3–24) shows that the fillet weld factored resistance (E480XX electrodes) is 0.152 kN per mm of weld leg size.

Use weld length $L = 200$ mm.   (nominal size of HSS)

Required fillet weld size is $560/(4\,(200)\,0.152) = 4.6$ mm

Since there could be a cutting gap of 1 mm to bridge, use 6 mm fillet welds.

Shear Lag Fracture in the Plate (section 3–3 of Fig. 7.25)

Clause 12.3.3.4 in CAN/CSA-S16.1-94 provides for shear lag in plates that are connected by a pair of welds parallel to the load, and the length of the welds L should generally not be less than the distance w between them. The effective net area is reduced if the length of the welds is less than $2w$.

Distance between welds $w = 203$ mm

For portion of plate between welds

A'_{ne} = shear lag factor $\times A_g$ (CAN/CSA-S16.1-94, Clause 12.3.3(b))

$= 0.75\,(203)\,12 = 1\,830\,\text{mm}^2$

The portion of the plate that is beyond the welds is $(285 - 203)/2 = 41$ mm on each side of the HSS member. Since the length of the weld is more than 4 times this plate dimension, it is suggested that this width is fully effective. This is consistent with CAN/CSA-S16.1-94, Clause 12.3.3.3 where plate is taken as fully effective when the weld length is at least 2 times the width of plate between the welds. Therefore, plate may be taken as fully effective out to a distance from the weld of 1/4 the weld length.

Area beyond HSS $= (285 - 203)\,12 = 984$ mm^2.   (all effective)

Therefore, total $A'_{ne} = 1\,830 + 984 = 2\,810$ mm^2

Resistance of the plate at the welds (section 3–3)

$T_r = 0.85\,\phi\,A'_{ne}\,F_u$ (CAN/CSA-S16.1-94, Clause 13.2(a) (iii) and [7.14] in this book)

$= 0.85\,(0.9)\,2\,810\,(0.450) = 967$ kN ≥ 560 ∴ OK

Shear Lag Fracture in the HSS (section 4 of Fig. 7.25)

Shear lag requirements apply only when unconnected elements of a shape must be included in the critical cross section for strength purposes (Clause 12.3.3.2 in CAN/CSA-S16.1-94). The principle applies to elements connected either by welding or by bolting.

In this example, the connected elements (the top and bottom walls of the HSS member) alone have sufficient "fracture on net area" resistance by [7.13] (as seen below) for the 560 kN load. Therefore, the unconnected side walls are not required for connection resistance, and shear lag is not a consideration.

$$T_r = 0.85 \phi A_{ne} F_u \quad ([7.13])$$

$$A_{ne} = 4\,250 \frac{152}{152+203} - 2\,(14)\,6.35 = 1\,640 \text{mm}^2$$

(if one assumes 14 mm wide slots for the 12 mm plate)

$$\therefore T_r = 0.85\,(0.9)\,1\,640\,(0.450) = 565 \text{ kN} \quad \geq 560 \quad \therefore \text{OK}$$

Therefore the connection meets the requirements of CAN/CSA-S16.1-94.

If the load were 650 kN, the unconnected side walls of the HSS would be needed for strength and shear lag would become a consideration. The distance between the welds measured along the developed perimeter of the HSS member can be taken as 69 + 203 + 69 = 341 mm compared with a weld length of 200 mm. That gives a L/w ratio of 0.59, which is less than the lowest value of 1.0 used in CAN/CSA-S16.1-94, Clause 12.3.3.3 for identifying shear lag factors.

Experimental work by Korol *et al.* (1994) confirmed that a shear lag factor of 0.62 as proposed by the authors of this Guide may be used for values of L/w less than 1.0. The work of Korol *et al.* also showed that when L/w is less than about 0.6, the failure mode shifts from shear lag-induced tension rupture to either weld failure or shear rupture of the material along the welds.

Let us look at the current example from both perspectives:

Shear lag,

$$A'_{ne} = 0.62\,(4\,250 - (2\,(14)\,6.35)) = 2\,530 \text{ mm}^2$$

$$T_r = 0.85\,(0.9)\,2\,530\,(0.450) = 871 \text{ kN} \quad \geq 650 \quad \therefore \text{OK}$$

Tube material shear at welds,

$$V_r = 0.5 \phi L_n t F_u \quad (\text{CAN/CSA-S16.1-94, Clause 13.4.4})$$

$$= 0.5\,(0.9)\,(4\,(200 \times 6.35))\,0.450 = 1\,030 \text{ kN}$$

Shear lag governs, but a shorter weld length (i.e., smaller L/w) would result in material shear governing.

Of course other elements of the connection would need to be resized for the 650 kN load.

An examination of the interface between shear lag tension failures and base metal shear failure along the welds is shown in Fig. 7.26. The transition from the former to the latter occurs when the ratio L/w is between 0.6 and 0.7 in this presentation.

Shear resistance (CAN/CSA-S16.1-94, Clause 13.4.4) of base material along weld is $V_r = 0.5 \phi L_n t F_u$

Therefore, shear strength is $V_u = 0.5 L_n t F_u$
$= 0.5 (2L t F_u)$

When $L/w = 0.5$, shear strength $= 0.5 w (t F_u)$
i.e., coefficient for w is 0.5, as plotted below

Slope of equivalent effective w for base metal shear strength

Slope of effective w for tension strength (shear lag)

1.0 *

0.87 *

0.75 *

0.62 †

* = S16.1-94, Clause 12.3.3.3
† = Projected

Coefficient to obtain effective w for A'_{ne}

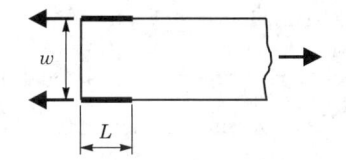

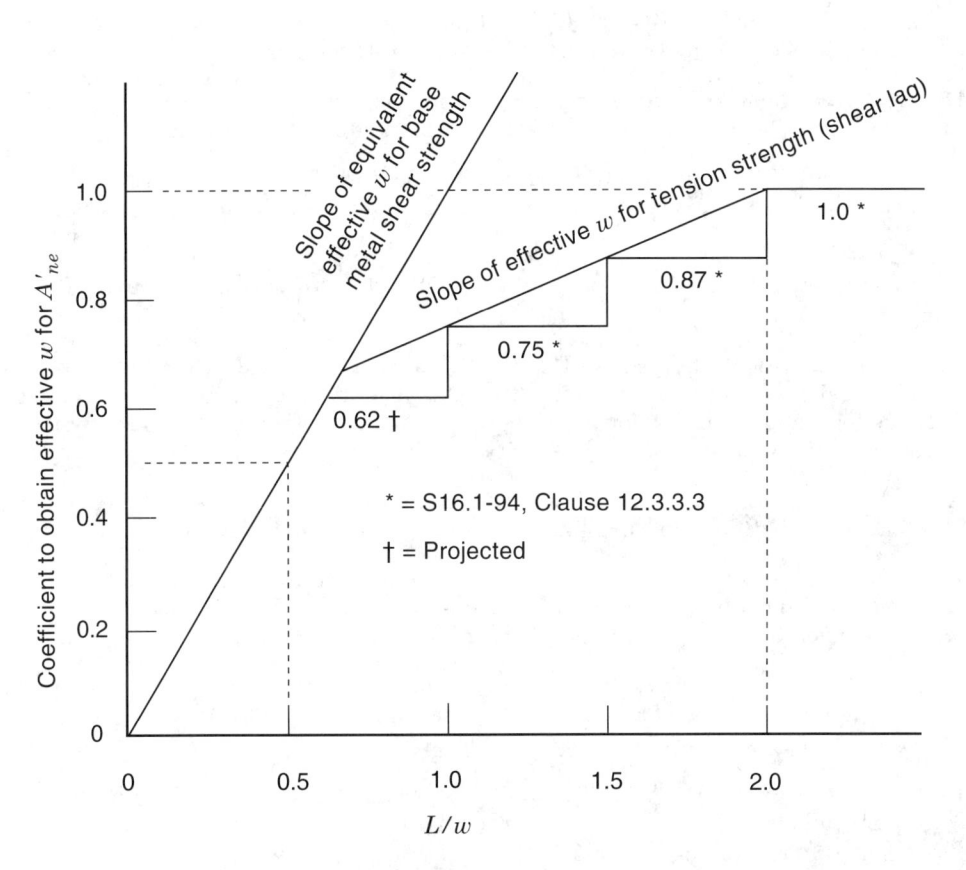

L/w

FIGURE 7.26

Examination of shear lag for L/w ratios less than 1.0

(An alternative label for the vertical axis would be, "Coefficient by which to multiply net area (A_{ne}) to get effective net area (A'_{ne})")

REFERENCES

BANKS, G. 1993. Flowdrilling for tubular structures. Proceedings, 5th. International Symposium on Tubular Structures, Nottingham, England, pp. 117–124.

BALLERINI, M., BOZZO, E., OCCHI, F., and PIAZZA, M. 1995a. The Flowdrill system for the bolted connection of steel hollow sections. Part I—the drilling process and the technological aspects. Costruzioni Metalliche, No. 4, July-August, Italy, pp. 13–23.

BALLERINI, M., BOZZO, E., OCCHI, F., and PIAZZA, M. 1995b. The Flowdrill system for the bolted connection of steel hollow sections. Part II—experimental results and design evaluations. Costruzioni Metalliche, No. 5, September-October, Italy, pp. 1–12.

BIRKEMOE, P.C., and PACKER, J.A. 1986. Ultimate strength design of bolted tubular tension connections. Proceedings, Steel Structures—Recent Research Advances and their Applications to Design, Budva, Yugoslavia, pp. 153–168.

CSA. 1994. Limit states design of steel structures, CAN/CSA-S16.1-94. Canadian Standards Association, Rexdale, Ontario.

CARAVAGGIO, A. 1988. Tests on steel roof joints for Toronto SkyDome. Master of Applied Science thesis, University of Toronto, Ontario.

DEKKERS, G.J. 1993. Flowdrill, technical guide. Available from Robert Speck Ltd., Little Ridge Whittlebury Rd., Silverstone, Northants, NN12 8UD, England.

DUTTA, D., and WÜRKER, K. 1988. Handbuch hohlprofile in stahlkonstruktionen. Verlag TÜV Rheinland GmbH, Köln, Federal Republic of Germany.

HENDERSON, J.E. 1996. Bending, bolting and nailing of hollow structural sections. Proceedings, International Conference on Tubular Structures, Vancouver, B.C., pp. 150–161.

IGARASHI, S., WAKIYAMA, K., INOUE, K., MATSUMOTO, T., and MURASE, Y. 1985. Limit design of high strength bolted tube flange joint—Parts 1 and 2. Journal of Structural and Construction Engineering Transactions of AIJ, Department of Architecture Reports, Osaka University, Japan.

KATO, B., and MUKAI, A. 1982. Bolted tension flanges joining square hollow section members. CIDECT Report 8B-82/3-E, University of Tokyo, Japan.

KATO, B., and MUKAI, A. 1985. Bolted tension flanges joining square hollow section members. Supplement—bolted at two sides of flange. CIDECT Program 8B Report, University of Tokyo, Japan.

KITIPORNCHAI, S., and TRAVES, W.H. 1989. Welded-tee end connections for circular hollow tubes. Journal of Structural Engineering, American Society of Civil Engineers, **115**(12): 3155–3170.

KOROL, R.M., GHOBARAH, A., and MOURAD, S. 1993. Blind bolting W-shape beams to HSS columns. Journal of Structural Engineering, American Society of Civil Engineers, **119** (12): 3463–3481.

KOROL, R.M., MIRZA, F.A., and MIRZA, M.Y. 1994. Investigation of shear lag in slotted HSS tension members. Proceedings, 6th. International Symposium on Tubular Structures, Melbourne, Australia, pp. 473–482.

MANG, F. 1980. Investigation of standard bolted flange connections for circular and rectangular hollow sections. CIDECT Report 8A-81/7-E, University of Karlsruhe, Federal Republic of Germany.

PACKER, J.A., BRUNO, L., and BIRKEMOE, P.C. 1989. Limit analysis of bolted RHS flange plate joints. Journal of Structural Engineering, American Society of Civil Engineers, **115**(9): 2226–2242.

ROCKEY, K.C., and GRIFFITHS, D.W. 1970. The behaviour of bolted flange joints in tension. Proceedings, Conference on Joints in Structures, Sheffield, England.

SADRI, S. 1994. Blind bolting. Modern Steel Construction, American Institute of Steel Construction, February, pp. 44–46.

SHERMAN, D.R. 1989. Evaluation of flow-drilled holes for bolted connections in structural tubes. CE Department Report, University of Wisconsin-Milwaukee, Wisconsin, U.S.A.

SHERMAN, D.R. 1995. Simple framing connections to HSS columns. Proceedings, American Institute of Steel Construction, National Steel Construction Conference, San Antonio, Texas, U.S.A., pp. 30.1–30.16.

STELCO. 1981. Hollow structural sections—design manual for connections, 2nd. ed. Stelco Inc., Hamilton, Ontario.

STRUIK, J.H.A., and de BACK, J. 1969. Tests on bolted T-stubs with respect to a bolted beam-to-column connection. Stevin Laboratory Report 6-69-13, Delft University of Technology, The Netherlands.

TANAKA, T., TABUCHI, M., FURUMI, K., MORITA, T., USAMI, K., MURAYAMA, M., and MATSUBARA, Y. 1996. Experimental study on end-plate to SHS column connections reinforced by increasing wall thickness with one side bolts. Proceedings, 7th. International Symposium on Tubular Structures, Miskolc, Hungary, pp. 253–260.

THORNTON, W.A. 1984. Bracing connections for heavy construction. Engineering Journal, American Institute of Steel Construction, 3rd. Quarter: 139–148.

TIMOSHENKO, S., and WOINOWSKY-KRIEGER. 1959. Theory of plates and shells, 2nd. ed. McGraw Hill.

WARDENIER, J., KUROBANE, Y., PACKER, J.A., DUTTA, D., and YEOMANS, N. 1991. Design guide for circular hollow section (CHS) joints under predominantly static loading. CIDECT (ed.) and Verlag TÜV Rheinland GmbH, Köln, Federal Republic of Germany.

WHITMORE, R.E. 1952. Experimental investigation of stresses in gusset plates. Bulletin 16, University of Tennessee Engineering Experiment Station, U.S.A.

YEOMANS, N.F. 1996. Flowdrill jointing system. Part I—mechanical integrity tests. CIDECT Report No. 6F - 13A/96, British Steel – Tubes & Pipes, Corby, England.

8
FABRICATION, WELDING AND INSPECTION

8.1 Fabrication

Generally the fabrication of structures from HSS requires only conventional shop practices, but with a few special considerations. Holes in HSS must be drilled, since physical constraints preclude punching them. Welding is done with traditional procedures for CAN/CSA-G40.21-M92 Grade 350W material. Burning, sawing, milling and other operations are applicable. Special profiling of weld preparations (particularly for connections involving circular HSS) is discussed in Section 8.2.

Section identification may require extra care (e.g., the use of temporary labelling or colour coding) since wall thicknesses cannot be measured with a tape once the sections are permanently closed. However, ultrasonics can be used if necessary. Table 8.1 shows the colour code developed by the Canadian Steel Service Centre Institute (CSSCI) for HSS wall thicknesses.

WALL THICKNESS (mm)	COLOUR	WALL THICKNESS (mm)	COLOUR
2.54	Pink	5.84	Dark Green
2.79	Yellow	6.35	White
3.18	Dark Blue	7.95	Brown
3.40	Gold	9.53	Red
3.81	Silver	11.1	Light Blue
4.78	Orange	12.7	Pink

TABLE 8.1
CSSCI standard colour coding for HSS wall thickness

8.1.1 Flattening the Ends of HSS Members

The use of HSS for bracing members offers visually pleasing components which are efficient, since HSS have a high degree of stiffness about all axes. End-flattened HSS are frequently used for light bracing or even for lightly loaded primary members. They have been used successfully connected to round, square or rectangular hollow sections and to other structural shapes. Connections can be welded or they can be bolted by following normal bolting practices.

Section 4.2 describes welded flattened connections, while Section 7.5 illustrates typical bolted flattened connections.

8.1.1.1 Flattening Procedures

Stelco (1981) suggested that the HSS ends be preheated prior to flattening. Cold flattening is possible, though the results are less predictable.

Preheating to a dark red temperature (540°C – 650°C) and immediately flattening with a high speed press gives the best results. Small cracks that may develop along the edges or on the flattened portion are structurally acceptable provided they do not extend beyond the last line of fasteners. However, the cracks should be welded closed, as shown in Fig. 8.1. If the HSS is to be used outside or in a corrosive atmosphere, the flattened end should be seal welded to prevent internal corrosion or the entry of water.

Generally, flattening is most successful with sections that have a large ratio of outside dimension to wall thickness, but HSS with ratios over 25

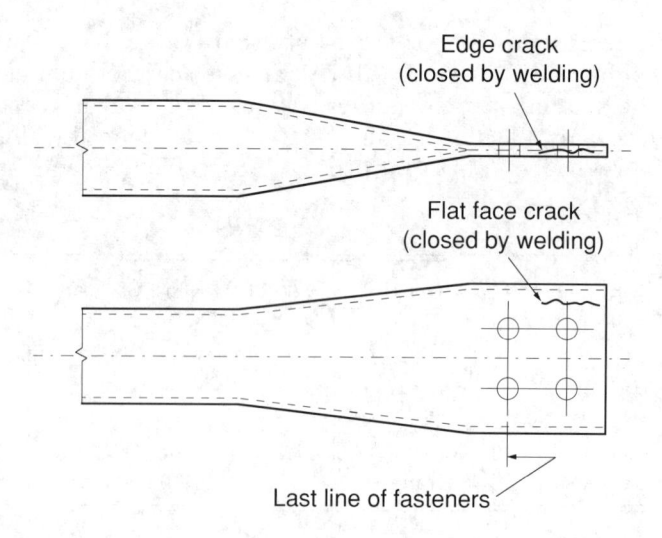

FIGURE 8.1
Possible cracking in end-flattened HSS

have reduced compressive strength (Wardenier *et al.* 1991). The slope of the taper should not be greater than 1:4 for maximum structural performance (Wardenier *et al.* 1991), but may be as much as 1:2 for secondary members (Stelco 1981).

8.1.2 Simplified End Cuts for Circular HSS Connections

Complexities regarding circular HSS connections are discussed in Section 8.2.4.1 under *Partial joint penetration groove welds—circular HSS*. However, in some instances simpler preparations are feasible:

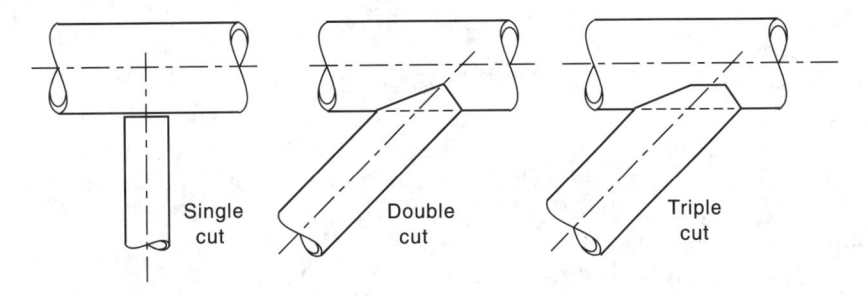

(a) Examples of webs with single, double and triple-cut ends

Chord d_0 (mm)	Web d_i (mm)
73	27
89	27
102	33
114	33
141	33
168	42
219	48
273	48
324	60
356	60
406	60

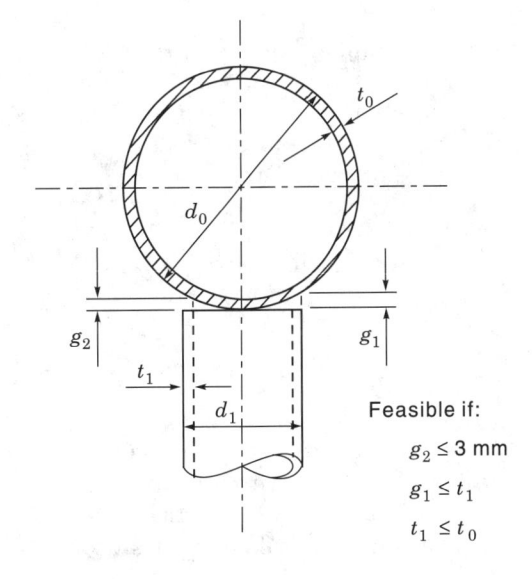

Feasible if:

$g_2 \leq 3$ mm

$g_1 \leq t_1$

$t_1 \leq t_0$

(b) Maximum size web with a single-cut end on a given chord size

(c) Feasibility criteria for single-cut ends

FIGURE 8.2
Simplified end cutting for circular HSS connections

Figure 8.2 provides information from Wardenier *et al.* (1991) pertaining to end preparations on web members that consist of one, two or three straight cuts as a function of the relative dimensions of the web and chord sections. The connection design engineer still needs to define the weld preparation and sizes.

8.1.3 Bending HSS Members

Hollow structural sections can be bent or rolled into curved lengths using either cold or hot processes. Cold working is relatively inexpensive, but smaller radii are obtainable with hot bending techniques.

8.1.3.1 Cold Bending HSS Members

Kennedy (1988) examined the limits of HSS cold rolling for CIDECT by using three-roller machines (Fig. 8.3(a)) to cold curve 27 different HSS sizes to four radii each. Deformations were measured on the resulting 108 samples and the results used to calibrate the machines in terms of HSS sectional dimensions and radius of curvature. These calibrations were then used to calculate deformations for all ISO (International Standards Organization) HSS sizes at various radii.

Two distinct forms of deformation were observed. The first was an inward bowing of the compression face of the section along its length, and the second was an outward bulging of a side face of the section (typically the side away from the supporting surface). Representative "before" and "after" cross sections are shown in Fig. 8.3(b). Deformations can be expressed as percentages:

$$\text{compression face bowing, } P_h = \frac{h - h_d}{b} \, 100\%$$

$$\text{side face bulging, } P_b = \frac{b_d - b}{h} \, 100\%$$

As the curvature is increased, either form of deformation may develop and limit the curvature obtainable without excessive distortion. Relatively thin wall sections exhibit early compression face bowing, while thicker wall sections tend towards earlier side wall bulging.

Arbitrary values of maximum P_b and P_h of 1% and 2% have been used to select the calculated data in Tables 8.2 and 8.3 for square and rectangular sections. The listings are minimum radii that result in either P_b or P_h, whichever governs, reaching the distortion values in the headings. Only radii less than 100 metres are listed. As shown in Fig. 8.4, 1% is just noticeable, while 2% is becoming pronounced. A wall deformation of 1% is

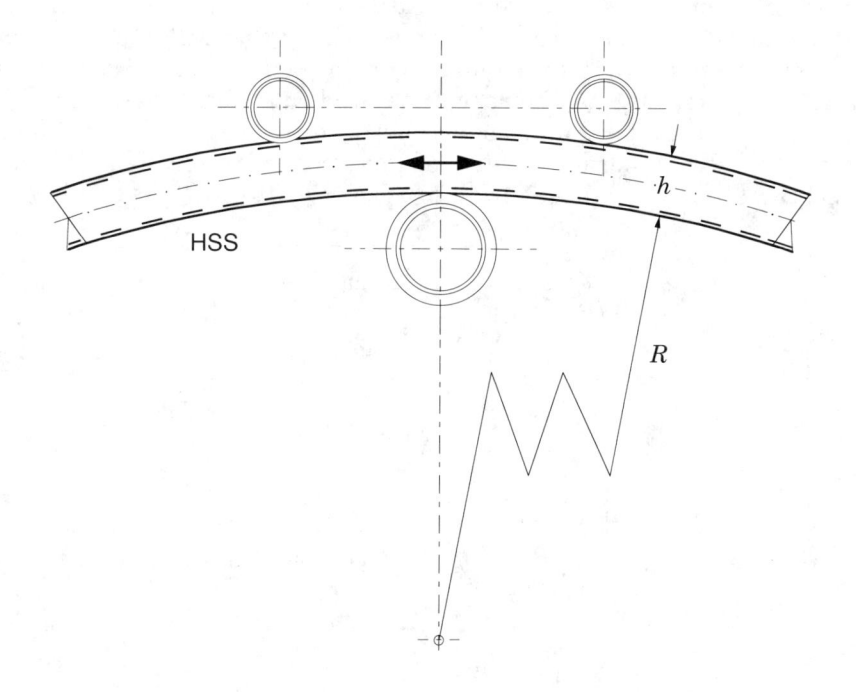

(a) Typical rolling arrangement for 3-roll cold bending

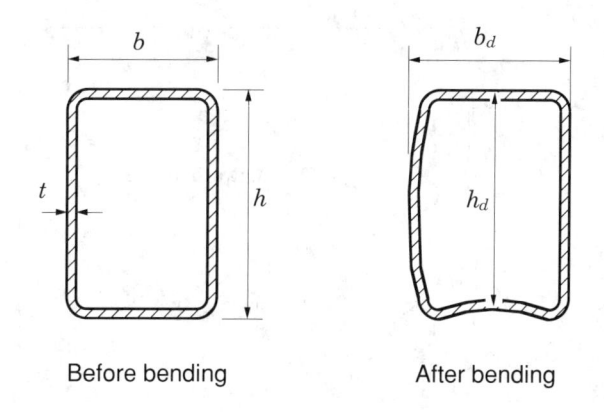

(b) Typical HSS sections

Before bending

After bending

FIGURE 8.3
Cold rolling of HSS to obtain curved members

commonly used by CIDECT as an acceptable level of distortion at connections under specified (service) load levels.

Some cautions need to be stated at this point. The experimental program employed ISO hot-formed HSS sections of 350 MPa specified minimum yield strength; many CSA sizes are close equivalents, some are not. Kennedy emphasized that the results showed some dependence upon both the rolling machines used and the actions of the operator. Even so, the data presented gives a general overview of the range of cold-rolled HSS curvatures obtainable by non-specialty fabricators.

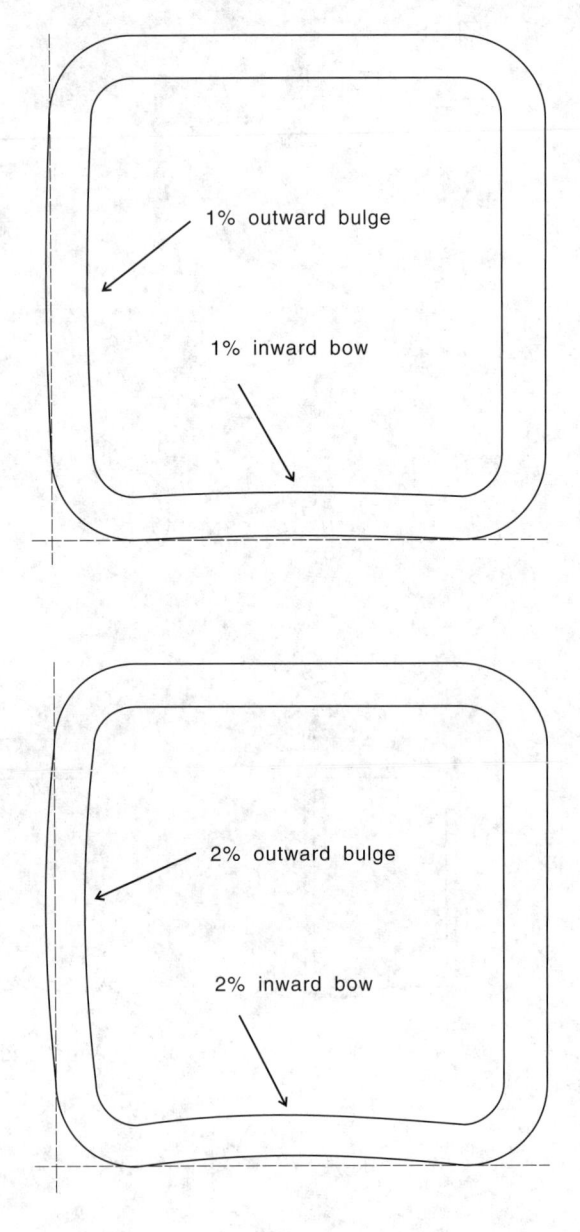

FIGURE 8.4
Illustration of distortions (1% and 2%)

HSS SECTION	1% P_b	1% P_h	2% P_b	2% P_h	HSS SECTION	1% P_b	1% P_h	2% P_b	2% P_h
30 x 30 x 2.0	0.71		<0.22		120 x 120 x 3.2				55.8
x 2.6	0.66		<0.22		x 4.0		80.6		30.0
x 3.2	0.63		<0.22		x 5.0		43.3		16.1
					x 6.3	38.7		12.0	
40 x 40 x 2.6	1.61		0.50		x 8.0	36.4		11.3	
x 3.2	1.53		0.47		x 10.0	34.4		10.7	
x 4.0	1.44		0.45						
					140 x 140 x 5.0				43.3
50 x 50 x 3.2	3.05		0.95		x 6.3	62.4			22.7
x 4.0	2.88		0.90		x 8.0	58.8		18.3	
x 5.0	2.72		0.85		x 10.0	55.5		17.3	
60 x 60 x 3.2	5.36		1.67		150 x 150 x 5.0				67.3
x 4.0	5.07		1.57		x 6.3		95.2		35.4
x 5.0	4.79		1.49		x 8.0	72.8		22.6	
					x 10.0	68.8		21.4	
70 x 70 x 3.2	8.65		2.69						
x 3.6	8.39		2.61		160 x 160 x 6.3				58.9
x 4.0	8.17		2.54		x 8.0		97.5		29.5
x 5.0	7.72		2.40		x 10.0		51.1		15.5
80 x 80 x 3.2	13.1			4.14	180 x 180 x 6.3				76.2
x 3.6	12.7		3.94		x 8.0				38.2
x 4.0	12.4		3.84		x 10.0		66.1		20.0
x 5.0	11.7		3.63						
x 6.3	11.0		3.42		200 x 200 x 6.3				95.9
					x 8.0				48.0
90 x 90 x 3.2		23.7		8.82	x 10.0		83.1		25.2
x 3.6	18.3			6.35					
x 4.0	17.8		5.53		220 x 220 x 8.0				59.1
x 5.0	16.8		5.23		x 10.0				31.0
x 6.3	15.9		4.93						
x 8.0	14.9		4.64		250 x 250 x 8.0				78.2
					x 10.0			42.5	
100 x 100 x 3.2		46.6		17.3					
x 4.0		25.1		9.31	260 x 260 x 8.0				85.2
x 5.0	23.3		7.25		x 10.0			48.5	
x 6.3	22.0		6.83						
x 8.0	10.7		6.43		300 x 300 x 10.0			78.8	
x 10.0	19.6		6.08						

TABLE 8.2

Minimum cold-rolled radii for square HSS members
(using conventional practice)

| HSS SECTION | MINIMUM RADIUS (Metres) FOR PERCENT DISTORTION |||| HSS SECTION | MINIMUM RADIUS (Metres) FOR PERCENT DISTORTION ||||
| | ≈ 1% || ≈ 2% || | ≈ 1% || ≈ 2% ||
	P_b	P_h	P_b	P_h		P_b	P_h	P_b	P_h
50 x 30 x 2.6		0.87		0.32	120 x 80 x 8.0		13.9		4.86
x 3.2	0.72		<0.22		x 10.0	8.58		2.67	
x 4.0	0.68		<0.22		140 x 80 x 4.0				64.5
60 x 40 x 3.2		1.97		0.73	x 5.0		93.2		34.6
x 4.0	1.26			0.39	x 6.3		49.0		18.2
x 5.0	1.19		0.37		x 8.0		25.2		9.36
70 x 40 x 3.2		3.79		1.41	x 10.0		13.5		5.03
x 4.0		2.04		0.76	150 x 100 x 5.0				75.3
x 5.0	1.77		0.55		x 6.3				40.0
80 x 40 x 3.2		6.69		2.49	x 8.0		54.7		20.3
x 4.0		3.59		1.34	x 10.0		29.4		10.9
x 5.0	2.49		0.77		160 x 80 x 4.0				73.9
90 x 50 x 3.2		17.9		6.65	x 5.0				38.8
x 3.6		12.9		4.79	x 6.3		65.6		19.9
x 4.0		9.61		3.57	x 8.0		32.8		9.94
x 5.0		5.16		1.92	x 10.0		17.2		5.21
100 x 50 x 3.2		28.0		10.4	180 x 100 x 5.0				70.9
x 3.6		20.1		7.49	x 6.3				36.3
x 4.0		15.0		5.58	x 8.0		60.1		18.2
x 5.0		8.07		3.00	x 10.0		31.5		9.53
100 x 60 x 3.2		41.5		15.4	200 x 100 x 5.0				63.1
x 3.6		29.9		11.1	x 6.3				32.3
x 4.0		22.3		8.29	x 8.0		53.5		16.2
x 5.0		11.9		4.45	x 10.0		28.0		8.48
x 6.3		6.30		2.34	200 x 120 x 6.3				58.9
120 x 60 x 3.2		90.0		33.5	x 8.0		97.4		29.5
x 3.6		64.8		24.1	200 x 120 x 10.0		51.1		15.5
x 4.0		48.4		18.0	220 x 140 x 6.3				88.0
x 5.0		26.0		9.66	x 8.0				44.1
x 6.3		13.7		5.07	x 10.0		76.3		23.1
x 8.0	7.76			2.61	250 x 150 x 6.3				95.9
120 x 80 x 3.2				62.4	x 8.0				48.0
x 4.0		90.1		33.5	x 10.0		83.1		25.2
x 5.0		48.4		18.0					
x 6.3		25.5		9.46	300 x 200 x 10.0				53.0

TABLE 8.3

Minimum cold-rolled radii for rectangular HSS members for bending about the major axis (using conventional practice)

Some firms that specialize in cold shaping HSS have developed techniques and equipment to better support the walls during bending, thus reducing deformation and achieving much smaller radii than those reported by Kennedy. Examples recently reported to the authors are listed in Table 8.4.

SECTION	RADIUS (m)	PROCESS	SECTION	RADIUS (m)	PROCESS
ROUND HSS			**RECTANGULAR HSS (bent about y-y axis)**		
60 x 4.8	0.4	Rolling	152 x 51 x 6.4	1.8	Mechanical
114 x 6.4	0.7	Rolling	152 x 102 x 6.4	2.1	Mechanical
168 x 9.5	0.9	Rolling	203 x 51 x 6.4	3.1	Mechanical
219 x 13	1.1	Rolling	203 x 102 x 6.4	2.4	Mechanical
			254 x 102 x 9.5	2.4	Mechanical
SQUARE HSS			305 x 102 x 9.5	2.4	Mechanical
50 x 50 x 5	0.6	Rolling	406 x 102 x 9.5	3.7	Mechanical
76 x 76 x 6.4	1.2	Rolling	406 x 203 x 9.5	10.4	Mechanical
100 x 100 x 6.3	1.1	Rolling	**RECTANGULAR HSS (bent about x-x axis)**		
102 x 102 x 6.4	1.8	Mechanical	102 x 51 x 6.4	1.8	Mechanical
102 x 102 x 9.5	1.5	Rolling	152 x 51 x 6.4	1.8	Mechanical
102 x 102 x 9.5	2.8	Mechanical	152 x 102 x 9.5	1.8	Mechanical
127 x 127 x 9.5	1.8	Rolling	203 x 51 x 6.4	2.4	Mechanical
152 x 152 x 9.5	2.0	Mechanical	203 x 152 x 6.4	2.3	Mechanical
152 x 152 x 9.5	2.1	Rolling	203 x 152 x 9.5	2.9	Mechanical
150 x 150 x 10	1.4	Rolling	254 x 51 x 6.4	3.1	Mechanical
150 x 150 x 12.5	3.0	Rolling	254 x 150 x 13	9.0	Rolling
152 x 152 x 12.7	12.1	Mechanical	254 x 102 x 9.5	3.8	Mechanical
203 x 203 x 6.4	4.9	Mechanical	254 x 203 x 9.5	7.5	Mechanical
203 x 203 x 9.5	3.1	Mechanical	305 x 102 x 9.5	3.5	Mechanical
203 x 203 x 9.5	4.9	Rolling	305 x 203 x 9.5	4.9	Mechanical
200 x 200 x 12.5	2.0	Rolling	305 x 203 x 13	9.2	Rolling
203 x 203 x 13	3.7	Rolling	406 x 102 x 9.5	8.6	Mechanical
254 x 254 x 9.5	7.0	Mechanical	406 x 203 x 13	19.9	Rolling
254 x 254 x 9.5	15.3	Rolling			
254 x 254 x 13	9.2	Rolling			
305 x 305 x 13	12.2	Rolling			
356 x 356 x 9.5	23.5	Mechanical			

TABLE 8.4

Some representative radii of curvature for cold-bent HSS
(using specialized techniques and equipment)

Processes such as draw bending (the HSS is forced around a mandrel), ram bending (hydraulic jack with die pushes the HSS against preset stops), and other jacking arrangements are all used in addition to roll bending for cold shaping HSS. The member is deformed through the yield point and becomes work hardened so that the material is somewhat less ductile and has lower toughness. That however is not usually a problem for most applications, and the stress-strain relationship is not altered in the elastic range.

Riviezzi (1984) suggests that the reduction in notch ductility becomes significant only when the amount of cold working produces a strain exceeding about 5% in the outermost fibres (when the radius is around 10 times the member depth). For fatigue or impact situations, he recommends stress relieving at 650°C for approximately 15 minutes when the strain is greater than about 2%. Galvanizing after cold bending is not advised unless the material has been stress relieved.

Depending on wall thickness, circular HSS can be rolled to a radius of about six times the outside diameter D, and draw bent to about three D, but stress relieving becomes increasingly desirable with very small radii. As with square and rectangular HSS, the thicker the wall, the smaller the achievable radius.

8.1.3.2 Hot Bending HSS Members

The most common method of hot curving HSS involves the use of induction heating by specialty firms. The process is particularly advantageous in applications where the use of heavy sections, tight radii or close geometrical tolerances are beyond the scope of cold bending. Bends are produced at over 820°C and water quenched, so that the radius is well "set" in the member. Precise control of temperature helps to assure accuracy and dimensional stability, with minimum risk of major metallurgical changes. Tensile strength, yield strength and toughness values are generally enhanced during the process (Riviezzi 1984).

The method uses an induction coil to apply a localized heat band in the material as the member is slowly pushed through the equipment, and hydraulic rams on a bend arm force the emerging section into the desired curvature. The bending takes place progressively in the heat band width, which is controlled by the water quench jets. This enables the temperature of the heat band to be at red heat while the adjacent zones are close to ambient temperature. Because of this precise heat control, the process is suitable for either light or heavy sections.

Industrial equipment varies from region to region, and designers would be well advised to confirm local capabilities when planning to use curved HSS.

8.1.4 Power Nailing HSS Connections

Circular HSS can be nailed together, as an alternative to bolting or welding, to form reliable structural connections (Packer and Krutzler 1994). At present, the method has been verified only for splices between two co-axial tubes where one tube is fitted reasonably snugly into the other. Nails are then driven through the two wall thicknesses and arranged symmetrically around the tube perimeter.

The nailing process itself is quick and inexpensive, and can be performed safely with minimal training. This involves a powder-actuated "gun" (no external power supply needed) that drives a high strength ballistic point pin (or "nail") into the steel. As this method avoids all shop fabrication and field bolting procedures, erection can be performed by relatively unskilled workers on site, at an extremely fast rate. This nailing procedure is shown in Fig. 8.5.

FIGURE 8.5
Nailing two tubes together for laboratory verification of the structural connection
(University of Toronto)

The appropriate gun settings for the type of nail used should be such that the nail tip penetrates the inside of the inner tube wall. This can be established by practice before working on site, and suppliers of the equipment (such as Hilti) can also provide assistance. Combined steel thicknesses up to 13 mm have been easily connected.

As an alternative, two tubes of the same outside diameter can be joined to each other by means of a tubular collar over both tube ends; or, two tubes of the same diameter can be mated together by flaring the end of one tube to fit over the other. The inside weld bead should be removed from the outer tube in the connection region, but testing did include specimens with zero to three mm local gaps between tubes, when slightly-out-of-round 406 mm HSS were used.

Research to date has covered tube diameters from 61 to 406 mm, tube wall thicknesses from 3.1 to 6.5 mm, diameter-to-thickness ratios from 13 to 64, and a variety of fits between the tubes. Over 40 tube connections have been tested to failure statically under axial tension or compression loading (Packer and Krutzler 1994, Packer 1996). The observed failure modes were by nail shear and by tube bearing, for which simple prediction formulae, derived from bolted connections, have been verified.

The nail shear failure mode resisted loads beyond the shear strength of the nails, about 20% more for loose fitting specimens and 30% more for tight fitting specimens. This additional or secondary strength resulted from a "nipple-dimple" effect at the interface between the tubes. A nail emerging from the inner face of the outer tube created a nipple protruding from that surface that interlocked with a matching dimple created in the outer face of the inner tube.

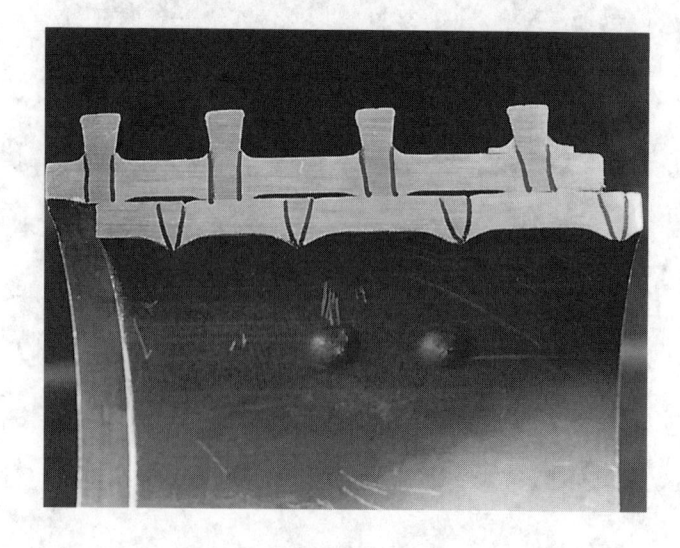

FIGURE 8.6
Shear failure of nails

Resistance for the failure mode of nail shear (Fig. 8.6) is given by

$$V_r = \phi_b \text{ (single shear strength of one nail) } n \qquad [8.1]$$

where $\phi_b = 0.67$

n = number of nails

This conservatively ignores the secondary contribution from the nipple-dimple effect.

Resistance for the failure mode of tube bearing (Fig. 8.7) is given by

$$B_r = 3\, \phi_b\, tdn\, F_u \qquad \text{(same as Clause 13.10(c)} \qquad [8.2]$$
$$\text{in CAN/CSA-S16.1-94)}$$

where t = tube wall thickness

d = diameter of the nail

n = number of nails

F_u = ultimate tensile stress of the tube material

This equation is subject to two conditions. The edge distance measured along the axis of the tube from the centre of the nail to the tube end should

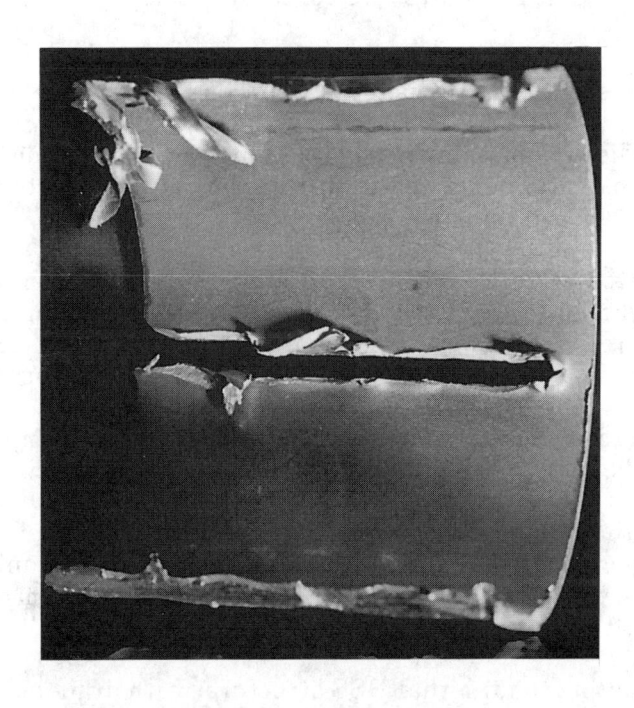

FIGURE 8.7
Bearing failure of tube wall

be at least $3d$. This should always be the case as the nail diameter is only 4.5 mm (for the Hilti ENPH2-L15). The nail pitch measured along the axis of the tube must also be at least $3d$.

Designers and detailers should also bear in mind that the nailing technique is ideal for fastening light gauge secondary members or fixings to HSS. For example, Z purlins can be nailed directly to square/rectangular HSS truss top chord members. Hangers or straps to HSS can likewise be easily nailed into place.

Concern has been expressed that the nails in lap splice tube connections might "work loose" under fatigue loading. Numerous constant-amplitude fatigue tests on co-axial nailed tube connections have now been undertaken, on a variety of tube sizes and nailing patterns, and it has been found that the nails do not "work out". Instead, cracks eventually develop in the tube wall or nails causing fatigue failure, but the fatigue performance is actually better than that for a bolted lap splice connection (Kosteski 1996).

8.1.5 Drainage and Interior Corrosion

Interior corrosion of HSS sections is not an issue when the sections are sealed by the connection details or by cap plates. Cap plates may conveniently be slightly larger or smaller than the outside dimensions of the tube, and need not be structural. Sealing HSS sections precludes the supply of fresh oxygen that is required for continuing corrosion.

There have been instances where moisture has accumulated inside apparently sealed HSS structures. Minute openings (for example at weld ends or connection details) can allow unexpectedly large volumes of water to enter. The action seems to be accelerated by contraction of the enclosed air as it cools after being heated during the day. Rain water on the surface tends to be sucked into any openings.

If allowed to freeze, enclosed water can split HSS members along their corners. Therefore, HSS structures exposed to precipitation and freezing temperatures must have the members thoroughly sealed or, alternatively, provided with drain holes at locations where water might accumulate.

It is recommended that drain holes be at least 10 mm diameter and that they be no more than half the tube width above the low point (Stelco 1989). Alternatively, a torch cut slot 10 mm wide by 20 mm high at the bottom of the HSS would be acceptable (Stelco 1989). Smaller holes may be prone to clogging with oxides or other build-ups. In galvanized structures that have vent holes for galvanizing, the hole at the bottom of each member could be left open.

Evidence has shown that HSS structures with drain holes develop only a light interior surface corrosion in normal environments. The exchange of air is insufficient to replenish the oxygen supply necessary for significant oxidation.

8.2 Welding and Inspection

Electric arc welding is the primary means of fabricating nearly all HSS connections. Even bolted connections generally have detail components that are welded to at least one of the members. Although the processes have been essentially standardized for many years, welding technology continues to evolve with increasing experience, better understanding and more sophisticated equipment. Some major welding codes and standards are revised and updated every two years or so.

On the one hand, the myriad of technical considerations that enter into the creation of a welded joint is so extensive that professional careers are devoted entirely to the subject. Yet on the other hand, the basic fundamentals for designing a welded joint are relatively manageable. The knowledge and understanding required of the practising design engineer in order to specify effective, practical and economical welded joints is not the same as that required of the welding specialist.

The incentive for the structural designer to master the elements of welding technology is the avoidance of unnecessary problems that can arise when inappropriate information or instructions are included in drawings and specifications. Such problems have the potential to assume major proportions as the parties involved (engineers, fabricators, inspection firms, building officials, owners) struggle with needless difficulties.

This section is therefore intended to give a brief overview of the technology behind the design of welded joints in HSS connections.

8.2.1 Codes and Standards

CAN/CSA-S16.1-94 *Limit States Design of Steel Structures*, is the current edition of the standard for steel structures built in Canada. Other standards intended for specific applications (e.g., bridges, antenna towers and offshore structures) are also available.

Within CAN/CSA-S16.1-94, Clause 13.13 presents expressions for the resistance of welded joints loaded in shear, tension or compression. The resistances given for fillet welds, partial joint penetration groove welds and complete joint penetration groove welds are examined in Section 8.2.3 of this chapter. Sub clauses in the Standard give compressive resistances for joints with partial joint penetration groove welds, and for plug and slot welds.

W59-M1989 *Welded Steel Construction (Metal Arc Welding)*, is the detailed Canadian standard which defines the specific requirements for welded joints. It applies to all welded steel construction specified to Standard CAN/CSA-S16.1-94 above. Structures such as pressure vessels or those governed by special codes such as the American Petroleum Institute are welded in accordance with other welding standards. Major chapters in W59 include:

- design of welded connections
- electrodes, workmanship, and technique
- radiographic and ultrasonic examination of welds
- details and welding procedure requirements for prequalified joints
- statically loaded structures—design and construction
- dynamically loaded structures—design and construction.

These chapters, along with others, furnish comprehensive information and instructions for the production of welded joints.

W47.1-1992 *Certification of Companies for Fusion Welding of Steel Structures*, by the Canadian Standards Association (CSA), outlines the requirements for a company to become certified to perform structural steel welding. The Canadian Welding Bureau is an independent organization that administers W47.1 for CSA, and also issues the certifications. Company personnel (engineers for design and for shop welding practices, and shop supervisory staff) must meet requirements of the Standard. Company Welding Engineering Standards, Welding Procedure Specifications, and Welding Procedure Data Sheets must be approved by CWB, and welders must be qualified by it. The details of welder qualification tests and the examination of weld specimens are all spelled out in the W47.1 Standard.

W48.1 to W48.6 from CSA, deal with welding consumables such as stick electrodes, solid and flux cored wire electrodes, and wire electrodes and flux for submerged arc welding.

Combined, the above Canadian standards constitute a complementary set of documents which have provided envied control, guidance and quality to the welding industry.

ANSI /AWS D1.1 *Structural Welding Code—Steel* (produced by the American Welding Society) is in many respects the American counterpart to W47.1 plus W59 from CSA. It includes requirements for procedure and welder qualifications, the design of welded connections, and a complete explanatory commentary. It is extremely comprehensive (the 1996 edition has about 480 pages), and is updated every two years.

Clauses pertaining to tubular structures originally evolved from a background of practice and experience with fixed offshore steel platforms. The connection capacities were therefore expressed with much greater confidence for circular tubes than for "box sections" (square and rectangular HSS). In 1992, the clauses dealing with square and rectangular HSS underwent major revisions to be in accord with the IIW/CIDECT recommendations, which are also presented in this Design Guide.

8.2.2 Welding Processes

The vast majority of welding in structural steel fabrication shops is done by the use of three distinct welding processes.

Shielded Metal Arc Welding (SMAW) is the conventional manual process that employs stick electrodes coated with a layer of shielding, alloying and purifying flux chemicals. The welder has a high degree of freedom, since only a hand-held electrode clamp connected by a single electrical cable to the source of power is manipulated. Stick electrodes about 300 mm long are replaced in the clamp as they burn off. The weld metal, which may be deposited in any position (including overhead), is laid down as required for each joint configuration. Deposition rates vary depending on many factors, but are in the order of one half to one kilogram an hour. SMAW welding continues as the general purpose process with a wide range of applications.

Flux Cored Arc Welding (FCAW) is known as a semi-automatic process. The consumable electrode, in the form of a continuous hollow wire containing the flux chemicals, is fed from a spool on the welding machine. Usually shielding gases are also delivered to the operator's "gun" to ensure protection of the arc and the molten metal from harmful elements like oxygen and nitrogen. The welder manipulates the gun by hand similar to the SMAW process, but there is a bundle attached to it consisting of the electrical cable, the wire feeding liner and the gas hose.

The cost of the more expensive equipment for FCAW is offset by a rate of deposition two to three times faster than with SMAW, so that substantial economies are possible when sizable welds are required with a minimum of moving around by the welder. The FCAW process is the main workhorse in many shops.

Submerged Arc Welding (SAW) is a fully mechanized process. Electric controls, spooled solid wire electrode(s), and wire and granular flux feeders are part of a powered mobile unit which travels relative to the work piece. Flux is deposited around the end of the wire electrode which is directed into the joint configuration of the weld. The electric arc is concealed by the flux build up, some of which is fused to create a protective slag (later removed), while the rest is vacuumed up for reuse. Size of the deposited nugget is controlled by a combination of the wire feed speed and the carriage travel speed. The process provides particularly smooth and regular welds of uniformly high quality.

Generally SAW is restricted to flat or horizontal straight joints, and the travel path must be unobstructed, with adequate clearances for the welding head. Higher amperages are possible with SAW, and sometimes multiple electrode wires are used to achieve deposition rates several times higher than with semi-automatic processes. Flange to web welds of bridge girders are typically made with SAW equipment.

Gas Metal Arc Welding (GMAW), also known as MIG welding, is similar to the flux core process, but uses a spooled, solid wire, continuous electrode. Since no flux chemicals are provided in the wire, shielding is accomplished by gas mixtures (some proprietary) provided as with the FCAW process. Training for GMAW is generally more demanding than for FCAW, but productivity is good. Use of the process is increasing, and GMAW weld joints may be prequalified in the next edition of CSA Standard W59.

While other processes are occasionally used for particular applications, the above four offer versatility adequate for the needs of most structural fabricating shops.

8.2.3 Types of Welded Joints

Three basic types of welds account for practically all structural weld joints: fillet welds, partial joint penetration groove (PJPG) welds and complete joint penetration groove (CJPG) welds. All can be made with any one of the processes discussed in the previous section. The nature and structural resistance of each type are now considered.

8.2.3.1 Fillet Welds

The traditional fillet weld, usually an isosceles triangle, is illustrated in Fig. 8.8, where the size is given by S, measured perpendicular from the plates. The fusion face f is the same as the leg length which, in the case of 90° fillets, is the same as the size. For any direction of force, resistance V_r is proportional to the throat area defined as $A_w = t_w L$. If more than one component of force exists, the vector resultant is used for the load; the resistance value is

$$V_r = 0.67 \, \phi_w \, A_w \, X_u \, (1.00 + 0.50 \sin^{1.5} \theta) \qquad [8.3]$$

where

    0.67 relates the shear resistance to the tensile resistance

    ϕ_w = the conservative resistance factor used for welds (also 0.67)

    X_u = the ultimate tensile strength of the weld metal.

    θ = angle between the axis of the weld and the line of action of the force (0° for a longitudinal weld and 90° for a transverse weld)

The expression $(1.00 + 0.50 \sin^{1.5} \theta)$ was introduced in the 1994 edition of CAN/CSA-S16.1 (Clause 13.13.2.2(b)) to reflect the greater strength of a fillet weld loaded transversely relative to one loaded longitudinally. Many designers use this increased resistance only in difficult circumstances, preferring the more conservative basic expression for all orientations of weld axis (by setting $\theta = 0°$).

Resistance of fillet welds = least of:

$$V_r = 0.67 \phi_w A_w X_u \qquad A_w = t_w L$$
$$V_r = 0.67 \phi A_{m1} F_{y1} \qquad A_{m1} = f_1 L$$
$$V_r = 0.67 \phi A_{m2} F_{y2} \qquad A_{m2} = f_2 L$$

where,
L = length of weld
f = width of fusion face
S = size of fillet weld
 (smaller S if legs unequal)
t_w = throat of fillet weld

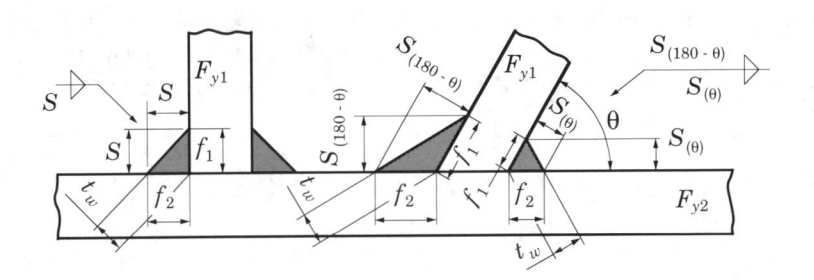

Relationship between t_w and S (for equal leg fillets):

when $\theta = 90°$, $t_w = 0.707 S_{(90)}$
when $\theta = 60°$, $t_w = 1.00 S_{(60)}$
when $\theta = 120°$, $t_w = 0.577 S_{(120)}$

FIGURE 8.8
Fillet weld information

Strength of the base metals (materials joined) rather than that of the weld metal might govern, especially for welds loaded transversely to their axes, and is given by the lesser of

$$V_r = 0.67 \phi_w A_{m1} F_{u1} \qquad [8.4a]$$

or

$$V_r = 0.67 \phi_w A_{m2} F_{u2} \qquad [8.4b]$$

where

$A_{m1} = f_1 L$, the fusion area on base metal 1
$A_{m2} = f_2 L$, the fusion area on base metal 2
F_{u1} = the specified minimum tensile strength of base metal 1
F_{u2} = the specified minimum tensile strength of base metal 2.

Since most fillet welds are symmetrical, $f_1 = f_2$; then [8.4a] = [8.4b]. The value of angle θ (for a 90° fillet weld) above which [8.4] governs rather than [8.3] is 49° for HSS material 350W from CAN/CSA-G40.21.

Fillet welds may vary from 60° to 120°. As can be seen in Fig. 8.8, the ratio between weld size and throat size varies with the weld angle. For the same resistance, larger welds are required for obtuse angles than for acute angles.

When loaded in tension or compression parallel to the weld axis, a fillet weld is taken to have the same resistance as the base metal, and the cross sectional area of a continuous fillet weld can be considered to contribute to the cross-sectional properties A, S, Z and I of the section (CAN/CSA-S16.1-94, Clause 13.13.4.2). If the tension or compression is normal to the weld axis, the resistance is assumed to be the same as for shear loads.

8.2.3.2 Partial Joint Penetration Groove Welds

Figure 8.9 shows typical cross sections of partial joint penetration groove (PJPG) welds. Edge preparation of one plate of the joint is usually necessary, but the required configuration is sometimes provided naturally by the angle between the members as in Fig. 8.9(c). The effective throat t_w and the fusion width f (here regarded as the smaller width of the two fusion faces) are the same for the joints in Figs. 8.9(a), (b) and (c), and are

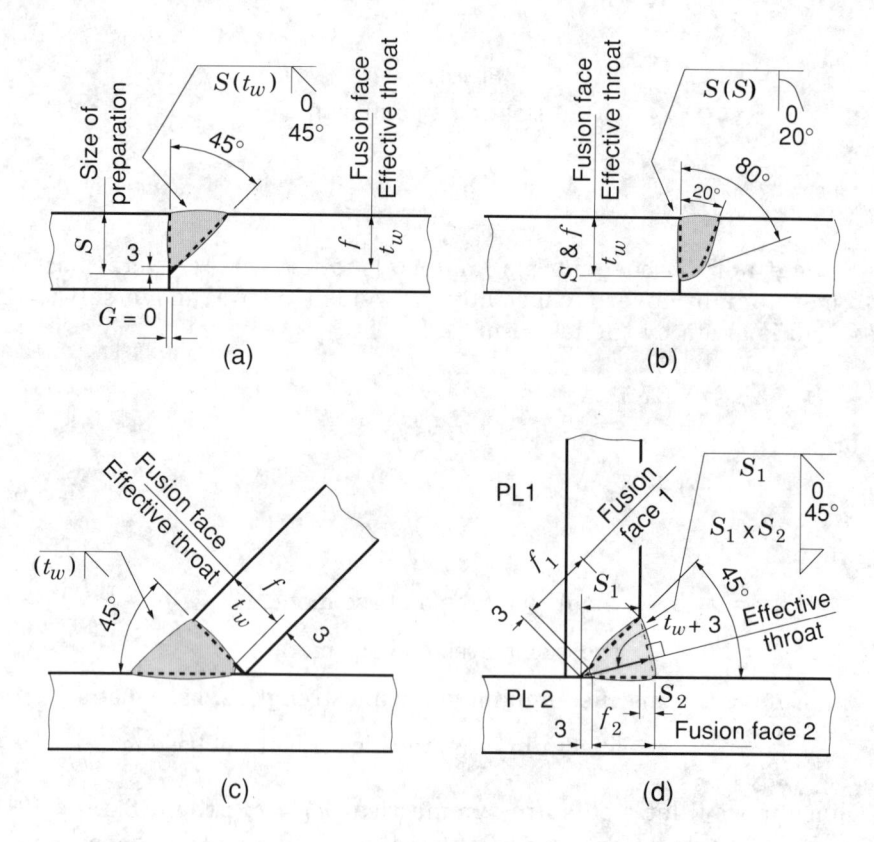

FIGURE 8.9
Partial joint penetration groove welds

generally a function of the edge preparation; therefore, the throat area A_w ($t_w L$) and the fusion area A_m (fL) are equal in these instances. Shear forces parallel and normal to the weld axis are combined vectorially to represent the load, and the shear resistance is given by the least of

$$V_r = 0.67\,\phi_w A_w X_u \qquad [8.5]$$

or

$$V_r = 0.67\phi_w A_{m1} F_{u1} \qquad [8.6a]$$
$$V_r = 0.67\phi_w A_{m2} F_{u2} \qquad [8.6b]$$

These equations are similar to those for fillet welds, but the values of V_r often vary from those of fillet welds because t_w (in A_w) is measured differently.

Fig. 8.9(d) provides an example of a partial joint penetration weld where t_w and f are not the same. The reinforced weld shown has two fusion widths and an effective throat, all of which are different.

Starting with the 1989 edition, CAN/CSA-S16.1 has recognized that partial penetration welds, loaded in tension normal to the axis of the weld, have a resistance comparable to the tension resistance of the base metal piece, pro-rated to the percentage of the thickness effectively fused. The resistance equation provides a constant margin against ultimate strength independent of the F_y/F_u ratio

$$T_r = \phi_w A_n F_u \quad \le \phi A_g F_y \qquad [8.7]$$

where

A_n = fusion area of base metal normal to the tensile load
F_u = specified ultimate tensile strength of the base metal.
ϕ = 0.9, the resistance factor for structural steels
A_g = gross area of the joined element
F_y = specified minimum yield strength of the base metal

Earlier editions of the Standard stipulated that the shear resistance value of the weld be used for this tension load.

The cap $\phi A_g F_y$ ensures that the tensile resistance given by [8.7] (a fracture failure at the weld) does not exceed the tensile resistance of the gross section (a yield failure of the joined element).

It should be noted that [8.7] does not provide unit tension resistance identical to that for the base metal:

Unit tension resistance by [8.7] is $0.67(450) = 300\,\text{MPa} \quad - \;(\phi_w F_u)$.

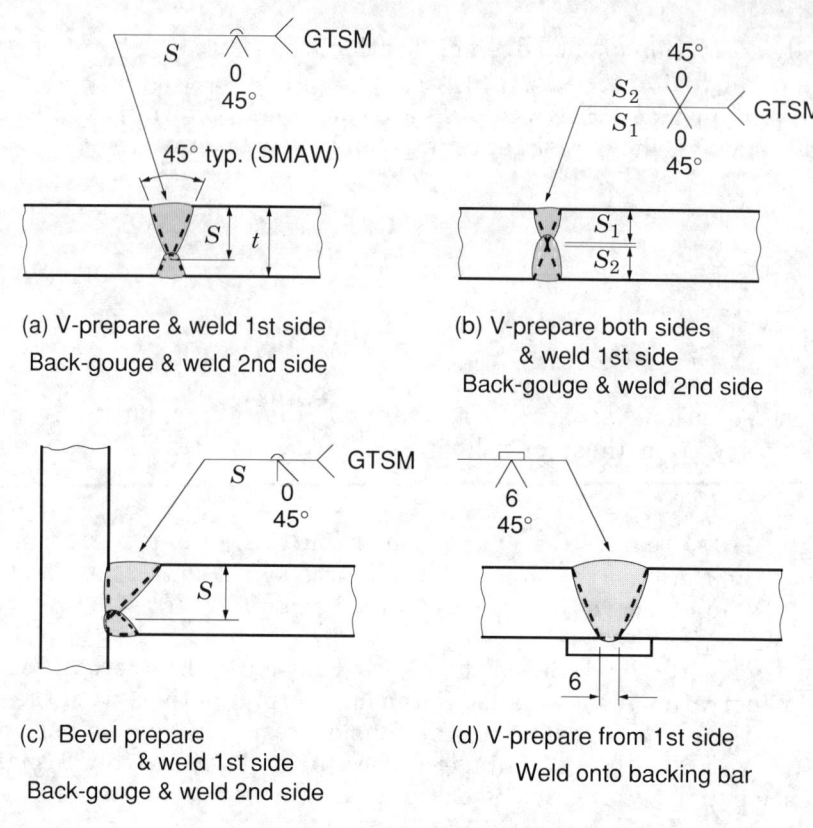

FIGURE 8.10
Complete joint penetration groove welds

(a) V-prepare & weld 1st side
Back-gouge & weld 2nd side

(b) V-prepare both sides
& weld 1st side
Back-gouge & weld 2nd side

(c) Bevel prepare
& weld 1st side
Back-gouge & weld 2nd side

(d) V-prepare from 1st side
Weld onto backing bar

GTSM = gouge to sound metal

For 350W HSS material, base metal unit tension resistance is

$$0.90\,(350) = 315\text{ MPa} \quad - \quad (\phi\,F_y).$$

The resistance for compression normal to the weld axis is taken to be the same as for the base metal, calculated on an area equal to the sum of A_n plus the area in contact bearing (CAN/CSA-S16.1-94, Clause 13.13.4.1).

When loaded in tension or compression parallel to the weld axis, PJPG welds are taken to have the same resistance as the base metal.

8.2.3.3 Complete Joint Penetration Groove Welds

Complete joint penetration groove (CJPG) welds, like those shown in Fig. 8.10, develop the full resistance of the base metal when they are made with matching strength electrodes (usually the case). The normal procedure for CJPG welds is to weld most of the joint from one side, to gouge out the root to sound metal from the second side, and then to complete the joint from that side as shown in Fig. 8.10(a), (b) and (c).

CJPG welds can be made against a backing strip, Fig. 8.10(d), which may be left in place for statically loaded structures. When the backing strips are removed (as required for CJPG welds transverse to the direction of stress in dynamically loaded structures), the joint must be ground or finished smooth, sometimes by grinding the root and depositing a weld pass.

A category of complete joint penetration groove welds, generally used in pressure piping, is made from one side without backing. They require specially qualified welders, meticulous joint preparation (bevels etc.) and precise fit-up. As discussed in Section 8.2.4.1, such joints are seldom justified in HSS structural construction.

8.2.3.4 Prequalified Weld Joints

Weld joints involve many variables: base material, joint configuration, weld process, welding position, electrodes, shielding gases, fluxes, preparation angle, root gap, root face, voltage, amperage, travel speed, possibly pre-heat and post-heat, number of passes and fit-up tolerances, to name some. All must be defined and controlled for a successful joint.

Fortunately, an extensive range of commonly used "prequalified joints" (fillet, partial joint penetration groove, and complete joint penetration groove) that use effective combinations of these variables has been codified in Standard W59. They specifically use the SMAW, FCAW and SAW processes discussed in Section 8.2.2, and can be applied without further procedure qualification testing. Joints that are not prequalified must be submitted to a qualification procedure that is documented and demonstrated by the fabricator, and tested and approved by the Canadian Welding Bureau, in accordance with the requirements of Standard W47.1 before they can be used in production work.

The expense and time to qualify a new joint naturally encourages utilization of the standard prequalified joints.

8.2.4 Weld Joint Applications to HSS Construction

Application of the preceding information to HSS joints will now be considered. In general, when welding joints between square or rectangular HSS members, it is recommended that weld start-stops be avoided at the corners of the sections, since they are the most highly stressed location in the joint. This is particularly the case with dynamically loaded structures.

8.2.4.1 Joint Types

Fillet welds

Fillet welds are the preferred type when the connection geometry will accommodate them. They are usually practical along the sides of square or rectangular web members connected to square or rectangular chords if the web member width is somewhat less than the flat width of the chord. Outside corner radii of HSS sections are considered to be twice the wall thickness, but Standard CAN/CSA-G40.20-M92, Table 34, (reproduced in Section 1.5 of this Design Guide) gives upper limits for the corner radius of currently produced sections that are up to three times the wall thickness.

A fillet weld which runs a little onto the radius can be built up as shown in Fig. 8.11(a), but this practice increases the cost of the joint. Fillet welds along the toe or heel of the web member are only applicable for enclosed angles between 120° and 60°, as shown in Fig. 8.11(b) and (c). Angles beyond these limits will most likely require partial joint penetration groove welds.

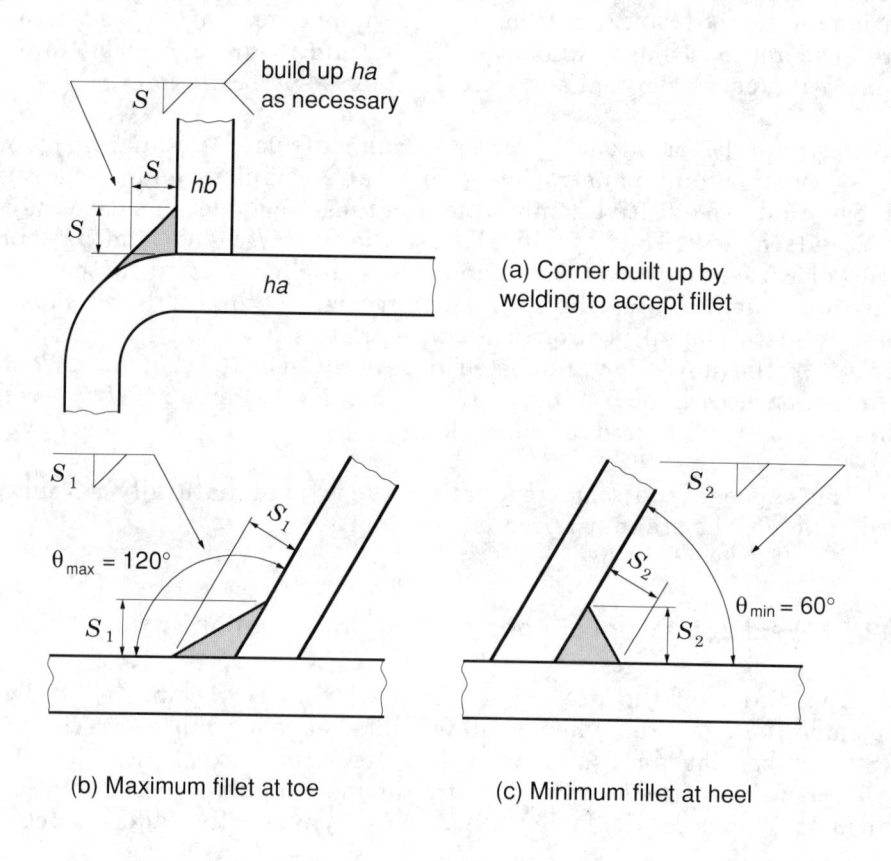

(a) Corner built up by welding to accept fillet

(b) Maximum fillet at toe

(c) Minimum fillet at heel

FIGURE 8.11
Fillet weld applications to HSS

Partial joint penetration groove welds—general

Almost all other HSS structural weld joints are PJPG welds. Many result from naturally occurring angles, as in Fig. 8.12(a); others are created by specific edge preparations on the material, as in Fig. 8.12(b). The latter has sometimes been regarded as a fillet weld, but it also qualifies as a 90° V-groove weld whose resistance, in tension transverse to the weld axis, is greater than that of a fillet weld (CAN/CSA-S16.1-94, Clause 13.13.3.2).

Partial joint penetration groove welds—flare bevel

A common PJPG weld is the flare bevel configuration which occurs when equal width sections are joined as illustrated in Fig. 8.13. The easiest arrangement for welding occurs with equal wall thickness sections as in Fig. 8.13(a) where backing can sometimes be avoided. However, if the corner radius approaches the upper limit of approximately three times the wall thickness, or a thin wall member joins a thick wall member, backing will likely be required.

One manner of reducing the gap is to install backing bars on the branch members as seen in Fig. 8.14 for the box HSS truss shown in Fig. 8.15. (Note the simple fixture employed to rotate the truss in order to avoid overhead or vertical welding.) Another method is to use a circular backing rod as in Fig. 8.13(b). An advantage of a solid round rod for backing is that it can be positioned somewhat more into or out of the gap to fit various gap sizes.

The joint looks like a CJPG weld, but it tends to be awkward at the corners of the connecting member where a transition occurs from flare bevel to fillet weld as one progresses around the corner. Corners are often the most highly stressed portions of an HSS joint, and welding complications there are undesirable.

(a) PJPG welds - without bevel preparations

(b) PJPG weld - with bevel preparation

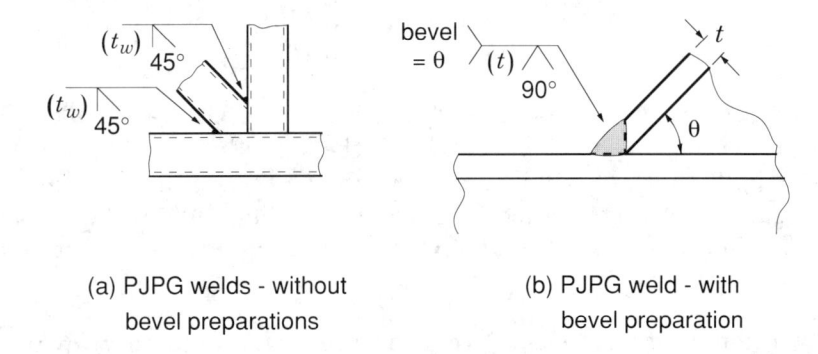

FIGURE 8.12
Partial joint penetration groove welds on HSS

(a) Flare-bevel PJPG Weld

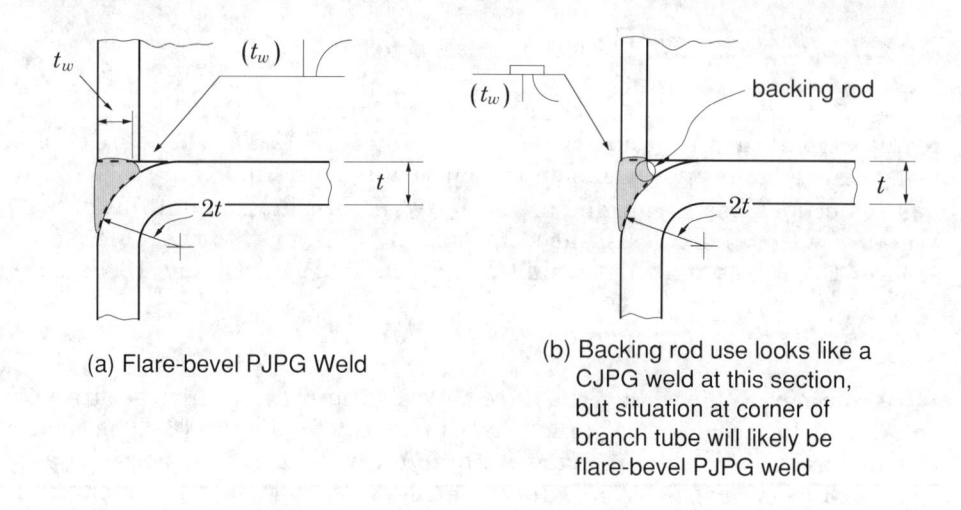

(b) Backing rod use looks like a CJPG weld at this section, but situation at corner of branch tube will likely be flare-bevel PJPG weld

FIGURE 8.13
Examples of equal-width HSS weld joints

Since flare bevel joints are not prequalified, Standard W59 requires that fabricators demonstrate the effective weld throats achieved by their HSS flare bevel weld procedures (for each situation) by preparing demonstration joints for sectioning and examination.

For all the reasons cited, it is best to minimize the use of expensive flare bevel welds.

Partial joint penetration groove welds—circular HSS

Circular HSS joints present considerable complexity, as is apparent in Fig. 8.16, which depicts joint geometry variations as one progresses around the perimeter of a member framing at 55°. Initially, the intersection is expressed as the line where the inner surface of the branch member meets the outer surface of the main member. Then bevels on the branch member are defined to permit clearances for the PJPG welds. (An exception to the bevel preparations occurs when small diameter branches join large diameter main members, as illustrated in Section 8.1.2.)

Usually there are four zones, as identified in Fig. 8.16(a). The toe, Fig. 8.16(b), in this case has to be bevelled 10°, the side, Fig. 8.16(c), at 20° and the heel, Fig. 8.16(d), at 35° in order to achieve the total groove angle of 45° necessary for a prequalified joint as per Standard W59. The shape of the intersection line is unique for each combination of tube diameters and framing angle.

Conventional practice was to produce a profile template to wrap around and mark the branch member for a cut made perpendicular to the tube axis. A second cut was then made with varying bevels for the weld preparation —skilled, somewhat tricky and expensive. Limited computer controlled cutting machinery is available for producing the final cut in a single

FIGURE 8.14
Pre-positioned backing bars for flare-bevel welds

FIGURE 8.15
Box Vierendeel truss fabricated from HSS

operation, but geometry ranges are restricted and few fabrication shops are so equipped. Obviously the complexities rapidly multiply when one gets into overlapping connections with multiple members of varying size.

Complete joint penetration groove welds

Complete joint penetration groove welds onto backing bars can be used to splice members as illustrated in Fig. 8.17. Accurate fitting of the backing bars is particularly important, and good inspection at that stage prior to welding is advised. CJPG welds without backing bars are best avoided because they are seldom justified economically or structurally. They are expensive for several reasons: welder qualifications are appreciably more demanding, the joints are not prequalified, and preparation and fit-up are more exacting. Inspection of such welds can be expensive and potentially disputatious when attempts are made to confirm acceptable complete penetration. Weld repairs are intricate.

Structurally, as seen in Chapter 3, HSS connections quite often cannot be configured to develop the full strength of the section. Thus, the need for CJPG welds would often seem to be questionable in any event.

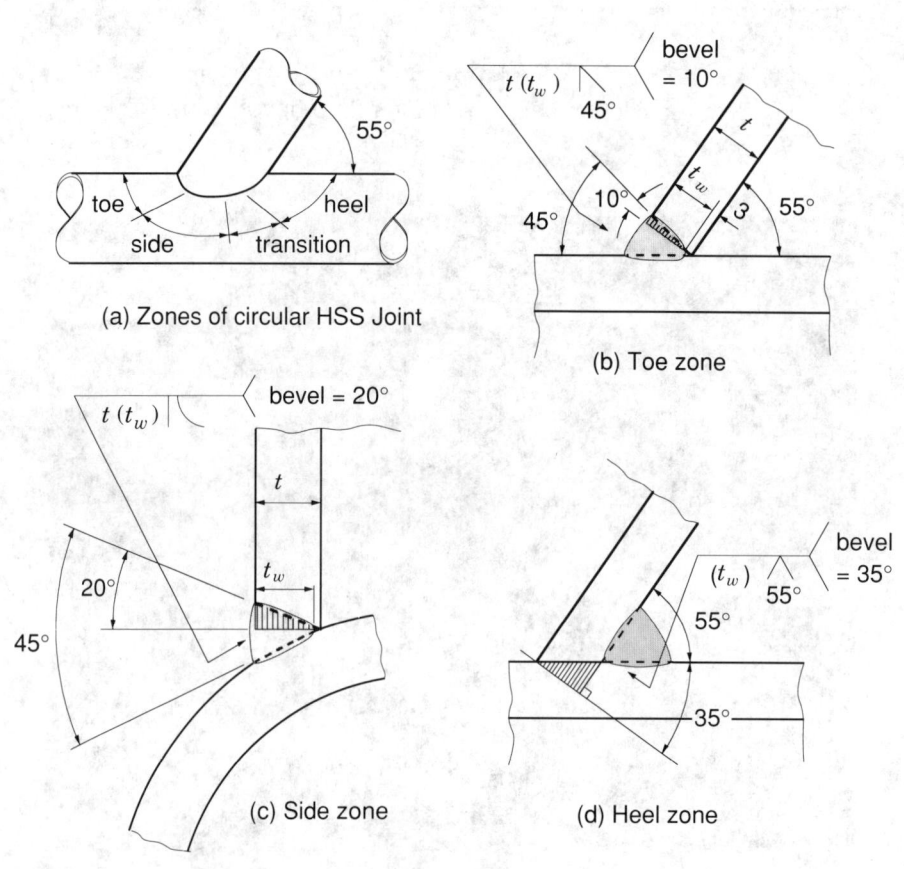

FIGURE 8.16
Bevel preparations for PJPG welds on circular HSS—55° branch angle

284

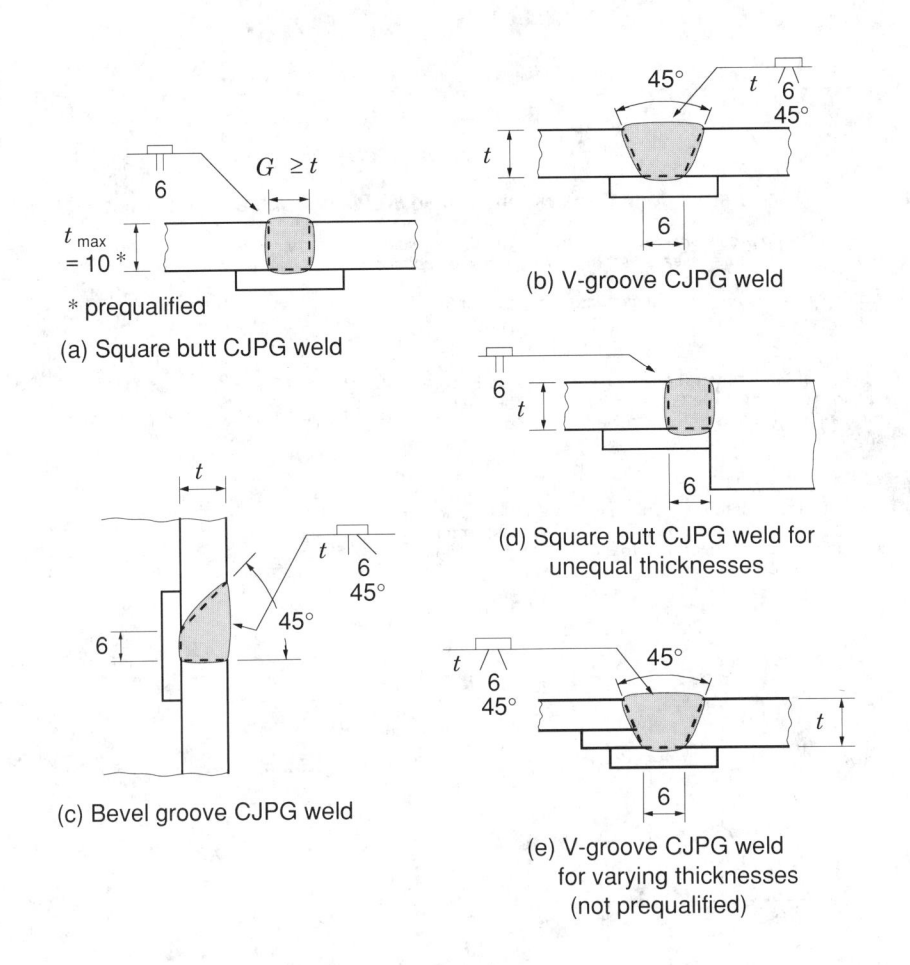

FIGURE 8.17
Examples of CJPG welds for splicing HSS sections

There is a provision in W59 for certain hollow structural section PJPG welds to be considered equivalent in strength to CJPG welds. Appendix L of W59, reproduced in Fig. 8.18, shows precisely reinforced joint configurations that may be made by welders qualified for conventional structural welding (rather than the more onerous one-sided CJPG welding without backing). However, the joint preparation and fit-up criteria are still substantially more stringent, and therefore more expensive, than for normal PJPG welds. In addition, since these joints are not prequalified, the fabricator would need to process a full procedure qualification with the Canadian Welding Bureau in order to use them.

Length of weld joints

The length of weld joints between intersecting square or rectangular HSS can be useful information for estimating, planning and weld design. Table 8.5 lists such perimeters calculated as recommended in Clause 10.8.5.1 of AWS D1.1, 1990.

Appendix L

Partial Joint Penetration Groove Welds in Tubular Construction*

*The following provisions deal with the strength of the weld only. The strength of the joint in case of this or any other type of weld must be assessed separately, taking into account the joint geometry and strain characteristics.

Note: This Appendix is not a mandatory part of this Standard.

L1. Partial joint penetration groove welds used in connections of hollow structural sections (HSS) when welded for the full thickness of material and when reinforced as shown in Figure L-1 may be considered as being equivalent in strength to complete joint penetration groove welds, provided that:

(a) Matching conditions required in Table 11-1 for complete joint penetration groove welds for the case of tension and compression normal to the axis of the weld and for shear are observed;

(b) The joint is statically loaded;

(c) The surface at which the HSS section is welded is wide enough to receive the welded reinforcement shown in Figure L-1.

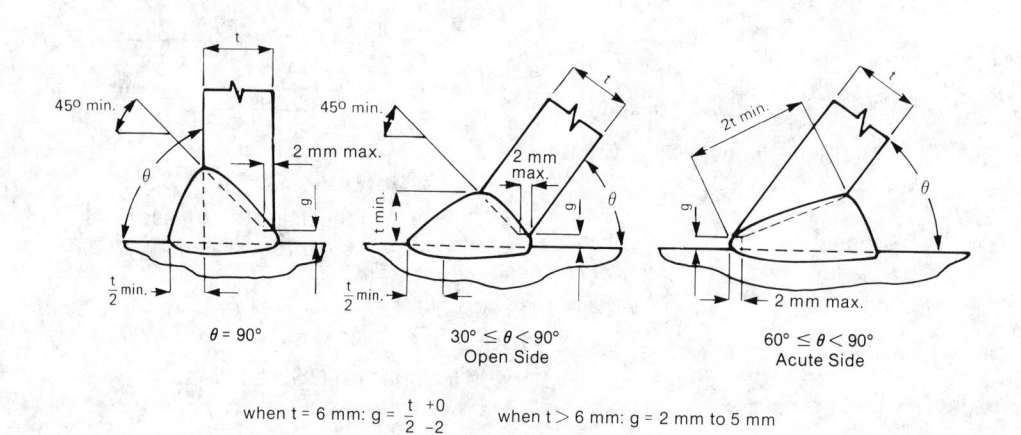

**Figure L-1
Partial Joint Penetration Groove Welds in Truss Connections of Hollow Structural Sections**

W59-M1989
January, 1989

FIGURE 8.18

Appendix L from Canadian Standards Association W59-M1989

HSS SECTION b x h x t (mm)	\multicolumn{9}{c}{Angle θ (degrees) Between Intersecting Members}
	30	35	40	45	50	55	60	65	70	90
25 x 25 x 3.8	133	121	113	107	102	98	95	93	91	89
32 x 32 x 3.8	171	157	146	138	132	127	123	120	118	114
38 x 38 x 4.8	204	187	174	164	157	151	147	143	140	136
51 x 51 x 6.4	272	249	232	219	209	201	195	191	187	181
64 x 64 x 6.4	348	319	297	280	268	258	250	244	240	232
76 x 76 x 8.0	416	381	355	335	320	308	299	292	286	278
89 x 89 x 9.5	484	443	413	390	372	359	348	340	333	323
102 x 102 x 13	544	498	464	438	418	403	391	382	374	363
127 x 127 x 13	697	637	593	561	535	516	500	488	479	464
152 x 152 x 13	849	776	723	683	652	628	610	595	584	566
178 x 178 x 13	1 000	916	853	806	770	741	719	702	689	668
203 x 203 x 13	1 150	1 060	983	928	887	854	829	809	794	769
254 x 254 x 13	1 460	1 330	1 240	1 170	1 120	1 080	1 050	1 020	1 000	972
305 x 305 x 13	1 760	1 610	1 500	1 420	1 360	1 310	1 270	1 240	1 210	1 180
51 x 25 x 4.8	181	170	161	155	150	146	143	141	139	136
25 x 51 x 4.8	227	203	186	174	164	156	150	145	142	136
76 x 51 x 8.0	317	294	277	264	254	247	241	236	233	227
51 x 76 x 8.0	363	328	302	283	268	257	248	241	235	227
102 x 51 x 8.0	370	346	329	316	306	298	292	287	283	278
51 x 102 x 8.0	463	415	380	354	334	318	306	297	289	278
89 x 64 x 8.0	393	363	342	325	313	303	295	289	285	278
64 x 89 x 8.0	439	398	367	345	327	313	303	294	288	278
102 x 76 x 9.5	461	426	400	380	365	353	344	337	332	323
76 x 102 x 9.5	507	460	425	399	379	364	351	342	335	323
127 x 51 x 9.5	415	391	374	361	351	343	337	332	329	323
51 x 127 x 9.5	554	494	451	418	393	374	359	347	338	323
127 x 64 x 9.5	464	435	413	396	384	374	366	360	356	348
64 x 127 x 9.5	580	521	477	444	419	400	384	372	363	348
127 x 76 x 13	499	464	438	419	404	393	384	377	372	363
76 x 127 x 13	590	531	489	457	432	413	398	386	377	363
152 x 102 x 13	650	602	568	541	521	505	493	484	476	464
102 x 152 x 13	743	672	619	580	549	526	507	493	482	464
178 x 127 x 13	802	741	697	664	638	618	602	590	581	566
127 x 178 x 13	896	811	749	703	667	639	617	600	587	566
203 x 102 x 13	755	706	671	644	624	608	595	585	578	566
102 x 203 x 13	943	847	776	722	681	649	624	605	590	566
203 x 152 x 13	954	880	827	786	755	731	712	697	686	668
152 x 203 x 13	1 050	951	880	826	784	752	727	707	692	668
254 x 152 x 13	1 060	984	929	889	857	833	814	799	788	769
152 x 254 x 13	1 250	1 130	1 040	968	916	875	844	819	800	769
305 x 203 x 13	1 360	1 260	1 190	1 130	1 090	1 060	1 030	1 010	997	972
203 x 305 x 13	1 560	1 410	1 300	1 210	1 150	1 100	1 060	1 030	1 010	972

TABLE 8.5

Length of welds for HSS square or rectangular members

1. Outside corner radius assumed equal to $2t$.

2. Perimeters shown in table are for the thickest HSS in each member outside dimension size. Therefore perimeters are conservative for smaller thicknesses.

3. Perimeters calculated by: $K_a (4\pi t + 2(b - 4t) + 2(h - 4t))$
   where $K_a = ((h/\sin\theta) + b)/(h + b)$.

8.2.4.2 Weld Effectiveness

A fundamental distinction between the resistance of weld joints (as discussed in Section 8.2.3) and the resistance of connections (as discussed in Chapter 3) must be kept in mind. As noted in the latter, the connection has a resistance (as a function of geometric parameters) that is often less than the capacity of the member. This is quite acceptable provided the connection resistances exceed the connection design loads.

The connection resistance cannot be increased by additional welding because the extra weld will not be effective in transferring load. Such superfluous weld is at least wasteful and possibly harmful in that unnecessary heating, shrinking, and restraint will have been introduced. Therefore the joint designer's task is one of determining how much weld is actually needed, where it should be placed, and in what configuration.

8.2.4.2.1 Fillet Weld Size Considerations

An approach that provides "pre-approved" weld sizes is taken by Eurocode 3 (EC3 1992), where it is recommended that web (branch) members be welded all around their perimeter, and that, when fillet welds are used, the weld throat be 1.10 times the member wall thickness for 350 MPa steels. Such welds are considered by Eurocode 3 to develop the resistance of the member.

The fact that CSA-W59, Clause 4.1.3.3.3 defines fillet welds as having angular limits of 60° to 120° is relevant to the following discussion. When the angle is less than 60°, weld fusion may not occur all the way to the root, and the rules for a partial joint penetration groove weld apply.

In CAN/CSA-S16.1-94, a fillet weld throat of 1.10 times the wall thickness usually (but not always) develops the yield resistance of an HSS wall. This is because the resistance of a fillet weld is a function of weld geometry and of the direction of the load.

Looking at weld geometry:

Axial load resistance per unit length of an HSS wall cross section is

$$T_r = \phi t F_y = 0.9\,(t)\,0.350 = 0.315t$$

Given an HSS with a wall thickness t and a fillet weld with a 90° cross section, resistance per unit length of the weld fusion face f is

$$V_r = 0.67\,\phi_w\,f\,F_u \qquad \text{(from [8.4])}$$
$$= 0.67\,(0.67)\,(\sqrt{2}\;1.10t)\,0.450 = 0.314t$$

which is essentially the same value as for the HSS wall.

However, when the fillet weld angle is less than 90° (therefore a shorter fusion face relative to the 1.10 t weld throat), the resistance of the fusion face is less than the yield resistance of the HSS.

Looking at the direction of the load:

When the direction of the load is at an angle of 49° to the axis of the weld, and the weld throat is 1.10 times the HSS wall thickness t, the resistance per unit length of the weld throat is

$$V_r = 0.67 \phi_w \, 1.10 t \, X_u \, (1.00 + 0.5 \sin^{1.5} \theta_{(weld)}) \quad \text{(from [8.3])}$$

where $\theta_{(weld)}$ is the angle between the direction of the load and the axis of the weld.

$$\therefore V_r = 0.67 (0.67) 1.10 t \, (0.480)(1 + 0.5 (0.656)) = 0.315 t$$

which is the same value as for the HSS wall.

However, when the angle between the direction of the load and the axis of the weld is less than 49°, the resistance of the weld throat is less than the yield resistance of the HSS wall.

But in the context of the welds on sides "a" and "b" (see Fig. 8.19) of a sloping HSS web member, the additional length of the weld (compared to the HSS face h_i) more than compensates for the reducing unit resistance of weld metal as the angle $\theta_{(web)}$ is decreased.

The result is that a fillet weld with a throat of 1.10 t now develops the resistance of an HSS face both by CAN/CSA-S16.1-94 and by Eurocode

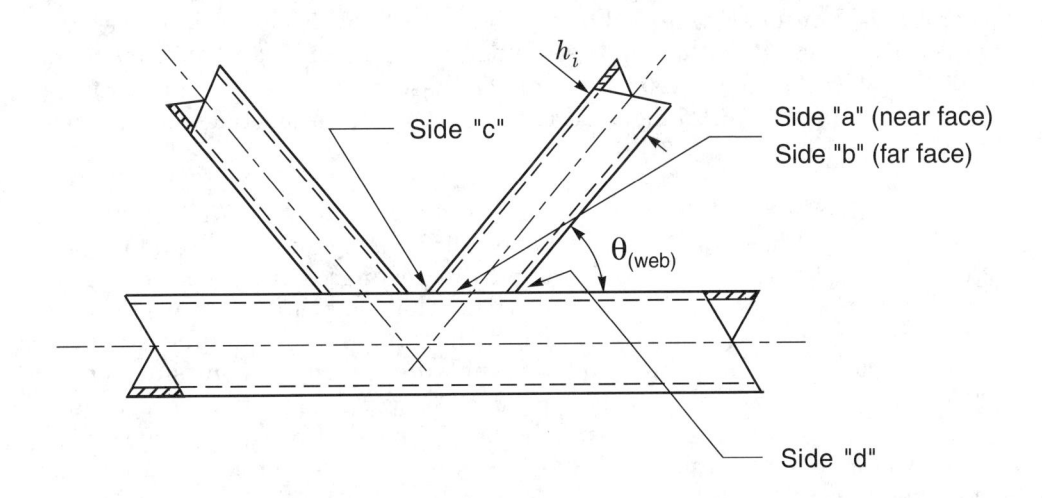

FIGURE 8.19

Identification of welds connecting a web (branch) member to a chord (main) member

3—with the S16.1 exception of the heel weld, side "d" in Fig. 8.19, (because the weld cross section is less than 90°, with the result that the fusion face of the weld governs). The resistance of the toe weld on side "c" exceeds that of the HSS wall because the obtuse cross section provides an increased fusion face, and the weld axis is at 90° to the load.

Cases where the designer should develop the yield resistance of the connected HSS wall include: frames in which significant plastic load redistribution must occur (such as plastically-designed portal frames or Vierendeel trusses), and frames designed for seismic loading. In the case of the latter, the Japanese go one step further and require welds (and connections) to develop at least 1.2 times the capacity of the connected member. This is done to allow yield stresses up to 20% higher than nominal in the connected member, yet still guarantee that plasticity (and hence energy dissipation) occurs in the member before the connection and welds.

8.2.4.2.2 Effective Weld Length Considerations

Under normal predominantly static loading conditions, the above approaches to weld sizing (based solely on web member thickness) are likely to be excessively punitive for many applications. By proportioning welds on the basis of actual loads, a designer may likely justify smaller weld sizes, but the effects of local connection deformations must be taken into account.

The advantages of this weld design approach can be significant in situations where aesthetics have controlled the member selection, or only a restricted number of web member sizes have been chosen for reasons of cost optimization, or a long compression web member causes the member to be loaded well below its full section squash load, or the web to chord member angle is low, thereby producing a much longer weld around the web member. When designing welds to resist actual web member forces, one must use effective weld lengths, however an upper limit on fillet weld throat size will still be 1.10 t (for weld cross-sections $\geq 90°$) because such a weld develops the unit axial load resistance of the connected web member wall.

By designing for web member forces and using the effective weld length method, (which has only been developed sufficiently for square and rectangular HSS members), the weld around the perimeter of a web member is treated as four separate welds, and the actual length of each weld along the four sides must initially be calculated. It may not be adequate to treat the web member "footprint" as a rectangle because of the outside corner radii (usually taken as $2t$). That approach may significantly *over*estimate the actual length of the weld, particularly for small hollow sections with thick walls. It is recommended that the contact perimeter of a sloping web member be calculated by

$$K_a (4\pi t + 2(b - 4t) + 2(h - 4t)) \qquad [8.8]$$

where $K_a = ((h/\sin\theta) + b)/(h + b)$, as was done for Table 8.5. [8.9]

Once the total contact perimeter of the web member has been determined, and the length apportioned among the four sides, the resistance of each of these four welds can be calculated separately, then added. In computing the resistance of a weld along one side, one takes into account:

(a) the amount of that weld length that can be assumed effective, and

(b) the orientation of the axis of that weld to the direction of loading in the web member, so that one can allocate a higher resistance to a weld that is transversely-loaded (or partially transversely) rather than longitudinally-loaded.

Effective Weld Lengths

Following is a summary of web member effective weld lengths for various truss-type, planar connections under predominantly static loading.

The effective lengths are given in terms of sides "a, b, c and d" and the angle θ between the web member and the chord (Fig. 8.17). For connections with inclined web members, location "c" is at the web member "toe" (also adjacent to the gap in a K or N gap connection), and location "d" is at the web member "heel".

T, Y and X connections

For $\theta_1 \leq 50°$: Effective length = a + b + d = $\dfrac{2h_1}{\sin\theta_1} + b_1$ [8.10]

For $\theta_1 \geq 60°$: Effective length = a + b = $\dfrac{2h_1}{\sin\theta_1}$ [8.11]

A linear interpolation is recommended between 50° and 60°.

K and N gap connections

For $\theta_i \leq 50°$: Effective length = a + b + c + d = $\dfrac{2h_i}{\sin\theta_i} + 2b_i$ [8.12]

For $\theta_i \geq 60°$: Effective length = a + b + c = $\dfrac{2h_i}{\sin\theta_i} + b_i$ [8.13]

A linear interpolation is recommended between 50° and 60°.

K and N 50% overlap connections

Limited research (127x127x13 HSS web members used at 60° angles relative to a 203x203x13 HSS chord) suggested that all sides are fully effective for at least that case.

Research results from full scale Warren trusses with gap and overlap K connections (Frater and Packer 1990, 1992a, 1992b) (Fig. 8.20) and from isolated T, Y and X connections (Packer 1995, Packer and Cassidy 1995) were used to validate [8.10] to [8.13], above. Figures 8.21 and 8.22 show fractured weld joints tested to failure during the truss program.

A design example that illustrates the application of effective weld lengths to a Warren truss is given in Section 13.2.8 of this Design Guide.

On occasion, information in Tables 3.2 and 3.3 might be helpful when assessing weld effectiveness. Therein, for square and rectangular chord HSS, effective widths of *web member faces* are defined for determining the member connection resistances. Table 3.2 identifies that portions of the transverse faces of overlapping web members in K or N connections may be partly ineffective. Table 3.3 additionally identifies that the heel face of web members in gap K and N connections, and both transverse faces of web members in T, Y and X connections with $\beta > 0.85$, may be partly ineffective. It follows that welds to the ineffective lengths of *faces* probably contribute little to the transfer of forces from one member to the other.

It is suggested that consideration be given to a uniform strength weld all the way around a web member, even in the regions which may not be fully effective. This will provide structural continuity for secondary effects.

It is convenient and conservative to calculate the weld length as the perimeter of the web member at the intersection with the chord. Standard W59, Clause 4.3.2.3 (and also AWS D1.1-96, Clause 2.4.2.2) allows the length to be based on the calculated length along the mid-point of the weld throat, an approach which is more practical for analysis than it is for design.

FIGURE 8.20
Typical rectangular HSS truss tested at the University of Toronto to develop rules for effective weld lengths

As was mentioned in Section 3.4, the concealed toe of a partially overlapped web member of a K or N connection need not be fully welded to the chord when the web force components, perpendicular to the chord, are substantially in balance. Otherwise, the full circumference should be welded.

For fatigue loaded T, Y, X, K or N connections, it was recommended by IIW (1985) that fillet weld throat thicknesses be at least equal to web member wall thicknesses. This is slightly less than the requirement for statically loaded connections (IIW 1989), that throat thicknesses be at least 1.07 (later adjusted to 1.10) times the wall thicknesses in order for welds to be considered pre-approved. The authors tentatively suggest that $t_w \geq 1.0t$ therefore be used for fatigue-critical connections, as still recommended by IIW (1985), although there is evidence that smaller fillet welds

FIGURE 8.21
Fillet weld fracture at a gap connection (University of Toronto)

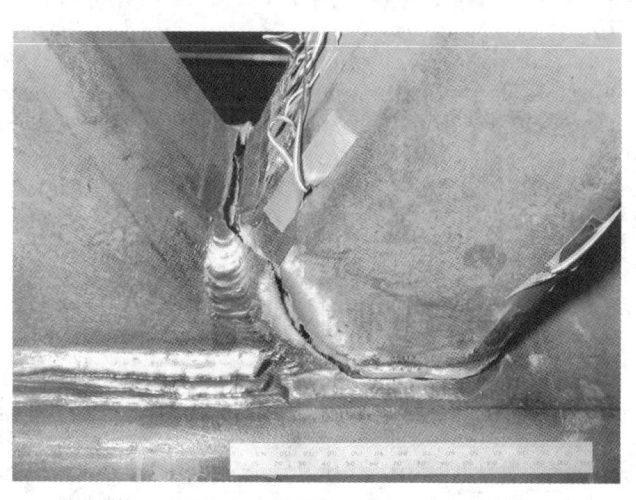

FIGURE 8.22
Fillet and groove weld fracture at an overlap connection (University of Toronto)

may still give satisfactory fatigue performance in some situations (van Wingerde et al. 1996).

8.2.5 Inspection

A wide variety of techniques is available for both destructive and non-destructive inspection of welds. Each has specific advantages and disadvantages; therefore each has appropriate applications. Five methods of non-destructive examination are in common use: visual, magnetic particle, liquid penetrant, ultrasonic, and radiographic. All are used to search for discontinuities such as porosity, lack of fusion, slag inclusions, and cracks in welded joints. The presence of some discontinuities in a weld does not necessarily mean that the weld is unacceptable to code. It is the nature and frequency of a discontinuity that determines whether it constitutes an unacceptable defect (i.e., that the weld is not "fit for purpose").

8.2.5.1 Methods of Inspection

Very often close visual scrutiny, both after fit-up (prior to welding), and upon completion of the weld, is all that is needed. An examination before welding will confirm that the geometric parameters (root gap, width of face, angle of bevel, regularity of preparation, alignment, etc.) are as prescribed. When these variables are correct, the probability of a sound weld is high. After completion, the examiner should check the size and profile of the weld, and should also look for surface discontinuites such as porosity, lack of fusion, and cracking. If everything appears good after fitting, and again after welding, if the welders have proper qualifications and materials are as specified, the weld likely meets all requirements for a sound structural weld of normal specification quality.

Magnetic particle examination provides an easy and convenient method for investigating steel (including welds) for surface discontinuities like fine cracks that have not opened up enough to be visually obvious. Fine magnetic particles are sprayed onto the surface in the presence of a magnetic flux field produced by a portable AC yoke (DC current is better when searching for flaws just below the surface). Structural discontinuities distort the magnetic field and the particles line up accordingly to produce a distinct indication. Figure 8.23 shows an HSS connection under test by this widely used method.

Liquid penetrant has long been used to explore and confirm the extent of discontinuities that intersect the surface of weldments. The surface is carefully cleaned and a penetrating red dye solution applied. Capillary action draws the dye into even minute discontinuities. After a few minutes, the excess dye is removed with a solvent and the surface allowed to dry. Thereupon a quick-drying white developer solution is sprayed on. It blots the dye from any discontinuity into which it was drawn, thus marking a clear outline of the imperfection, red on white.

Ultrasonic examination (whereby high frequency sound waves are directed into a weldment and their echo displayed electronically on a screen) is increasingly popular. Interior discontinuities distort the echo, which reveals the location and approximate size of the potential defect. The method is very sensitive and will detect any type of discontinuity. However, interpretation of the nature of the flaw is to a great extent subject to the judgement, experience and skill of the operator. The lack of a permanent record of the findings has been considered a disadvantage of the process in some situations, but electronic recording equipment is increasingly available.

Radiographic inspection is carried out by directing either X-rays (produced by electrical equipment), or far more likely gamma-rays (from radio isotopes like cobalt or iridium) through the specimen onto a photographic film. The process records an image of most types of weld discontinuities. Radiography is excellent at revealing internal voids such as those caused by slag inclusions or porosity. Critical butt joints formed of complete penetration groove welds are generally radiographed to permanently record that the specified quality was achieved.

A significant limitation of radiography is the fact that laminations or cracks perpendicular to the direction of the rays will not be shown because the rays pass through the same total thickness of material with or without the discontinuity. Ultrasonic examination detects those flaws, and it complements radiography well in that respect. Irregular shapes or thicknesses are not generally suitable for radiography because the pattern developed on the film is determined by the thickness through which the rays pass. For that reason, partial joint penetration groove welds with their uneven root profiles are seldom radiographed.

FIGURE 8.23
Magnetic particle inspection of a connection
(University of Toronto)

8.2.5.2 Application of Inspection Methods to HSS Construction

As discussed above, timely and knowledgeable visual inspection is the best all-around method of achieving specified regular structural quality for HSS fabrication.

Magnetic particle examination is useful for specific instances where problems may be suspected, where absence of cracking in areas of high restraint needs to be confirmed, or for routine spot checks.

Liquid penetrant examination tends to be messy and slow, but can be particularly helpful when determining the extent of a discontinuity. This is especially true when gouging and grinding are being done to remove a defect for the repair of a weld.

Ultrasonic examination has a limited role in HSS fabrication, since small fillet welds and partial joint penetration groove welds are usually not suitable objects for the method. Relatively thin sections with great variations in joint geometry make the signals difficult to interpret, but technicians with specific experience on weldments similar to those to be examined may be able to do so in some instances. Even the interpretation of readings from complete joint penetration groove welds (with or without backing bars) can be difficult.

Radiographic examination has limited application to HSS fabrication because of the irregular shape of the joints and the resulting variations in thickness of the material as projected onto a film. Butt splices provide some opportunities for radiography, but shots onto backing bars provide limited information about the condition of fusing to the bars at the root corners. The general inability to place either the radiation source or the film inside HSS means that exposures must usually be taken through both the front face and the back face of the section with the film attached to the outside of the back face. Several such shots progressing around the member are needed to examine the complete joint.

In summary, in the context of HSS fabrication, radiography is quite limited, ultrasonics has a little application, magnetic particle examination has a useful role, and liquid penetrant has certain specific functions. But the most practical and effective inspection technique is timely visual examinations by an experienced inspector.

Critical joints should be scrutinized prior to the commencement of welding. Fit-up, preparation bevels, gaps, alignment, etc. all need to be as per the CWB approved procedure data sheets (and W59 fabrication tolerances) for the joint. After welding is complete, how does the joint look? Is the weld size correct? Are there surface flaws? Maybe a magnetic particle examination is needed; an ultrasonic reading may be justified in certain instances, but most of the time a good visual inspection is sufficient.

The Gooderham Centre for Industrial Learning (a division of the Canadian Welding Bureau) administers a highly successful educational program that was initially developed by the no-longer-operating Welding Institute of Canada. It is available in a series of 39 home study modules (Modular Learning System 1996) arranged in various groupings to cover welding technology, design, inspection, etc. The reader is directed to that Centre for detailed technical information and assistance.

Reference can also be made to papers by Post (1989, 1990), a welding engineer consultant who has been a fitter, boilermaker, certified welder, and certified welding inspector. These papers were written from the practitioner's perspective, and they provide a knowledgeable source of practical information on the subject of tubular welding and inspection.

8.2.6 Avoiding Problem Areas

Loads

Provide *actual loads* to be transferred at the connections. Avoid "connect for capacity of the member"; often with HSS, that is not possible without local reinforcement. Even "connect for the full connection resistance of the members" may be more than is necessary, particularly when members have been sized for uniformity or for stiffness.

Amount of welding

Do not specify larger welds, or more extensive welding than is needed for the structural function. It is wasteful, costly, and possibly harmful.

Weld symbols

It is more important for the design engineer to provide actual loads to be transferred rather than elaborate weld symbols. The specifics of the joint depend on the weld process to be used and on the welding position (flat, horizontal, vertical or overhead). It is common practice for the fabricator to provide weld joint cross sections on the shop drawings to complement the weld symbols if necessary to clarify weld details.

Type of joint

Specify fillet welds rather than groove welds whenever possible, until the size of fillets becomes excessive (generally, when fillet legs are around 16 mm).

Use partial joint penetration groove welds when fillets are not feasible.

Minimize the use of flare bevel groove welds.

Avoid complete joint penetration groove welds—for other than butt splices on backing bars for similar size sections.

In particular, avoid CJPG welds without backing bars. The joints are not prequalified. Welder qualifications, joint preparation, fit-up, welding, and inspection are all much more difficult and expensive than for alternative welds.

Inspection

Use visual inspection extensively; concentrate on fit-up examination before welding is started.

Avoid ultrasonics unless it is known to be feasible on the joint, and then only for critical joints.

REFERENCES

AWS. 1990. Structural welding code—steel, 12th. ed. ANSI/AWS D1.1-90, American Welding Society, Miami, U.S.A.

AWS. 1996. Structural welding code—steel. ANSI/AWS D1.1-96, American Welding Society, Miami, U.S.A.

CSA. 1980. Bare mild steel electrodes and fluxes for submerged arc welding, W48.6-M1980. Canadian Standards Association, Rexdale, Ontario.

CSA. 1989. Welded steel construction (metal arc welding), W59-M1989. Canadian Standards Association, Rexdale, Ontario.

CSA. 1990. Carbon steel electrodes for flux-and metal-cored arc welding, W48.5-M1990. Canadian Standards Association, Rexdale, Ontario.

CSA. 1991. Carbon steel covered electrodes for shielded metal arc welding, W48.1-M1991. Canadian Standards Association, Rexdale, Ontario.

CSA. 1992. Certification of companies for fusion welding of steel structures, W47.1-92. Canadian Standards Association, Rexdale, Ontario.

CSA. 1992. Chromium and chromium-nickel steel covered electrodes for shielded metal arc welding, W48.2-M1992. Canadian Standards Association, Rexdale, Ontario.

CSA. 1992. General requirements for rolled or welded structural quality steel, CAN/CSA-G40.20-M92. Canadian Standards Association, Rexdale, Ontario.

CSA. 1992. Structural quality steels, CAN/CSA-G40.21-M92. Canadian Standards Association, Rexdale, Ontario.

CSA. 1993. Low-alloy steel covered electrodes for shielded metal arc welding, W48.3-M1993. Canadian Standards Association, Rexdale, Ontario.

CSA. 1994. Limit states design of steel structures, CAN/CSA-S16.1-94. Canadian Standards Association, Rexdale, Ontario.

CSA. 1995. Solid carbon steel filler metals for gas shielded arc welding, W48.4-M1995. Canadian Standards Association, Rexdale, Ontario.

EUROPEAN COMMITTEE for STANDARDIZATION. 1992. Eurocode 3: design of steel structures. Part 1.1—general rules and rules for buildings. ENV 1993-1-1:1992E, British Standards Institution, London, England.

FRATER, G.S., and PACKER, J.A. 1990. Design of fillet weldments for hollow structural section trusses. CIDECT Report No. 5AN/2-90/7, University of Toronto, Ontario.

FRATER, G.S., and PACKER, J.A. 1992a. Weldment design for RHS truss connections. I: Applications. Journal of Structural Engineering, American Society of Civil Engineers, **118**(10): 2784–2803.

FRATER, G.S., and PACKER, J.A. 1992b. Weldment design for RHS truss connections. II: Experimentation. Journal of Structural Engineering, American Society of Civil Engineers, **118**(10): 2804–2820.

IIW. 1985. Recommended fatigue design procedure for hollow section joints. Part I: hot spot stress method for nodal joints. International Institute of Welding Subcommission XV-E, IIW Doc. XV-582-85, IIW Annual Assembly, Strasbourg, France.

IIW. 1989. Design recommendations for hollow section joints—predominantly statically loaded, 2nd. ed. International Institute of Welding Subcommission XV-E, IIW Doc. XV-701-89, IIW Annual Assembly, Helsinki, Finland.

KENNEDY, J.B. 1988. Minimum bending radii for square & rectangular hollow sections (3-roller cold-bending). CIDECT Report 11C-88/14-E, University of Windsor, Ontario.

KOSTESKI, N. 1996. Nailed tubular connections. Master of Applied Science thesis, University of Toronto, Ontario.

MODULAR LEARNING SYSTEM. 1996. Home study courses in 39 modules. Gooderham Centre for Industrial Learning, 7250 West Credit Ave., Mississauga, Ontario, L5N 5N1.

PACKER, J.A. 1995. Design of fillet welds in rectangular hollow section T, Y and X connections using new North American code provisions. Proceedings, 3rd. International Workshop on Connections in Steel Structures, Trento, Italy, pp. 463–472.

PACKER, J.A. 1996. Nailed tubular connections under axial loading. Journal of Structural Engineering, American Society of Civil Engineers, **122**(8): 867–872.

PACKER, J.A. and CASSIDY, C.E. 1995. Effective weld length for HSS T, Y and X connections. Journal of Structural Engineering, American Society of Civil Engineers, **121**(10): 1402–1408.

PACKER, J.A. and KRUTZLER, R.T. 1994. Nailing of steel tubes. Proceedings, 6th. International Symposium on Tubular Structures, Melbourne, Australia, pp. 61–68.

POST, J.W. 1989. Gaining confidence with the fabrication, welding and inspection of tubular connections. Proceedings, American Institute of Steel Construction, National Steel Construction Conference, Nashville, Tennessee, U.S.A., pp. 22.1–22.30.

POST, J.W. 1990. Box-tube connections; choices of joint details and their influence on costs. Proceedings, American Institute of Steel Construction, National Steel Construction Conference, Kansas City, Missouri, U.S.A., pp. 22.1–22.26.

RIVIEZZI, G. 1984. Curving structural steel members. Steel Construction, Australian Institute of Steel Construction, **19**(3): 10–11.

STELCO. 1981. Hollow structural sections—design manual for connections, 2nd. ed. Stelco Inc., Hamilton, Ontario.

STELCO. 1989. Effects of freezing temperature on HSS containing water, 3rd. ed. Information Bulletin, Stelpipe, Welland, Ontario.

WARDENIER, J., KUROBANE, Y., PACKER, J.A., DUTTA, D., and YEOMANS, N. 1991. Design guide for circular hollow section (CHS) joints under predominantly static loading. CIDECT (ed.), Verlag TÜV Rheinland GmbH, Köln, Federal Republic of Germany.

van WINGERDE, A.M., PACKER, J.A., STRAUCH, L., SELVITELLA, B., and WARDENIER, J. 1996. Fatigue behaviour of non-90° square hollow section X-connections. Proceedings, 7th. International Symposium on Tubular Structures, Miskolc, Hungary, pp. 315–332.

9
BEAM TO HSS COLUMN CONNECTIONS

9.1 Introduction

With their superior compressive capacities, hollow structural sections are a natural choice for columns. Square, rectangular and circular HSS readily accommodate connections for simple beam supports, and either square or rectangular HSS provide convenient shapes for beam moment connections.

For simply supported behaviour to be realized, connections must provide a certain degree of flexibility in order to accommodate beam end rotations as the beam deflects under load. A full moment connection which prevents beam rotation relative to the column implies a contraflexure point some distance along the length of the beam. In practice, most connections being semi-rigid fit somewhere between the idealized extremes, with simple connections often having the contraflexure point only 100 mm or so from the column face. Increased moment transfer to the column with increasing connection stiffness should be kept in mind.

For heavily loaded square and rectangular HSS columns, it is preferable that connections to thin walled columns transfer the forces near the edges of the column face, rather than near its centre. Edge transfer is necessary for Class 4 HSS (CSA 1994) columns (Sherman 1995) in order to prevent excessive rotation of the column face, which could reduce the column capacity.

9.2 Rectangular HSS Beams to Rectangular HSS Columns

In general, bolted connections tend to be used for simply supported beams, and welded connections for moment connected beams. Both types will be considered.

9.2.1 Bolted Connections

Bolted HSS beam to column connections generally entail the use of detail material such as tees, angles or plates, in a manner similar to the variants of Fig. 9.1.

Figure 9.1(a) shows a double tee connection. The tees are either built up from plates or cut from rolled sections. Central alignment of the beam and column can be maintained by offsetting the tees. The column tee should be welded only along the vertical edges, and should include a short return around the top corners. Also, the width to thickness ratio of the tee flange, previously suggested by White and Fang (1966) to be 10 or more, is now recommended to be a minimum of 13 (Astaneh and Nader 1990, and Dawe and Mehendale 1995) in order to provide connection flexibility primarily through deformation of the tee flange, rather than the HSS wall.

Figure 9.1(b) shows a pair of angles that provide double shear loading on the bolts. Welding of the angles to the column is the same as for the tee in Fig. 9.1(a).

Figure 9.1(c) illustrates a relatively narrow beam framing into a wide HSS column. Angles, welded near the column corners, are used on either side of the beam that has the bottom surface cut back to provide access for bolting. Eccentric loading in the plane of the column face upon the angle welds may limit the capacity of this arrangement.

Figure 9.1(d) shows a better match of beam and column widths where two shear plates are used. This would be a relatively stiff connection with the plates welded near the column corners.

Figure 9.1(e) depicts equal width members connected with two gusset plates welded to the column. While convenient, this arrangement has some potential for difficult fitting at the site. It may be necessary to spread the plates slightly with jacks after the welds cool because welding contraction will tend to deflect and pull the plates together.

Figure 9.1(f) portrays an end plate example which affords sealing of the HSS beam while providing easy bolt installation.

$\dfrac{b}{t} \geq 10$

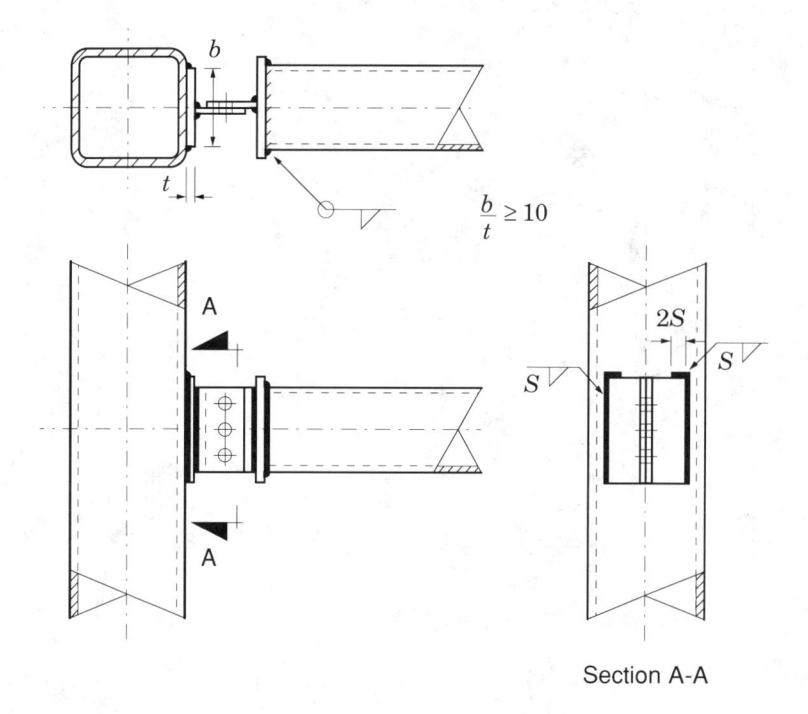

Section A-A

(a) Tees on column and beam

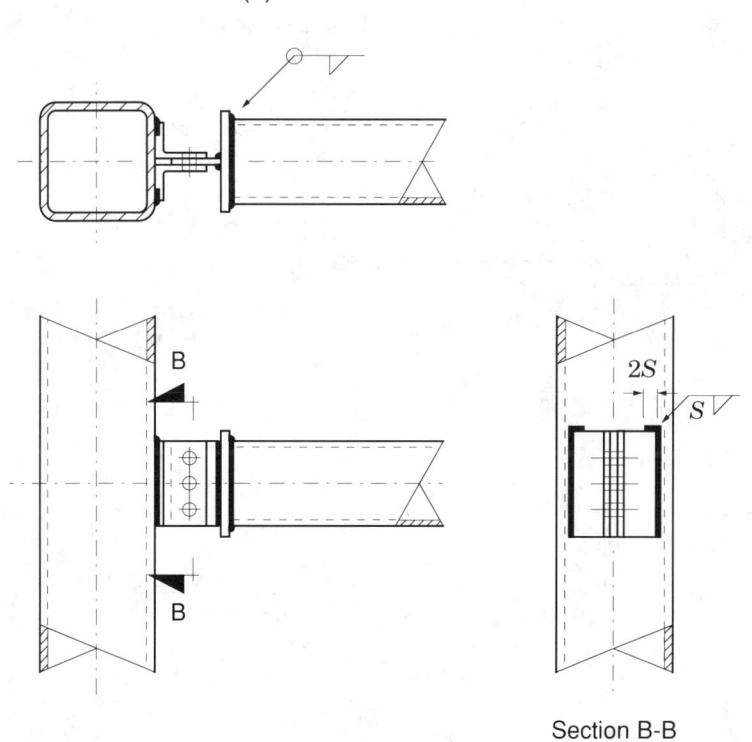

Section B-B

(b) Header angles on column with tee on beam

FIGURE 9.1 (start ...)
Simple connections

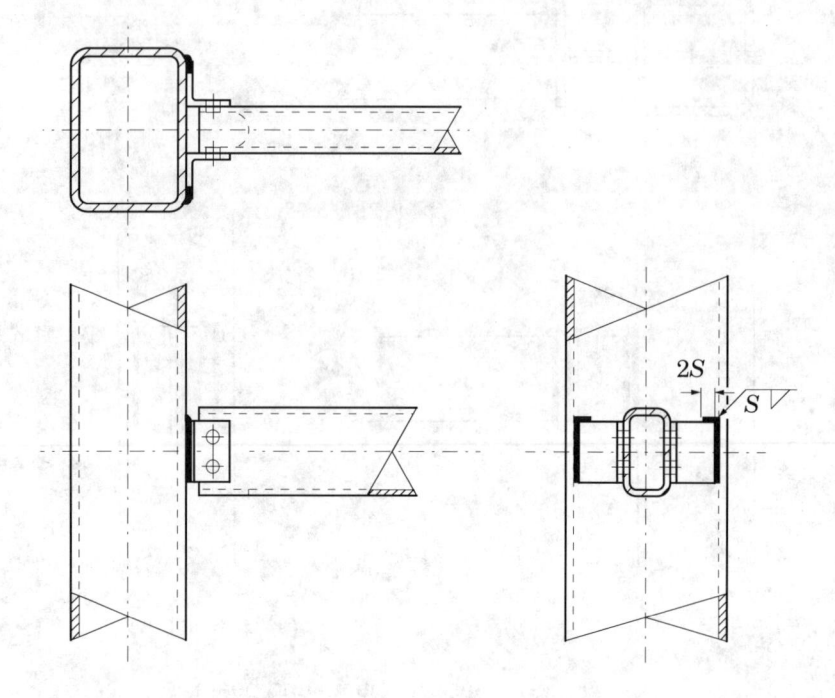

(c) Separated double angles on column

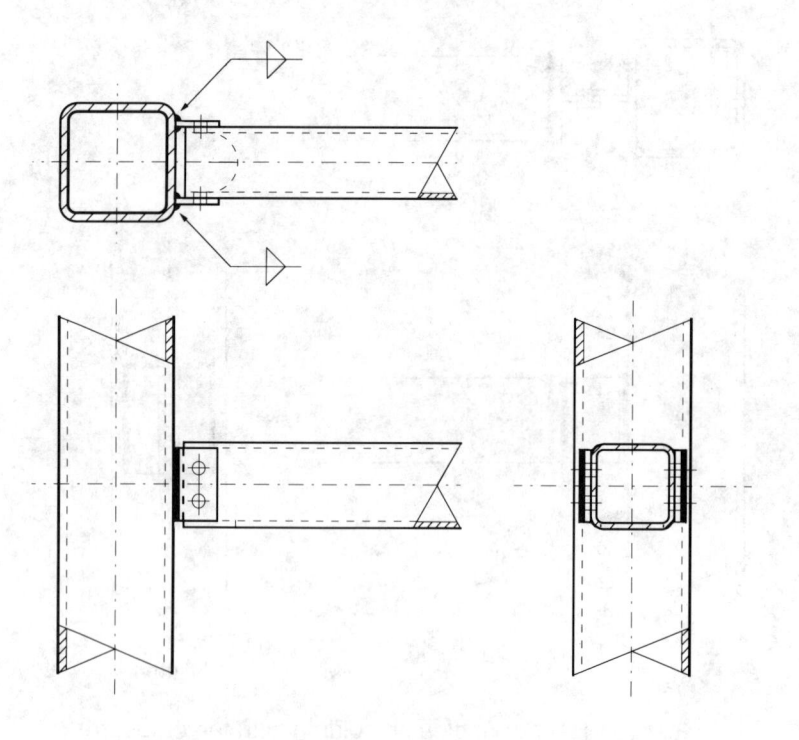

(d) Plates on column face

FIGURE 9.1 (... continued ...)

304

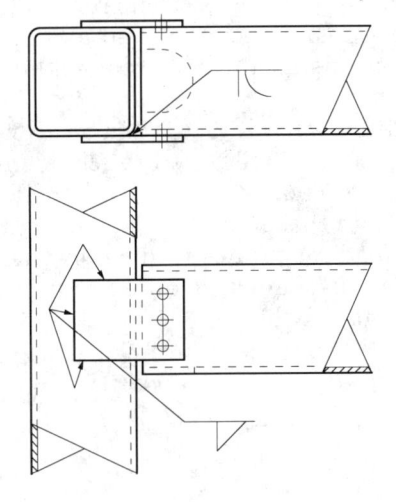

(e) Plates on sides of column

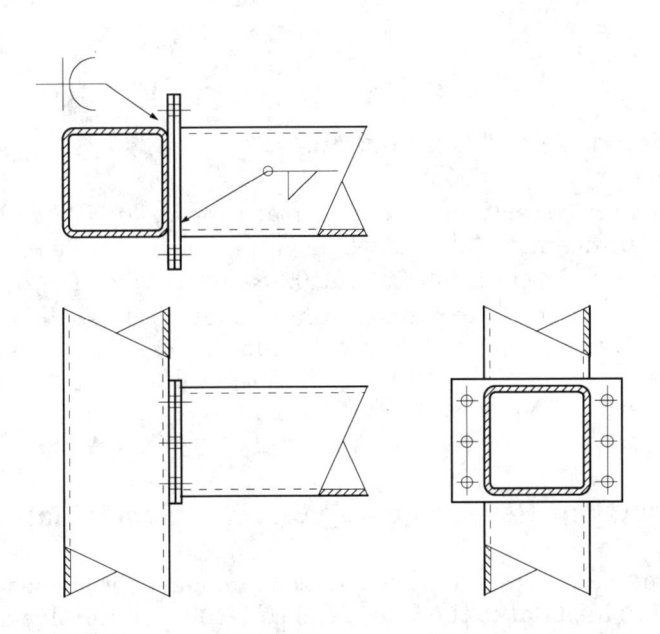

(f) Plate on column face with beam end plate

FIGURE 9.1 (... concluded)

9.2.2 Unreinforced Welded Connections

Section 6.1 on Vierendeel trusses presents a detailed discussion of the behaviour and strength of HSS to HSS welded moment connections. Unreinforced welded HSS beam to HSS column connections resemble Vierendeel connections, so their capacity to transmit moment and axial forces can by determined with the same expressions used for Vierendeel trusses.

Three modes of connection failure were developed in Section 6.1.2: Mode (a) chord (column) face plastification, Mode (c) branch member (beam) effective width failure and Mode (d) chord (column) side wall crippling—with the least value governing. (Note that identified Modes (b) and (e) were not considered pertinent.) Axial load resistance is given by the formulae in Tables 3.2 and 3.3 for T connections, and the interaction with moment resistance was expressed conservatively by

$$\frac{N_1}{N_1^*} + \frac{M_{f1}}{M_{r1}^*} \leq 1.0 \qquad [9.1]$$

Since HSS beam shear loads are carried primarily by the side walls of the section, the interaction of shear resistance with moment or axial loads is likely to be low and the shear resistance of the HSS beam can be taken as

$$V_{p1} = 2h_1 t_1 \frac{F_{y1}}{\sqrt{3}} \qquad [9.2]$$

9.2.3 Reinforced Welded Connections

Both the bending resistance and flexural rigidity of HSS to HSS connections are significantly reduced when values of β, the beam width to column width ratio, are less than about 0.85 and values of b_0/t_0, column width to column thickness, are greater than about 16. In such cases, local reinforcement may be an option. Figure 9.2 illustrates two means of applying effective reinforcement, a plate to stiffen the column face, or a haunch to increase the area of beam contact ("footprint").

9.2.3.1 Connections Reinforced with Chord (Column) Plate

The most common, most easily fabricated and least obtrusive reinforcement is a plate welded flat to the face of the HSS column. Korol *et al.* (1982) developed a yield line analysis which led to reinforcing plate parameters that result in sufficient strength to resist the bending or axial capacity of a member framing to the column. Fifty combinations of sizes were analyzed

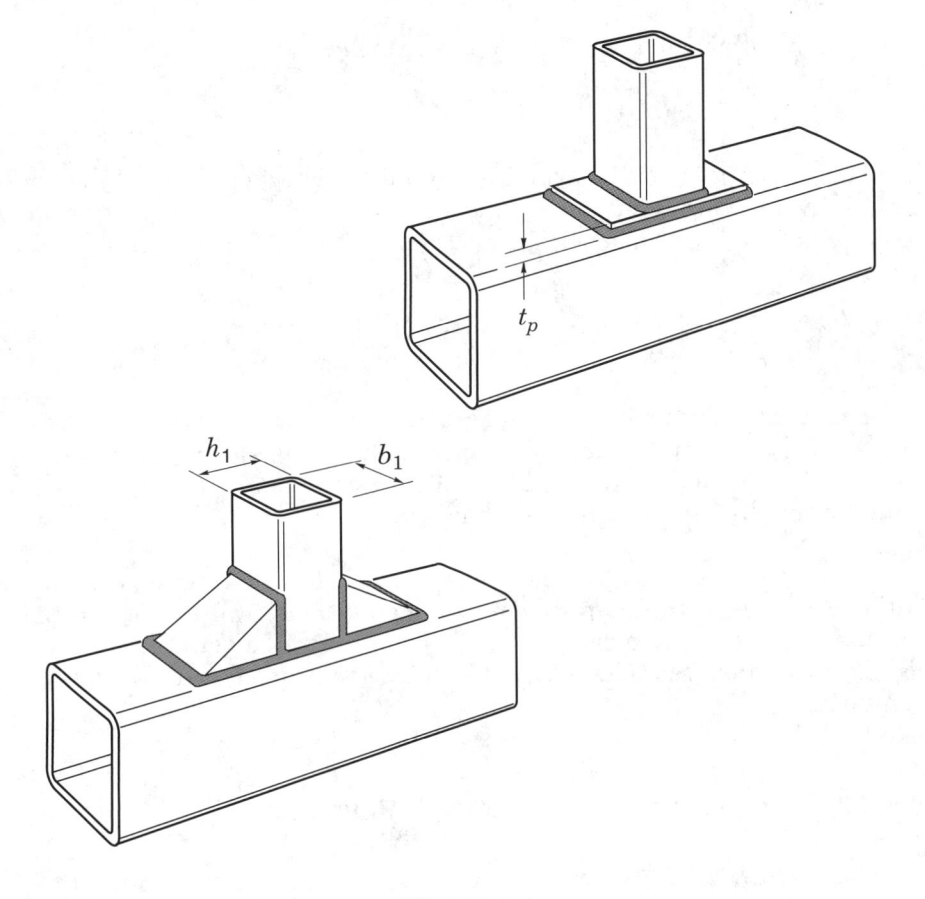

FIGURE 9.2

Effective reinforcement for HSS to HSS connections

for axial (punching) capacity and 73 combinations for moment capacity, with values of β from 0.25 to 0.80 and plate thickness t_p from t_0 to $2t_0$.

A comparison was also made with earlier (Korol *et al.* 1977) test results of seven full size specimens. They ranged from an HSS 152x152x4.8 beam on an HSS 254x254x9.5 column with a 9.5 mm thick plate to an HSS 254x254x8.0 beam on an HSS 305x305x9.5 column with a 19 mm plate.

Recommendations by Korol *et al.* (1977) to obtain full strength connections are

1. Plate width should be at least equal to the flat width of the HSS face (that is, taken to be $\geq b_0 - 4t_0$).
2. Plate length should be twice the HSS column width, i.e., $2b_0$.
3. Plate thickness depends on whether axial or bending loads dominate. For full axial compression capacity of the branch,

$$t_p \geq 4t_1 - t_0 \qquad [9.3]$$

For full moment capacity of the branch,

$$t_p \geq 0.63(b_1 t_1)^{0.5} - t_0 \qquad [9.4]$$

The side walls of the column should be checked for web crippling, especially for values of β larger than about 0.85. An adaptation of the expression used for Mode (d) failure with Vierendeel trusses [6.5] may be used:

$$M_{r1}^* = 0.5 F_k t_0 (h_1 + 5(t_0 + t_p))^2 \qquad [9.5]$$

The term F_k is the buckling stress of the column side walls, and can be taken as F_{y0} as explained in Section 6.1.2 for one-sided moment connections. For two-sided moment connecions, F_k can be taken as $0.8 F_{y0}$ as was done for axially loaded X connections.

Once again, welds need to be proportioned for the intended capacities. An alternative to the empirical recommendations by Korol *et al.* (1977) above would be to use a rational approach based on a set of failure modes, as illustrated in Section 9.3.2.3 for plate-reinforced wide flange beam connections.

9.2.3.2 Connections Reinforced with Haunches

Korol *et al.* (1977) demonstrated that another efficient and aesthetic form of connection reinforcement is to use 45° haunches. Cuttings from the branch (beam) member provide convenient haunches, and an edge dimension equal to the branch depth h_1 (which gives a total contact length of $3h_1$) works well. Then the equations for Mode (a), chord face plastification, and Mode (d), chord side wall bearing or buckling capacity (from Vierendeel truss connections in Section 6.1.2) may be applied with $3h_1$ rather than h_1.

Failure Mode (a), ($\beta \leq 85$):

$$M_{r1}^* = 3F_{y0} t_0^2 h_1 \left(\frac{1}{6h_1/b_0} + \frac{2}{\sqrt{1-\beta}} + \frac{3h_1/b_0}{1-\beta} \right) f(n) \qquad [9.6]$$

The function $f_2(n)$ allows for reduced connection moment capacity due to the axial column load.

$$f_2(n) = 1.2 + \left(\frac{0.5}{\beta} \right) n \qquad \text{but} \not> 1.0 \qquad [9.6a]$$

where n is the column load as a fraction of the column yield load, $N_0/A_0 F_{y0}$, and is negative for compression. Equation 9.6(a) was shown graphically in Fig. 6.5.

Failure Mode (d), $(0.85 < \beta \leq 1.0)$:

$$M_{r1}^* = 0.5 F_k t_0 (3h_1 + 5t_0)^2 \qquad [9.7]$$

The column side wall buckling stress F_k can again be determined as noted at [9.5]. However, since the ratio of bearing length to column depth $(1.5h_1 + 2.5t_0)/h_0$ is now becoming large, this assumption for F_k should only be used for $h_0/t_0 \leq 23$ (Davies and Packer 1987). For haunch-reinforced moment connections with $h_0/t_0 > 23$, F_k should be calculated using the Functions part of Table 3.3.

The third potential failure, Mode (c) effective width failure of the branch, is not applicable in this case.

9.3 Wide Flange Beams to Square and Rectangular HSS Columns

Research has been carried out on both simple shear connections and full moment connections for framing wide flange beams into square and rectangular columns.

9.3.1 Simple Shear Connections

Quite likely the most suitable connection for general purpose use is the traditional double angle arrangement illustrated in Fig. 9.3. It provides the strength of bolts in double shear combined with excellent flexibility and, being symmetrical, the connection avoids any lateral torsion. Fabricators can select standard detail angles from stock rather than prepare special components such as tees, and handbook capacity values may be used.

White and Fang (1966) examined five other types of simple wide flange beam to square HSS column connections, as shown in Fig. 9.4. That work formed the basis for general recommendations, but the limit on the crucial column face slenderness parameter was not determined.

Sherman and Ales (1991) looked more closely at characteristics of the first of these connections, the simple shear tab. More recently, Sherman (1995) published the results of a large number of experiments on realistically-loaded, simple framing connections between wide flange beams and square HSS columns, in which the load imposed on the column was predominantly shear. In all, nine different types of simple framing connections to HSS columns were considered:

Section A-A

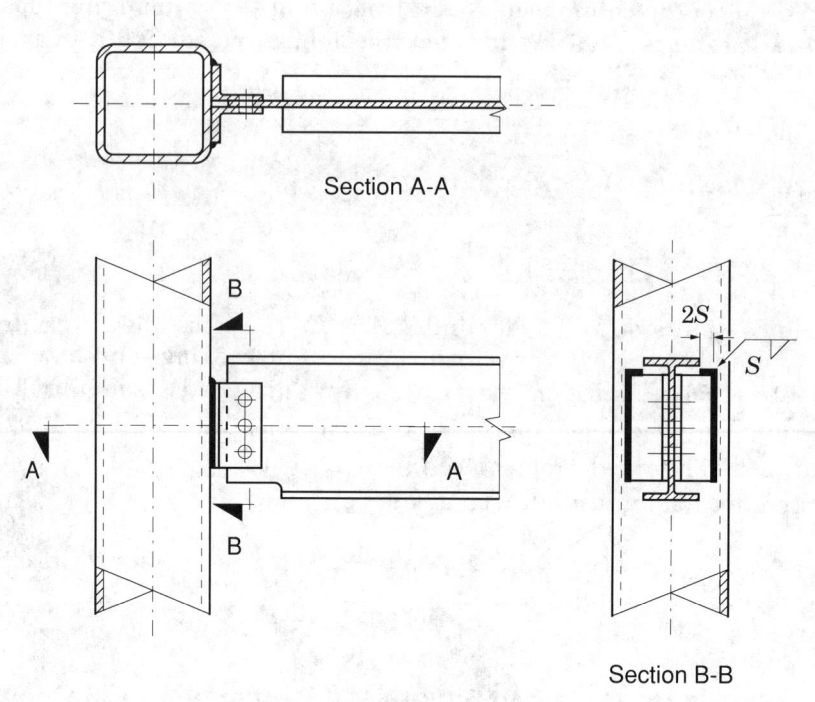

FIGURE 9.3
Double angle header connection

Section B-B

- » double angles
- » shear tabs
- » through-plates (similar to shear tabs but the plate passes through the column member by means of slots in the walls)
- » tees (narrower than the HSS) with vertical fillet welds
- » tees (wider than the HSS) with flare-bevel groove welds
- » unstiffened seated connections
- » single angles with L-shaped fillet weld (toe and bottom of angle)
- » single angles with two vertical fillet welds
- » web end plates

For all except the web end plate (where flowdrilling of the column face was used), the connecting elements were welded to the square HSS column. Then for all except the seat angle (where the flange bears on the outstanding leg) and the web end plate, the connecting elements were bolted to the web of the wide flange beam. The b/t ratio of HSS walls varied from 5 to 40.

Shear tab tests were performed with bolts both snug tight and fully pretensioned. The connections with snug tight bolts had the same ultimate

(a) Type A: Shear tab connection

(b) Type B: Tee connection

(c) Type C: Top angle and seat angle connection

(d) Type D: Shear tab to column corner

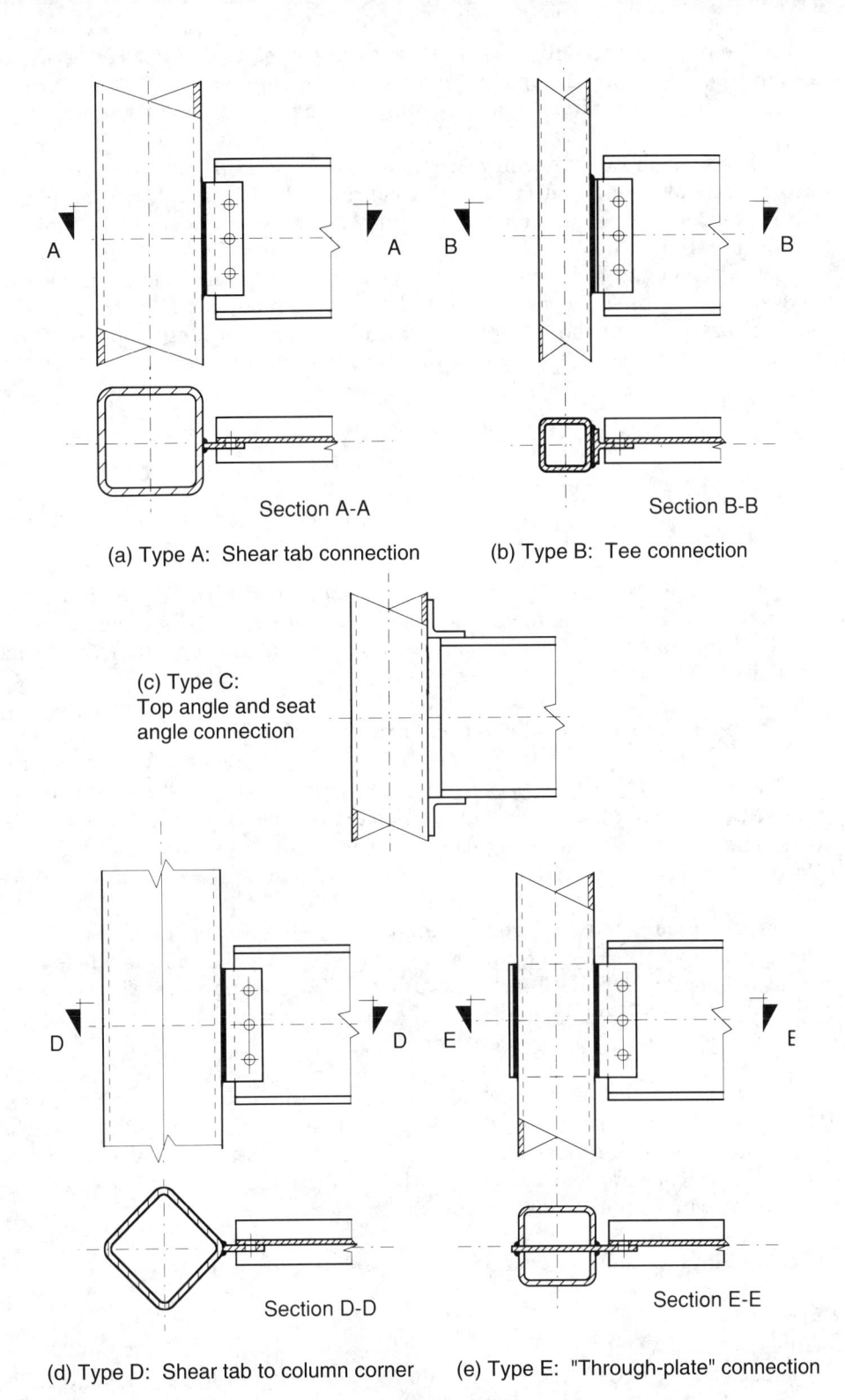

(e) Type E: "Through-plate" connection

FIGURE 9.4
Types of simple connections tested by White and Fang (1966)

capacities and eccentricities as those with pretensioned bolts. However, at working loads, pretensioned bolts produced larger eccentricities (to the contraflexure point where negative moment changed to positive moment).

Over the wide range of connection types and variables studied, only *one* limit state was identified for an HSS column. This was a punching shear failure related to end rotation of the beam when a thick shear tab was joined to a thin-walled (Class 4) HSS. Two connections failed when the shear tab pulled out from the HSS wall at the top of the tab around the perimeter of the welds. A simple criterion to avoid this failure mode is to ensure that the tension resistance of the tab under axial load (per unit length) is less than the shear resistance of the HSS wall along two planes (per unit length):

$$\phi F_{yp} t_p < 2 \phi \, 0.5 \, F_{u0} t_0$$

Thus, $\quad t_p < \left(\dfrac{F_{u0}}{F_{yp}}\right) t_0$ [9.8]

Yield line distortion (flexural failure associated with connection rotation) of the HSS face was never a critical limit state. This is because there is a limit to the end slope of a simply supported beam, and this restrains the distortion of the column face.

The local distortion that does occur in the HSS wall (for connections on one or both sides of the HSS) has negligible influence on the column resistance <u>provided the HSS is not thin-walled</u>, where the definition of thin-walled by Sherman (1995) amounts to a b/t ratio of 35.6 for 350 MPa steel. The 350 MPa steel b/t ratio (i.e., $(b_0 - 4t_0)/t_0$ value) maximum in CAN/CSA-S16.1-94 (CSA 1994) for Class 3 (non-compact) sections is 35.8.

There is also no advantage to using through-plates, as opposed to shear tabs, <u>providing the HSS is not thin-walled</u>; i.e., the column is Class 3 or better (Sherman 1995). For thin-walled HSS, through-plates would be necessary. In addition, special attention must be paid to the column resistance, which may be adversely affected.

Of all the types of simple (shear) connections studied by Sherman (1995), a relative cost review showed that shear tab and single angle connections were the least expensive. Double angle and fillet-welded tee connections were more expensive, while through-plate and flare-bevel-welded tee connections were among the most expensive.

Some further comments can be made about the connections in Fig. 9.4:

Shear tabs

A possible failure mode for the connection is warping of the tab due to twisting of the beam end. It is therefore recommended that long unbraced beams connected by shear tabs be provided with lateral support in the vicinity of the connection. Alternatively, avoid shear tabs in such situations.

Shear tab to column corner

Heavy welding in the corner of the HSS may cause problems if the cold-formed tube material has low ductility in the corners.

Tee Connections

White and Fang (1966) originally proposed that the width to thickness ratio of the tee flange be ≥ 10 in order to provide desired flexibility. Subsequent research by Astaneh and Nader (1990) on tee connections to heavy wide flange columns concluded that a tee flange width to thickness ratio ≥ 13 provides sufficient flexibility for the connections to be considered simple. This has been subsequently verified by shear tests on tee connections to square HSS columns by Dawe and Mehendale (1995). There is little difference in capacity, whether the tee is centred or offset (to allow the beam to be on the column centreline).

9.3.2 Moment Connections

A large number of moment connection tests between wide flange beams and HSS columns have been undertaken, but the number of connection types and connection variables is immense.

Researchers have all confirmed that moment connections on unstiffened (or unthickened) HSS columns behave in a semi-rigid manner, and that their design moment resistances are limited by a yield moment, produced by plastification in the column. This yield moment can be related to a column face deformation limit, such as $0.03b_0$ or $0.03d_0$ (Lu *et al.* 1994), which is acknowledged as a reasonable ultimate load deformation limit for design purposes (International Institute of Welding Subcommission XV-E).

However, in Japan the full beam moment capacity must be resisted under seismic design conditions. Consequently, of the many HSS moment connections tested there, predominantly on circular columns (Kamba and Tabuchi 1994), nearly all are reinforced (usually with diaphragms). Recent efforts in Japan have centred on avoiding diaphragm-reinforced connections by thickening the column wall locally in the connection region (Kamba *et al.* 1994, and Tanaka *et al.* 1996). This is done by induction heating of the intended connection region of the HSS column and then applying an axial compression force to the column with a hydraulic jack until the HSS wall thickness increases (by as much as a factor of two). The Japanese research (plus extensive CIDECT research in Europe) is generating complex, regression-fitted, yield moment expressions that it is hoped will be in an appropriate design format soon.

As semi-rigid connection design is still very rare, this section will concentrate on the design of stiffened, full-strength, moment connections. A number of concepts are used to transmit moments from wide flange beams to square or rectangular HSS columns. They range from the use of continuous beams with the column interrupted, to the provision of continuity from beam to beam across the column, to the reinforcing of the column face to accept a beam moment connection.

9.3.2.1 Continuous Beams

The continuous beam approach, Figure 9.5(a), avoids the task of transmitting moments into the HSS column by running a continuous beam through an interrupted column. Column continuity is provided by reinforcing the wide flange beam with a split length of the column section. This particular detail presupposes beam flanges that are as wide as the column section, and would not be intended to transmit major moments into the column. At the top of a column, the beam is landed on a cap plate as seen in Fig. 9.5(b). If web reinforcement is needed, partial stiffeners (usually located over the column walls) can be used.

If the critical design condition for the cap plate occurs when the HSS column is in tension due to uplift, the cap plate connection can be designed as a bolted flange-plate connection with bolts along two sides of the HSS, as discussed in Section 7.2.1.

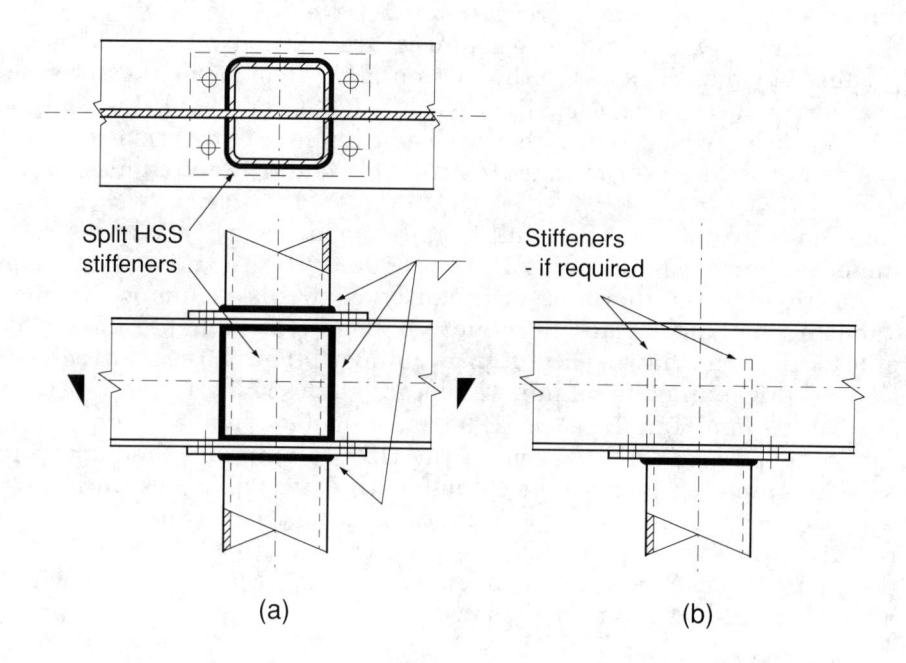

FIGURE 9.5
Continuous beam to column connections

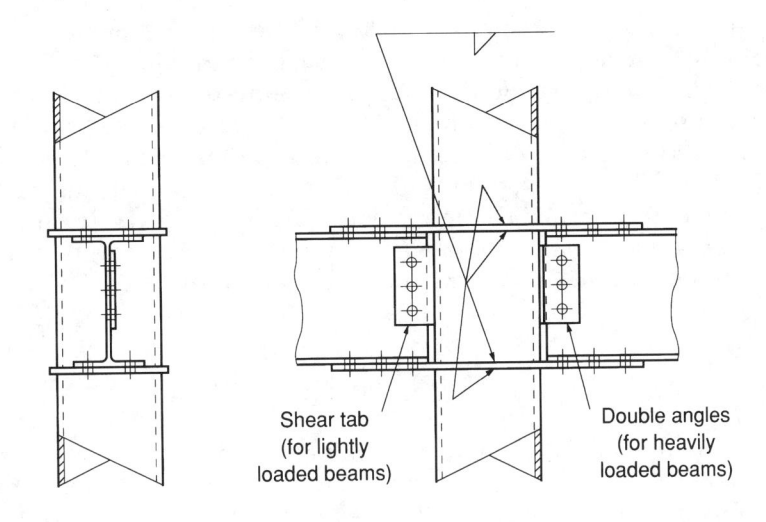

(a) At intermediate floors

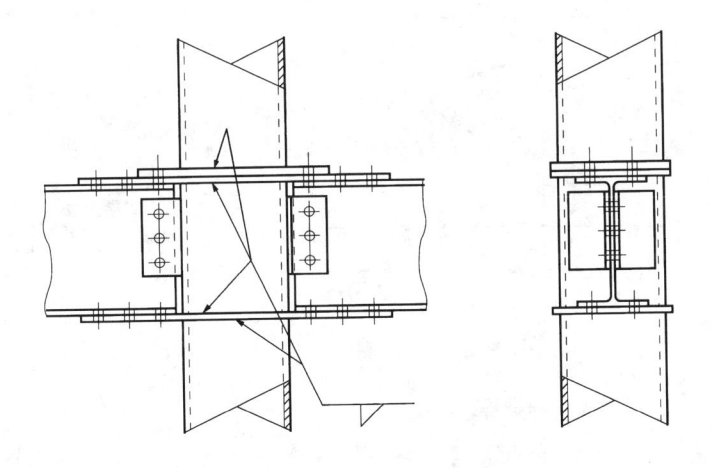

Shear tab (for lightly loaded beams)

Double angles (for heavily loaded beams)

(b) At roof or with column splice

FIGURE 9.6
Through-plate moment connections

9.3.2.2 Continuity from Beam to Beam

Through-plates

A simple arrangement which provides direct moment transfer from beam to beam across a column (or to a column) is shown in Fig. 9.6(a) where the column is interrupted to pass flange plates through it from one beam to the other. In this instance shear tabs can be used for beam webs to transfer the shear, with replacement by double angles for heavy beam reactions. The corresponding detail for a column top can be seen in Fig. 9.6(b) where another column tier can be bolted directly on top if desired. This arrangement can be modified for beams framing from three or four directions, but they all need to be within shimming range of the same depth.

Strap angles

One alternative when there are beams in only one plane and they are the same depth is to use strap angles which connect the beams to the column side walls parallel to the beam, as shown in Fig. 9.7. One leg of the straps is welded to the beam flanges and the other is welded to the column

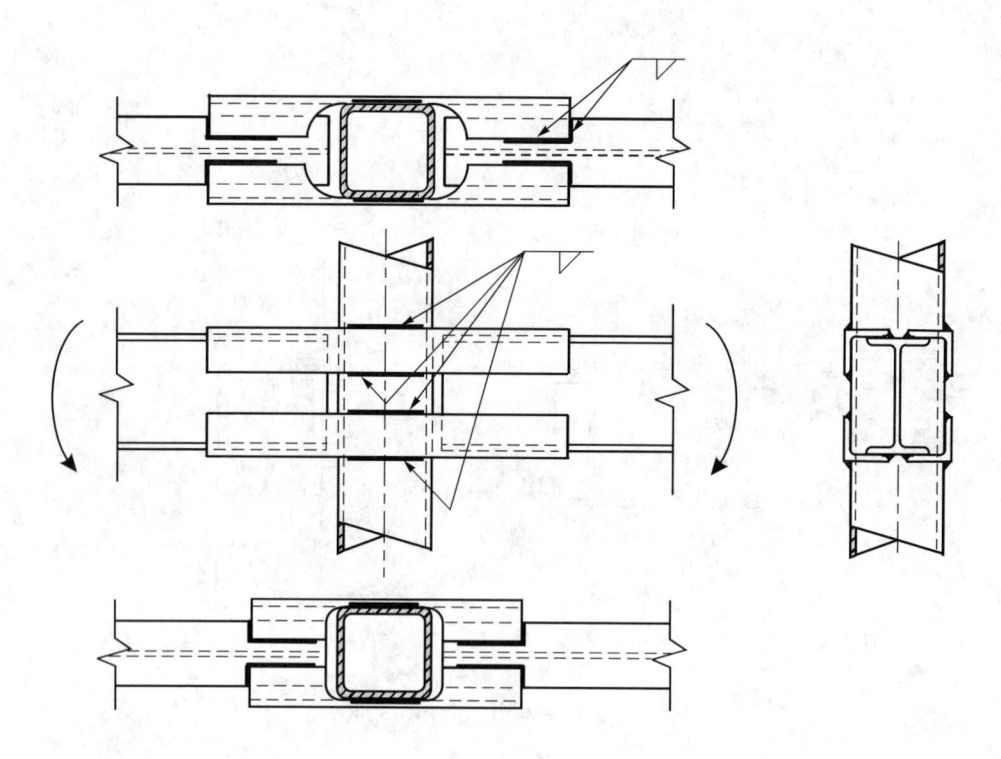

FIGURE 9.7
Strap angle moment connection
(web shear connections not shown for clarity)

side wall. Experimental programs on large-scale 203 square and 305 square columns (Picard and Giroux 1976, Giroux and Picard 1977) established satisfactory details to develop the plastic moment capacity of the tested beams.

When the beam width was the same as the column width the strap angles were able to carry both the moment and the shear loads from the beams without any beam web connections. When the beam width was 2/3 the column width shear connections for the full beam shear were necessary to avoid twisting the strap angles. One connection with a beam width equal to the column width had a beam connected to only one side of the column (unlike the other tests which were two-sided), and it performed almost identically to the two-sided connection (Picard and Giroux 1976, Giroux and Picard 1977).

Strap angles for the bottom (compression) flange can be shaped with the radius of the coped leg just a little larger than the corner radius of the HSS column. For the top (tension) flange a larger radius is necessary to avoid local fracture, the recommended radius being the leg dimension less the angle thickness. Also, a short length of the top (tension) strap angles at the cope should not be welded, as shown in Fig. 9.7. These factors result in somewhat longer strap angles for the tension flange than for the compression flange.

Flange diaphragms

Another method of connecting beam flanges to an HSS column and to each other was investigated by Kato *et al.* (1981), and is presented in Fig. 9.8. A plate diaphragm is fitted around the column for each of the beam flanges, and vertical web plates are located between them. The beams are connected by simple shear connections at the extremities of this assembly where beam contraflexure points are expected. The arrangement can be adapted to locations with two, three or four beams framing at a column. Six specimens were tested using Japanese sections from HSS 250x250 with 350 mm beams to HSS 350x350 with 470 mm beams. Moment, shear, and axial loads were combined.

The Architectural Institute of Japan (1990) has published allowable stress design recommendations for the moment resistance of reinforced welded connections between wide flange beams and interior HSS columns, where the connections are made with exterior diaphragms as shown in Table 9.1. These recommendations have been translated from the Japanese specification by Kamba and Kanatani (1993), with confirmation that the coefficients in the formulae were determined such that the allowable connection strength was almost half that obtained from test results. Thus, the limit states connection resistance can be taken as 1.5 times the P_a values given in Table 9.1. The AIJ specification also warns that all the corners of such exterior diaphragms should have a smooth finish to avoid excessive stress concentrations; this is especially good advice for dynamic loading conditions.

Exterior column
Section A-A

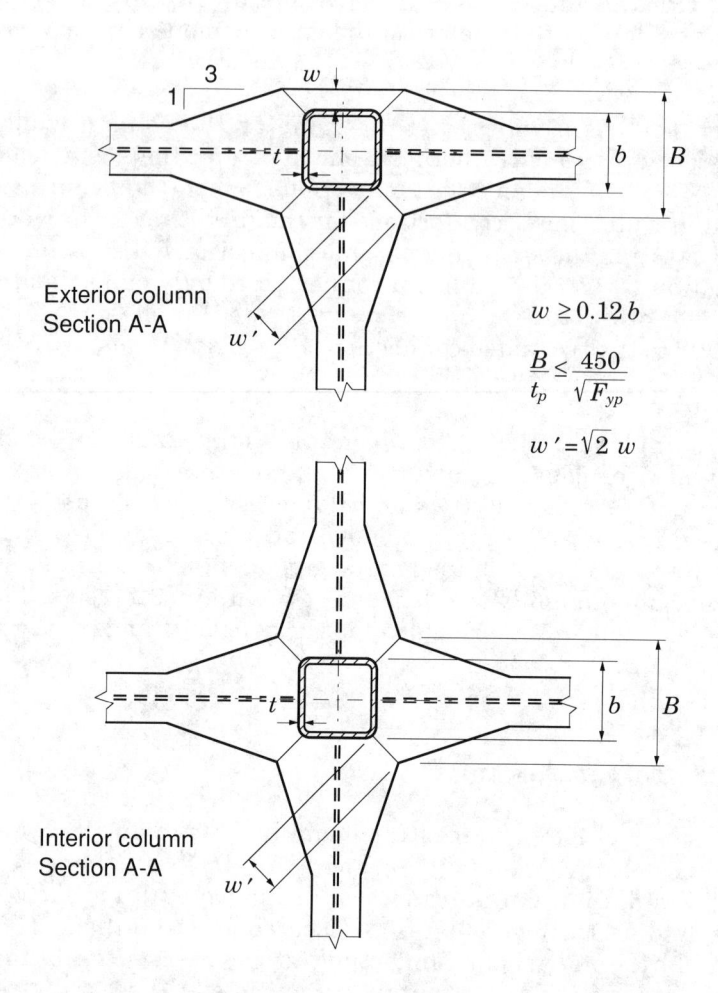

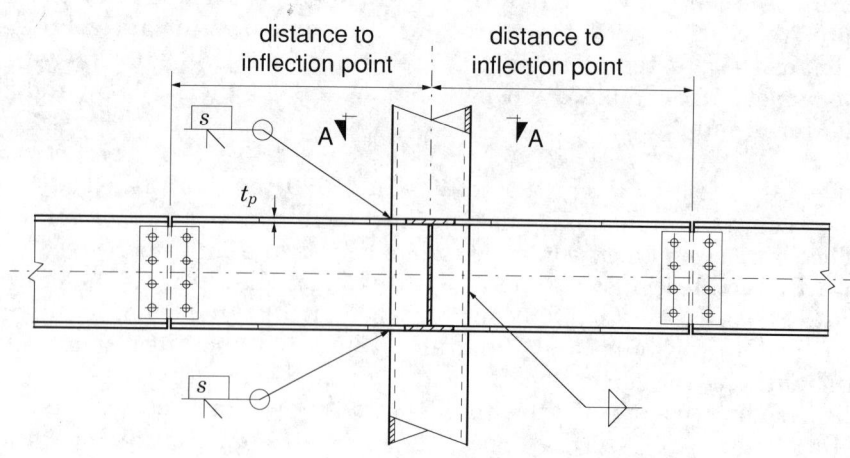

$w \geq 0.12\,b$

$\dfrac{B}{t_p} \leq \dfrac{450}{\sqrt{F_{yp}}}$

$w' = \sqrt{2}\,w$

Interior column
Section A-A

distance to inflection point

distance to inflection point

FIGURE 9.8
Flange diaphragm moment connections

CONNECTION RESISTANCE OF EXTERIOR DIAPHRAGM ON SQUARE HSS COLUMN

Shape of Exterior Diaphragm	Connection Resistance P_a
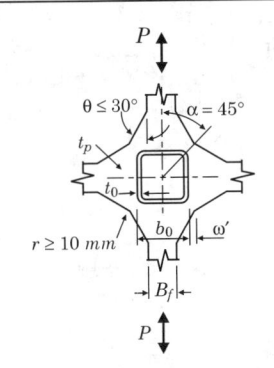 ($\theta \leq 30°$, $\alpha = 45°$, $r \geq 10$ mm)	$P_a = 1.48 \left(\dfrac{t_0}{b_0} \dfrac{t_p}{t_0 + \omega'} \right)^{2/3} \left(\dfrac{t_0 + \omega'}{b_0} \right) b_0^2 F_{up}$

RANGE OF VALIDITY

$17 \leq \dfrac{b_0}{t_0} \leq 67$ $\dfrac{\omega'}{b_0} \leq 0.4$ $0.75 \leq \dfrac{t_p}{t_0} \leq 2.0$ $\dfrac{(b_0/2) + \omega'}{t_p} \leq \dfrac{237}{\sqrt{F_{yp}}}$ $\theta \leq 30°$

CONNECTION RESISTANCE OF EXTERIOR DIAPHRAGM ON CIRCULAR HSS COLUMN

Shape of Exterior Diaphragm	Connection Resistance P_a
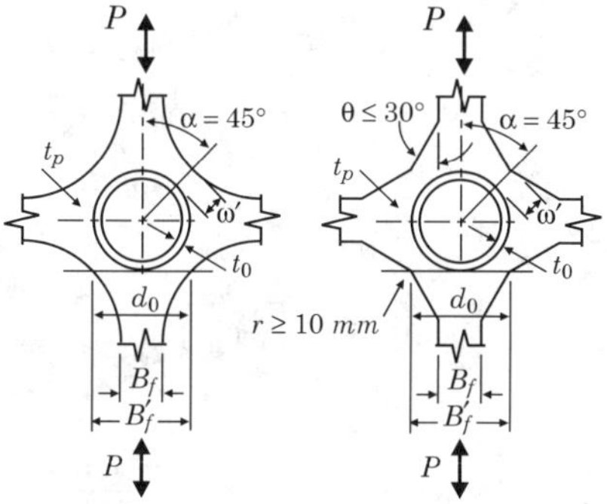 ($\alpha = 45°$, $\theta \leq 30°$, $r \geq 10$ mm)	$P_a = \left(3.28 \dfrac{B'_f}{d_0} + 1.43 \right) t_0 \sqrt{t_p (t_0 + \omega')} \, F_{y0}$ For $\sqrt{2} \left(\dfrac{d_0}{2} + \omega' \right) \geq d_0$, $B'_f = d_0$ For $\sqrt{2} \left(\dfrac{d_0}{2} + \omega' \right) < d_0$, B'_f to be determined from connection details as a tangent to the HSS

RANGE OF VALIDITY

$15 \leq \dfrac{d_0}{t_0} \leq 55$ $\dfrac{B'_f}{2 t_p} \leq \dfrac{237}{\sqrt{F_{yp}}}$

F_{y0} and F_{yp} in MPa B_f = flange width t_p = thickness of exterior diaphragm
ω' = edge distance of exterior diaphragm
P = connection resistance = beam moment over beam depth
P_a = connection resistance in the "long term" (allowable load)

TABLE 9.1

Allowable loads for welded beam to column moment connections reinforced with exterior diaphragms (AIJ 1990)
(Multiply P_a by 1.5 to get limit states connection resistances)

9.3.2.3 Column Face Reinforcement

Reinforcement by Doubler Plates

The most direct approach for moment-connecting a beam to an HSS column is to reinforce the column face to accept the flange forces from the beam. Dawe and Grondin (1990) have carried out an experimental and analytical programme on 10 specimens incorporating the two types of connections illustrated in Fig. 9.9. The results can be used to demonstrate four basic failure modes, expressed as

a) "effective width" rupture of the beam tension flange (or of the beam tension flange plate if one is used) where it is welded to the column doubler plate

b) punching shear failure of the doubler plate at the beam tension flange (or beam tension flange plate)

c) web crippling of the column side walls near the beam compression flange

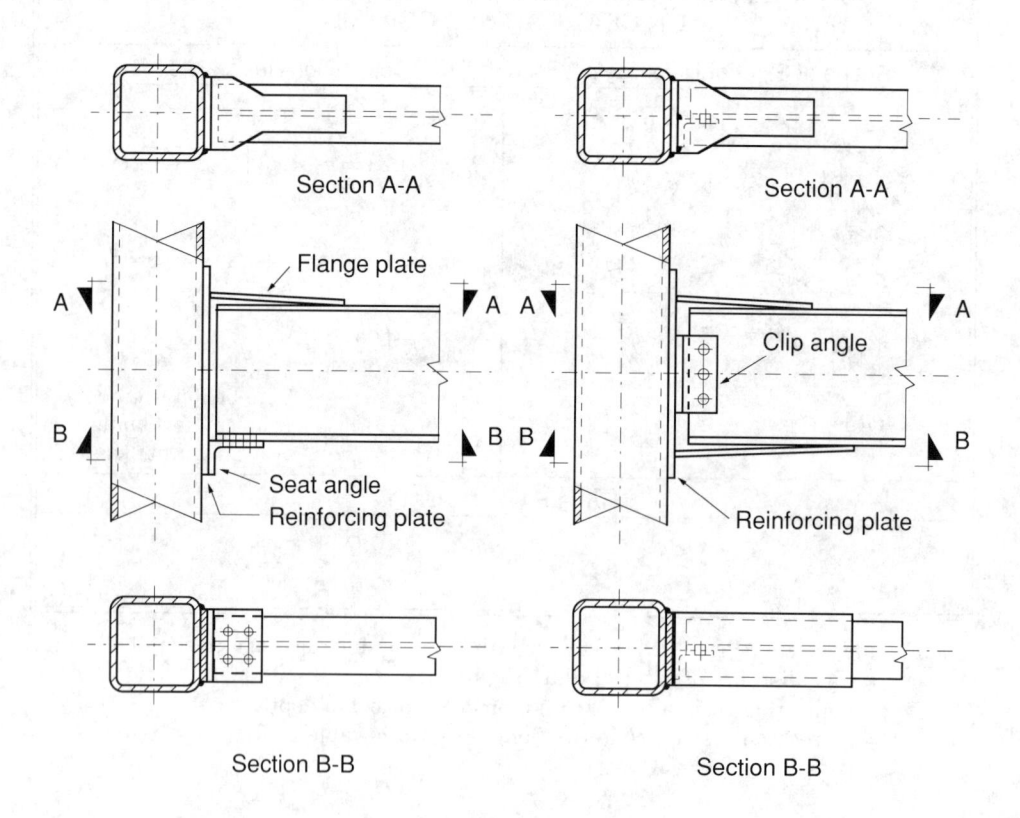

FIGURE 9.9
Column face doubler plate reinforcement for moment connections

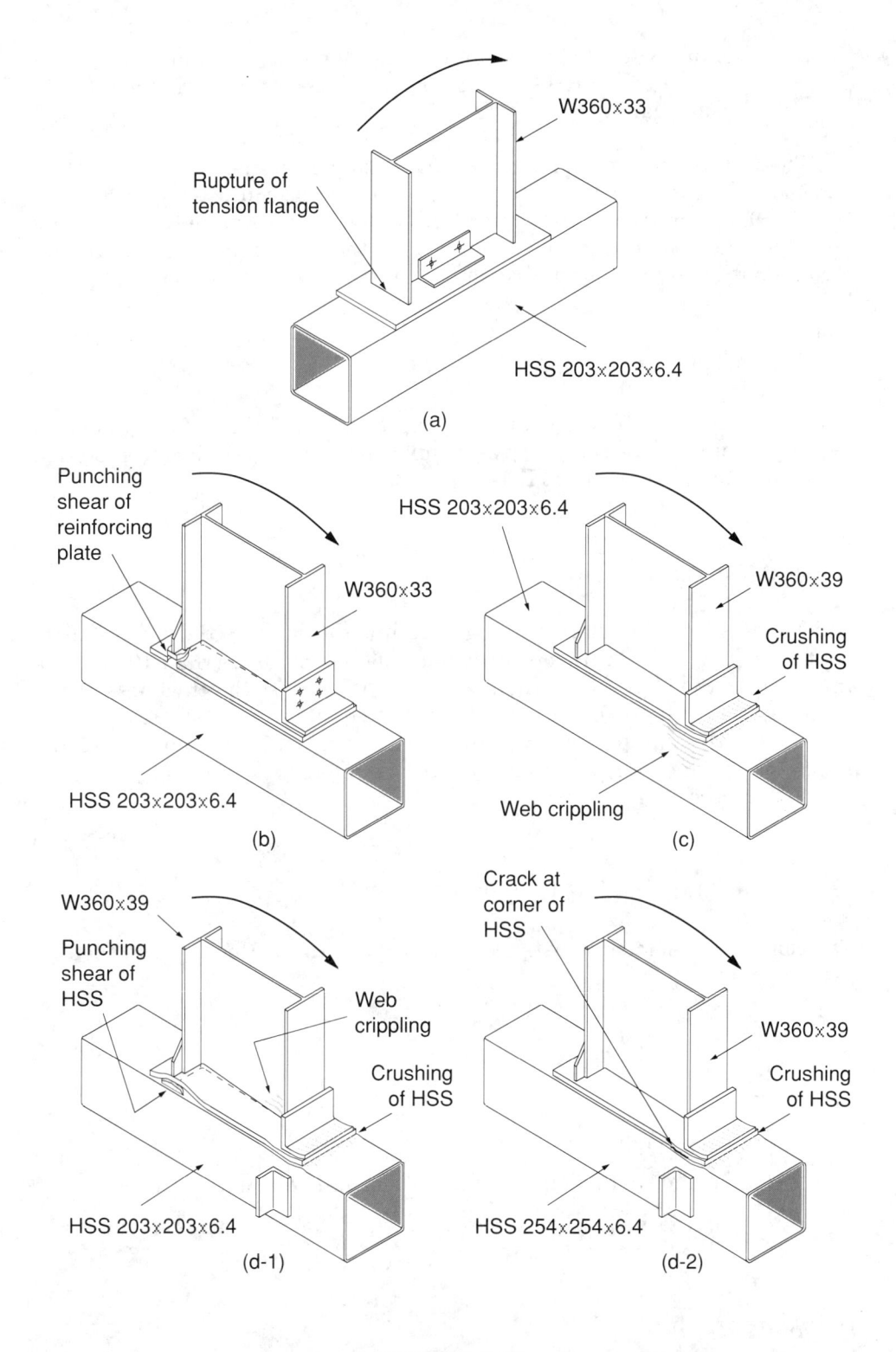

FIGURE 9.10
Moment connection failure modes for wide flange beams to plate-reinforced HSS columns (Dawe and Grondin 1990)

d) punching shear of the column face along the edge of the doubler plate, either near the beam tension flange or near the beam compression flange.

The failure modes are shown in Fig. 9.10, where the specimen section sizes are also identified. (These modes are not to be confused with those in Section 9.2.2 for HSS to HSS moment connections.) Applied loading consisted of a force on the beam 1160 mm from the column in order to produce a combination of moment and shear.

Failure Mode (a)

Although the effective width (b_e) expression in Table 3.2 is applied to overlap K connections there, it was developed from plate-to-HSS connection tests and is applicable here. It is

$$b_e = \left(\frac{10}{b_0/t_0}\right)\left(\frac{F_{y0} t_0}{F_{yi} t_i}\right) b_i \quad \text{but} \leq b_i$$

which states that the effective flange width is the same as the actual flange width b_i when the b/t ratio of the doubler plate is not more than 10, and the thickness of the "landed on" (doubler) plate is at least the thickness of the "landing" (beam flange) plate, provided that equal strength steels are used. Restated with appropriate subscripts for the present situation (where the subscript b refers to the beam and subscript p refers to the doubler plate), the expression becomes

$$b_e = \left(\frac{10}{b_p/t_p}\right)\left(\frac{F_{yp} t_p}{F_{yb} t_b}\right) b_b \quad \text{but} \leq b_b \qquad [9.9]$$

The connection moment resistance governed by failure Mode (a) can then be estimated by

$$M_{r1}^* = h_b F_{yb} t_b b_e \qquad [9.10]$$

where

b_e is given by [9.9]

t_b is the beam flange (or beam tension flange plate) thickness

h_b is the height of the beam between flange centres (or distance between flange plate centres).

Failure Mode (b)

The second guide to predicting behaviour is the effective punching shear width equation in Table 3.3. With adjusted subscripts, it is

$$b_{ep} = \left(\frac{10}{b_p/t_p}\right) b_b \qquad \text{but} \leq b_b \qquad [9.11]$$

which states that the effective width for punching shear b_{ep} is the same as the width of the beam flange b_b when the b/t ratio of the doubler plate is not more than 10.

The connection moment resistance governed by failure Mode (b) can then be estimated by

$$M_{r1}^* = 2h_b \frac{F_{yp}}{\sqrt{3}} t_p b_{ep} \qquad [9.12]$$

where

b_{ep} is given by [9.11]

t_p is the doubler plate thickness.

Failure Mode (c)

Web crippling of the column side walls is similar to loading from a transverse plate on an HSS member as shown in Table 11.1. The expression there for chord side wall failure is

$$N_1^* = 2 F_{y0} t_0 (t_1 + 5t_0)$$

When modified for the presence of a doubler plate on the HSS, this would become

$$N_1^* = 2 F_{y0} t_0 (t_b + 5 (t_0 + t_p)) \qquad [9.13]$$

Dawe and Guravich (1993) tested 13 plate-reinforced HSS specimens with compression-loaded transverse plates to determine the strength of this configuration. Their test results have indicated that [9.13] provides little, if any, margin of safety, thus suggesting that the doubler plate spreads the compression load out much further than assumed for an unreinforced connection (see Section 11.1). Hence, it would seeem prudent to consider the connection as behaving rather like a T connection (for a one-sided moment connection) or an X connection (for a two-sided moment connection), with the resistances being given by the chord side wall failure expressions in Table 3.3. Then, for a one-sided connection in compression

$$N_1^* = 2 F_k t_0 (t_b + 5 (t_0 + t_p)) \qquad [9.14]$$

where $F_k = C_r/\phi A$ as determined in Table 3.3. For a two-sided connection, $F_k = 0.8 C_r/\phi A$ should be used. Equation [9.14] is very conservative relative to the Dawe and Guravich (1993) test results.

The connection moment resistance governed by failure Mode (c) can then be estimated by

$$M_{r1}^* = N_1^* h_b \quad [9.15]$$

The column side walls are stabilized against buckling when there are beams framing into them with connecting material mounted on the vulnerable area of the walls. For that reason an angle was attached to the side walls of some specimens as shown in Fig. 9.10 to induce Mode (d) failures. Dawe and Grondin (1990) also conducted limited analytical studies to establish the thickness of reinforcing plates necessary on column side walls to provide stability there.

Failure Mode (d)

Punching shear of the column face at the edges of the doubler plate, either outwards at the beam tension flange or inwards at the compression flange would only be expected if the width of the doubler plate B_p is less than $b_0 - 4t_0$. One could speculate, by employing principles of virtual work, that the connection moment resistance for this failure mode (assuming uniform punching shear stress all around the doubler plate) might be

$$M_{r1}^* = 0.5 \frac{F_{y0}}{\sqrt{3}} t_0 (L_p^2 + 2L_p B_p)$$

However, since the punching shear stress can be far from uniform around the doubler plate, this expression overstates the moment resistance. An examination of results from Dawe and Grondin (1990) suggests actual moment resistance may be only about two thirds of this value for the specimens tested.

Until a practical solution is validated for failure Mode (d), it is suggested that

$$M_{r1}^* = 0.25 \frac{F_{y0}}{\sqrt{3}} t_0 (L_p^2 + 2L_p B_p) \quad [9.16]$$

can be used, but with the application of careful judgement.

Other failure modes

Dawe and Guravich (1993) observed two failures by fracture of the weld joining the doubler plate to the HSS, near the beam tension flange. They occurred when $b_b/B_p > 0.85$. The effective length of the weld that carries the load is a matter of some judgement, but Dawe and Guravich (1995) indicated in a discussion note that it appeared, in effect, to be about 10 times the thickness of the reinforcing plate for the two welds that failed.

Another potential failure mode, although not observed by Dawe and Grondin (1990) nor Dawe and Guravich (1993), would be a yield line failure mechanism in the doubler plate.

Based on limited test data, no advantage was gained when the length of a 13 mm thick doubler plate was increased beyond 80 mm from the outer face of a beam flange plate.

Generally, it should be possible to develop the moment capacity of beams by considering the above failure modes. As the connection resistance and stiffness are more a function of the doubler plate than of the column wall slenderness, that plate should be approximately the same width as the column.

Reinforcement by Angles

Hollow structural section column faces have been reinforced in Japan by wrapping the column with angles as shown in Fig. 9.11. In this case, four 200x200x25 angles (trimmed to fit) were welded on to HSS 300x300x16 columns and to each other at the toes. The assembly was then drilled for field bolting with Huck high strength blind bolts (see Section 7.4 of this Design Guide). Suitable moment connections have been configured from either tee stubs bolted to the beam flanges or end plates welded to the beams. The resistance of the connection in the tension region can be calculated by appropriate tee-stub prying models (Tabuchi *et al.* 1994).

Reinforcement by Local Thickening of the HSS Wall

As mentioned in Section 9.3.2, the Japanese have developed a means to thicken the walls of HSS columns in the connection region by the use of

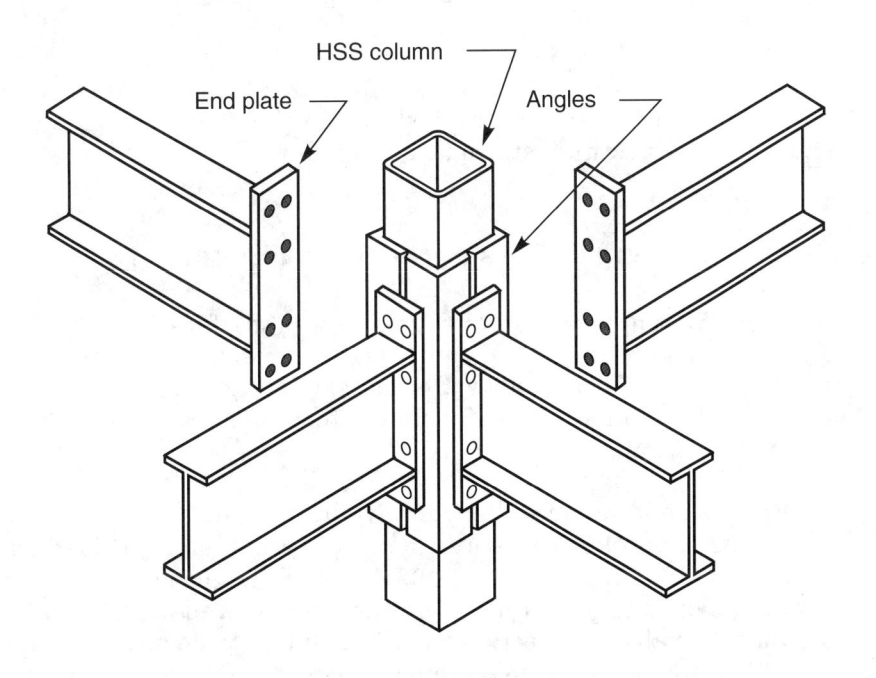

FIGURE 9.11

Column reinforcement with angles for moment connections

FIGURE 9.12
HSS columns with locally-thickened walls in Japan
(Photograph courtesy of Steel Structures Laboratory of Kobe University and Daiwa House Industry Co., Ltd., Japan)

heat and axial compression. A building frame that uses such columns and beams with welded end plates is shown in Fig. 9.12. The moment connections employ Huck high strength blind bolts.

9.3.2.4 Stiffening by Tees on the Beam Flanges

An interesting alternative to reinforcing the face of an HSS column is to broaden the beam flanges with tee stiffeners to deliver flange forces directly into the column side walls, as seen in Fig. 9.13. (Tees are generally cut from wide flange shapes or fabricated from plates, so their use may involve a cost consideration.)

Both tee and angle stiffeners have been studied, with tees proving to perform better under both static and cyclic beam moment loading (Ting *et al.* 1991, and Shanmugam *et al.* 1991). Beam plastic moments can typically be developed with this type of reinforcement and web crippling of the column side walls (see Section 9.3.2.3) is likely to be an influential failure mode. Requirements for the design of the tee stiffeners have been produced (Ting *et al.* 1993) that, after review for this Design Guide, could be best summarized as follows:

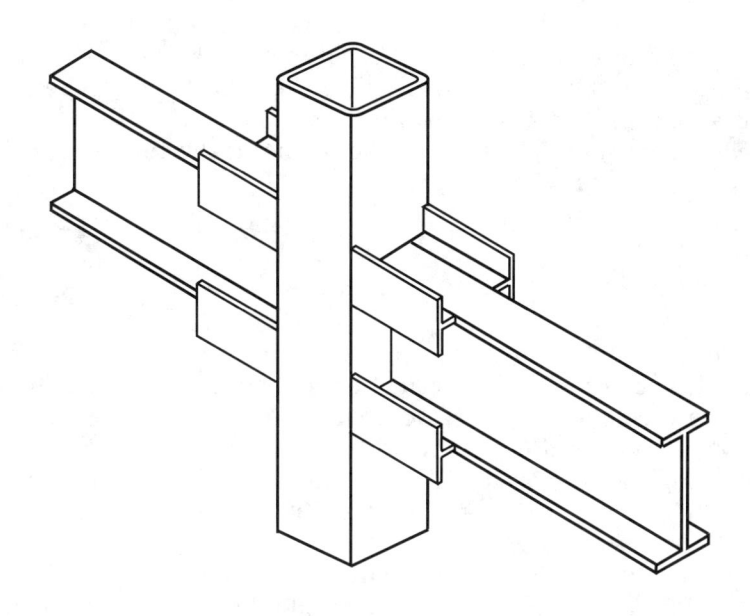

FIGURE 9.13

Stiffening by tees on the beam flanges

a) The connection should be designed to develop the plastic moment resistance of the beam.

b) The tee web thickness should be at least half the beam flange thickness.

c) A load dispersion angle of no more than 20°, from the beam flange/tee stiffener junction to the corner of the HSS, should be assumed. This defines a minimum tee stiffener length, L_{min}, as:

$$L_{min} \geq \frac{b_0 - b_b}{2 \tan 20°} = 1.37 (b_0 - b_b) \qquad [9.17]$$

where

b_0 = width of the column flange

b_b = width of the beam flange

In practice, the stiffener length should exceed this minimum because the flange force must disperse laterally to the full column width, then disperse longitudinally to the full width of the tee flange.

d) Research (Ting et al. 1993) has shown that at the column face junction the "effective width" of the beam flange plus tee webs is very low, so it would be reasonable to assume that full dispersion of the beam flange force into the tee flange has taken place prior to the intersection with the column. Assuming the same steel grades for the beam and tees, the tee flange area should then be greater than half the beam flange area, so that the beam moment capacity can be resisted by the forces in the tee flanges. This will control the selection of the tee section.

9.3.2.5 Concrete Filling of the Column

In a concrete-filled column, the bearing strength of the confined concrete is well above the normal crushing strength, and there should be no need for reinforcement in this region. The beam connection could be directly welded to the column. See Section 3.9.2.1 on concrete-filled compression-loaded rectangular (or square) HSS X connections for a method to quantify this concrete bearing strength (Packer 1995).

In the tension region, however, a directly-welded beam will still tend to pull the column face outwards, causing a plastic mechanism to form in the chord wall, as with an unfilled column. Under tension loading, the connection stiffness has been found to increase for rectangular and square HSS (Packer 1995) and for circular HSS (Yamamoto *et al.* 1994) concrete-filled members, relative to their unfilled counterparts. In most cases the yield load of the connection in the tension region also increased, due to the preservation of the hollow section shape, but this was not always the case (Packer 1995).

Thus, it is prudent to base the connection design in the tension region on unfilled HSS design rules, but with a further constraint. Transfer of tensile forces to the HSS face can result in separation of the tube from the concrete core, which will decrease the confining effect of the steel and hence possibly compromise the performance of the column, if designed as a composite member. It has thus been suggested that a tension region detail that transmits the forces to the HSS side walls would be preferable.

9.4 Wide Flange Beams to Circular HSS Columns

Simple shear connections for wide flange beams can be provided on circular HSS columns as shown in Fig. 9.14. Detail (a) is a simple shear tab for relatively lightly loaded beams. More heavily loaded beams on thin walled columns can be connected with a detail like (b), which consists of a channel welded to the column with a suitable shear connection attached to the channel.

Some moment connections to circular columns can be configured like those used for rectangular columns in the preceding section. Obvious exceptions are the direct use of strap angles to provide continuity from one

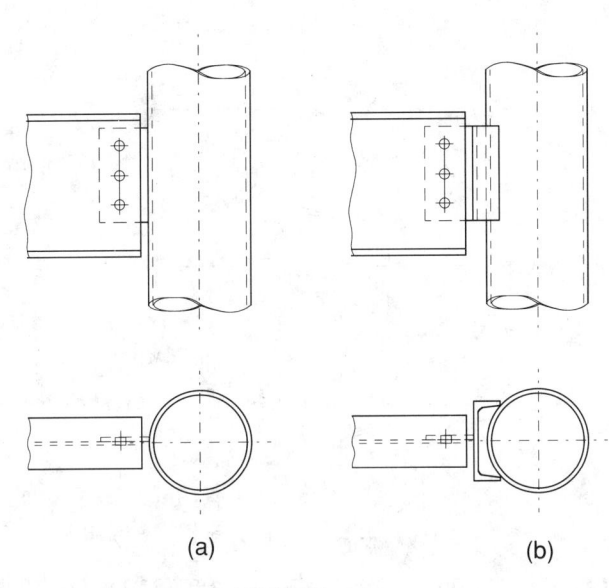

FIGURE 9.14

Shear connections for wide flange beams to circular HSS columns

beam to another and the concept of doubler plates to reinforce the column face.

An alternative moment connection is to contour the beam flanges so the beam can be fitted and welded directly to the column without any intermediate pieces as seen in Fig. 9.15. The connection moment resistance is given by Wardenier *et al.* (1991):

$$M_{r1}^* = h_b F_{y0} t_0^2 \left(\frac{5.0}{1 - 0.81\beta}\right) f_3(n') \qquad [9.18]$$

where

h_b = height of the beam between flange centres
F_{y0} = yield stress of the column
t_0 = wall thickness of the column
β = b_b/d_0
b_b = flange width of the beam
d_0 = diameter of the column

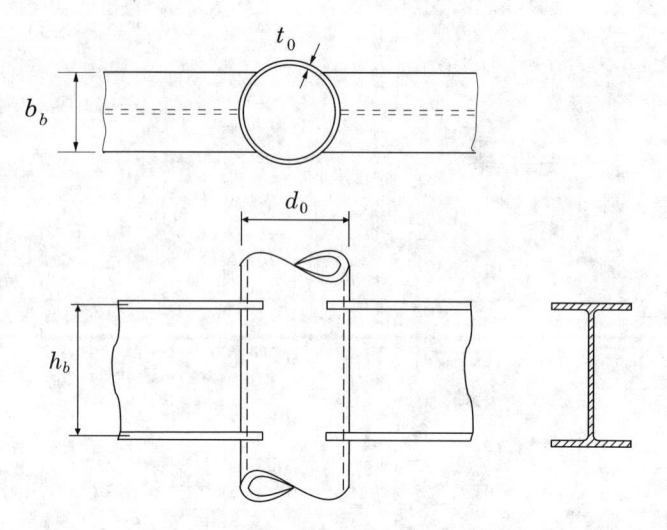

FIGURE 9.15

Moment connection for profiled wide flange beams to circular HSS columns

$f_3(n')$ = 1.0   (for column tension, n' positive)
         = $1 + 0.3n' - 0.3n'^2$ (for column compression, n' negative)
n' = $N_{0p}/(A_0 F_{y0})$
N_{0p} = axial load in the column above the connection.

This value of the connection resistance may be applied where beams frame into a column from both sides or from just one side. Both situations must be checked for punching shear (Wardenier *et al.* 1991):

$$\left(\frac{N_1}{A_1} + \frac{M_{f1}}{S_1}\right) t_1 \leq 1.16 F_{y0} t_0 \qquad [9.19]$$

where

N_1 = axial force in the beam
A_1 = cross sectional area of the beam
M_{f1} = bending moment in the beam at the column
S_1 = elastic section modulus of the beam.

A recent, more rigorous study of this connection type, which will typically be semi-rigid, has refined the function for the influence of column compression loading $f_3(n')$ (de Winkel and Wardenier 1996).

Within the validity range of

$$0.2 \leq \beta \leq 0.7$$
$$15 \leq 2\gamma \leq 45$$
$$0.2 \leq \eta \leq 1.3$$
$$0 \leq |n'| \leq 0.8$$

where

$$\beta = b_b/d_0 \qquad \gamma = d_0/2t_0 \qquad \eta = d_b/d_0$$

$$f_3(n') = 1 - n'^2 (\beta - \beta^2)(2\gamma)^{0.303}$$

The largest influence of the column compression load was found to occur for a β ratio equal to 0.50.

REFERENCES

AIJ. 1990. Recommendations for the design and fabrication of tubular structures in steel, 3rd. ed. Architectural Institute of Japan, Tokyo, Japan.

ASTANEH, A., and NADER, M.N. 1990. Experimental studies and design of steel tee shear connections. Journal of Structural Engineering, American Society of Civil Engineers, **116**(10): 2882–2902.

CSA. 1994. Limit states design of steel structures, CAN/CSA-S16.1-M94. Canadian Standards Association, Rexdale, Ontario.

DAVIES, G., and PACKER, J.A. 1987. Analysis of web crippling in a rectangular hollow section. Proceedings of the Institution of Civil Engineers, Part 2, **83**: 785–798.

DAWE, J.L., and GRONDIN, G.Y. 1990. W-shape beam to RHS column connections. Canadian Journal of Civil Engineering, **17**(5): 788–797.

DAWE, J.L., and GURAVICH, S.J. 1993. Branch plate to reinforced HSS connections in tension and compression. Canadian Journal of Civil Engineering, **20**(4): 631–641.

DAWE, J.L., and GURAVICH, S.J. 1995. Branch plate to reinforced HSS connections in tension and compression: Reply to discussion. Canadian Journal of Civil Engineering, **22**(1): 203.

DAWE, J.L., and MEHENDALE, S.V. 1995. Shear connections using stub tees between W beams and HSS columns. Canadian Journal of Civil Engineering, **22**(4): 683–691.

GIROUX, Y.M., and PICARD, A. 1977. Rigid framing connections for tubular columns. Canadian Journal of Civil Engineering, **4**(2): 134–144.

KAMBA, T., and TABUCHI, M. 1994. Database for tubular column to beam connections in moment-resisting frames. International Institute of Welding, Document No.IIW-XV-E-94-208, Kobe University, Kobe, Japan.

KAMBA, T., and KANATANI, H. 1993. Design formulae for CHS column-to-beam connections with exterior diaphragms. Proceedings, 5th. International Symposium on Tubular Structures, Nottingham, England, pp. 249–256.

KAMBA, T., KANATANI, H., and WAKIDA, T. 1994. CHS column-to-beam connections without diaphragms. Proceedings, 6th. International Symposium on Tubular Structures, Melbourne, Australia, pp. 325–333.

KATO, B., MAEDA, Y., and SAKAE, K. 1981. Behaviour of rigid frame sub-assemblages subject to horizontal force. Joints in Structural Steelwork, John Wiley and Sons, New York, N.Y., U.S.A., pp. 1.54–1.73.

KOROL, R.M., EL-ZANATY, M., and BRADY, F.J. 1977. Unequal width connections of square hollow sections in vierendeel trusses. Canadian Journal of Civil Engineering, **4**(2): 190–201.

KOROL, R.M., MITRI, H., and MIRZA, F.A. 1982. Plate reinforced square hollow section T-joints of unequal width. Canadian Journal of Civil Engineering, **9**(2): 143–148.

LU, L.H., de WINKEL, G.D., YU, Y., and WARDENIER, J. 1994. Deformation limit for the ultimate strength of hollow section joints. Proceedings, 6th. International Symposium on Tubular Structures, Melbourne, Australia, pp. 341–347.

PACKER, J.A. 1995. Concrete-filled HSS connections. Journal of Structural Engineering, American Society of Civil Engineers, 121(3): 458–467.

PICARD, A., and GIROUX, Y.M. 1976. Moment connections between wide flange beams and square tubular columns. Canadian Journal of Civil Engineering, 3(2): 174–185.

SHANMUGAM, N.E., TING, L.C., and LEE, S.L. 1991. Behaviour of I-beam to box-column connections stiffened externally and subjected to fluctuating loads. Journal of Constructional Steel Research, 20: 129–148.

SHERMAN, D.R. 1995. Simple framing connections to HSS columns. Proceedings, American Institute of Steel Construction, National Steel Construction Conference, San Antonio, Texas, U.S.A., pp. 30.1–30.16.

SHERMAN, D.R., and ALES, J.M. 1991. The design of shear tabs with tubular columns. Proceedings, American Institute of Steel Construction, National Steel Construction Conference, Washington, D.C., U.S.A., pp. 1.2–1.22.

TABUCHI, M., KANATANI, H., TANAKA, T., FUKUDA, A., FURUMI, K., USAMI, K., and MURAYAMA, M. 1994. Behaviour of SHS column to H beam moment connections with one side bolts. Proceedings, 6th. International Symposium on Tubular Structures, Melbourne, Australia, pp. 389–396.

TANAKA, T., TABUCHI, M., FURUMI, K., MORITA, T., USAMI, K., MURAYAMA, M., and MATSUBARA, Y. 1996. Experimental study on end plate to SHS column connections reinforced by increasing wall thickness with one side bolts. Proceedings, 7th. International Symposium on Tubular Structures, Miskolc, Hungary, pp. 253–260.

TING, L.C., SHANMUGAM, N.E., and LEE, S.L. 1991. Box-column to I-beam connections with external stiffeners. Journal of Constructional Steel Research, 18: 209–226.

TING, L.C., SHANMUGAM, N.E., and LEE, S.L. 1993. Design of I-beam to box-column connections stiffened externally. Engineering Journal, American Institute of Steel Construction, 4th. quarter: 141–149.

WARDENIER, J., KUROBANE, Y., PACKER, J.A., DUTTA, D., and YEOMANS, N. 1991. Design guide for circular hollow section (CHS) joints under predominantly static loading. CIDECT (Ed.) and Verlag TÜV Rheinland GmbH, Köln, Federal Republic of Germany.

WHITE, R.N., and FANG, P.J. 1966. Framing connections for square structural tubing. Journal of the Structural Division, American Society of Civil Engineers, 92(ST2): 175–194.

de WINKEL, G.D., and WARDENIER, J. 1996. Parametric study on the static behaviour of uniplanar I-beam-to-tubular column connections loaded with in-plane bending moments combined with pre-stressed columns. Proceedings, 7th. International Symposium on Tubular Structures, Miskolc, Hungary, pp. 229–235.

YAMAMOTO, N., INAOKA, S., and MORITA, K. 1994. Strength of unstiffened connection between beams and concrete-filled tubular column. Proceedings, 6th. International Symposium on Tubular Structures, Melbourne, Australia, pp. 365–372.

10

TRUSSES AND BASE PLATES TO HSS COLUMNS

10.1 Trusses to HSS Columns

Aesthetically pleasing and economical truss to column connections are readily possible between HSS trusses and HSS columns; many variations in detailing are feasible. Round or rectangular truss members are easily connected to either round or rectangular columns. As always, ease of fabrication and erection are important factors when designing the connections.

Usually such connections are configured for bolting, and various arrangements are illustrated in Fig. 10.1 for standard HSS trusses and in Fig. 10.2 for double chord HSS trusses. Normal design considerations and bolt values are applicable.

Worked examples are shown in Sections 13.2.9 and 13.3.9.

10.2 Base Plates

When steel columns bear on concrete footings, steel base plates are required to distribute the column load to the footing without exceeding the bearing resistance of the concrete. In general, the ends of columns are saw-cut or milled to a plane surface so as to bear evenly on the base plate (Fig. 10.3). Connection of the column to the base plate (and then to the footing) depends on the combinations of compression, shear, bending moment and tension loads that are expected. Horizontal shear forces can be resisted by the anchor bolts (nominal forces only), by friction between the base plate and the concrete footing, by a shear key (block welded to the underside of the base plate), or by recessing the base plate into the concrete

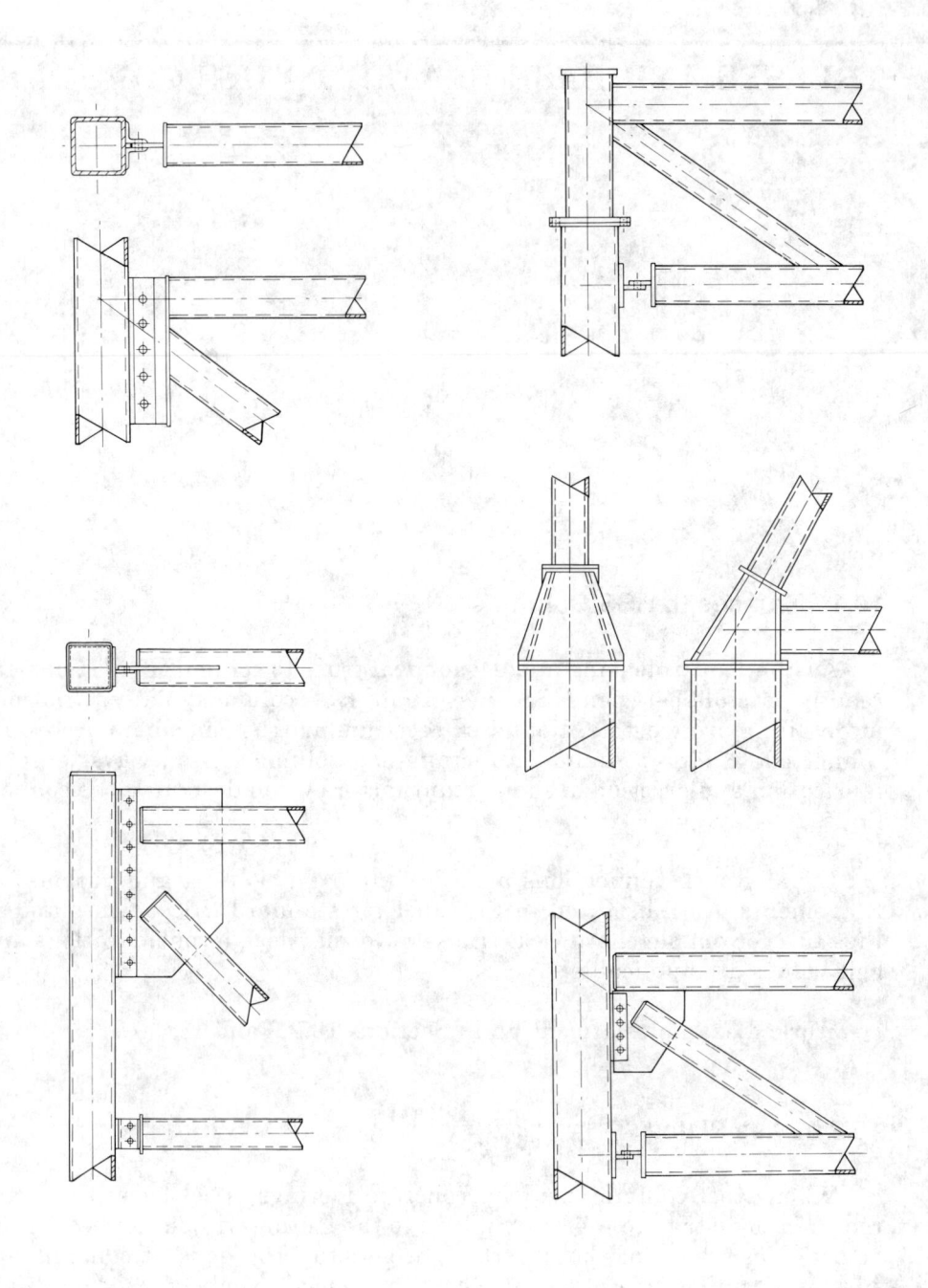

FIGURE 10.1
Typical HSS truss to HSS column connections

336

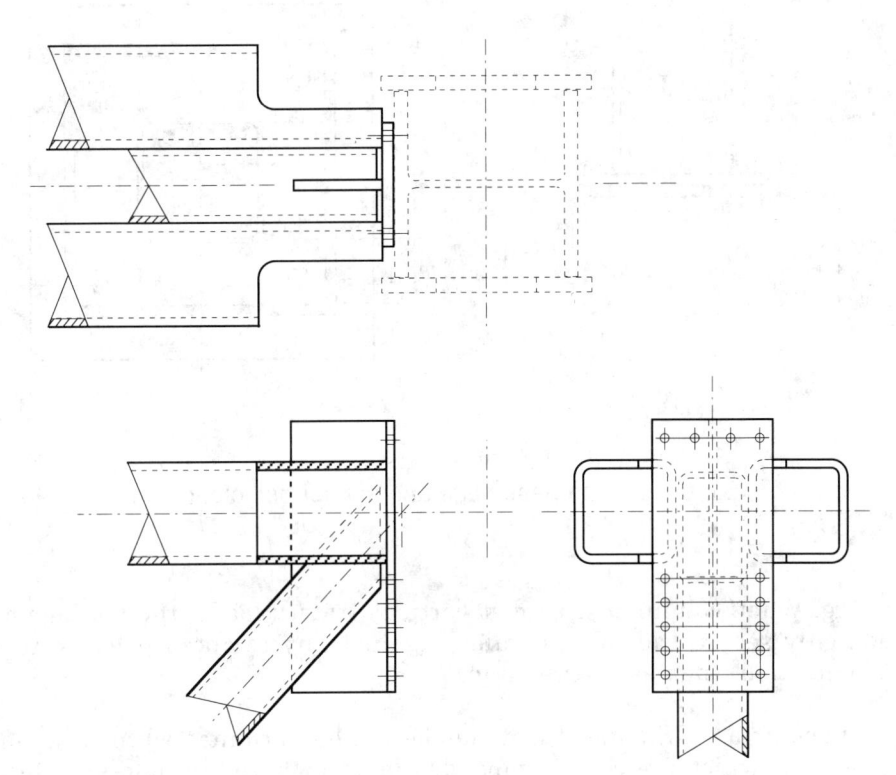

(a) Truss with double HSS chord separated by HSS web

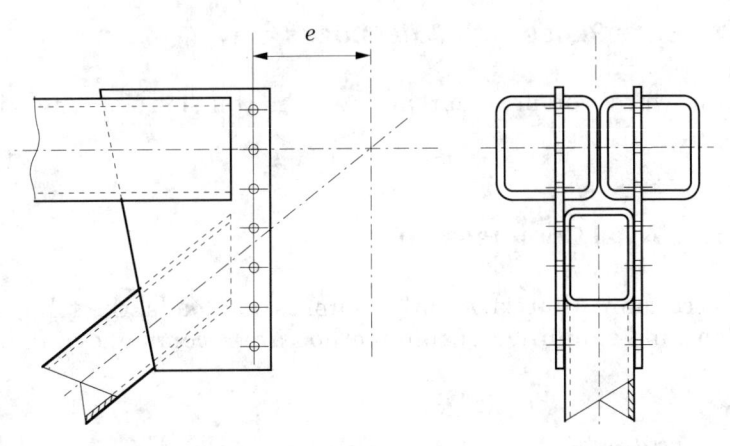

(b) Truss with back-to-back double HSS chord & HSS web

FIGURE 10.2
HSS double chord truss to HSS column connections

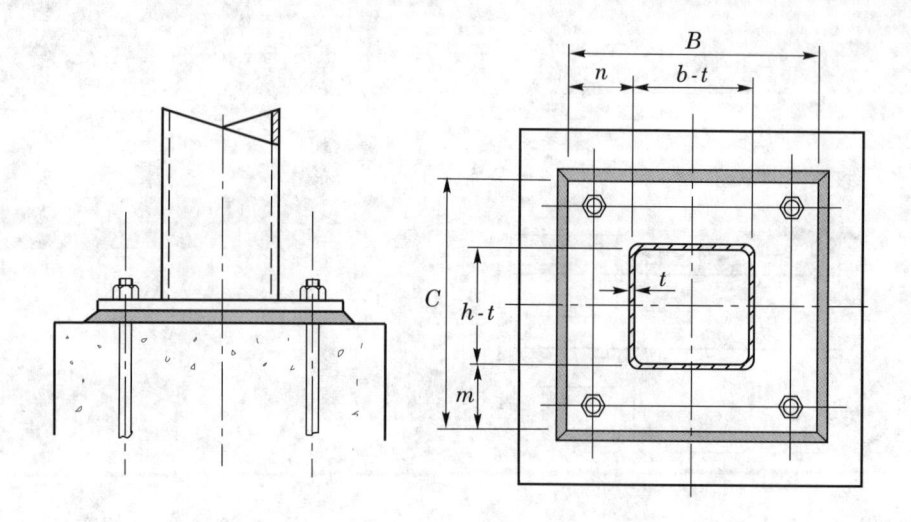

FIGURE 10.3

Simple column base plate and anchor bolt clearance

footing. When columns carry just vertical gravity loads, the arrangement need only secure the column safely against any temporary loads during steel erection, such as overturning.

On occasion, HSS structural members are pin connected to base plates or fixtures, which are either embedded in or anchored to concrete. Details of some suitable arrangements are presented in Fig. 10.4.

10.2.1 Base Plates with Axial Loads only

Downward (compression) and upward (tension) loading will be considered.

10.2.1.1 Axial Compression

For columns that carry only compression loads, the following assumptions and simplified design method are recommended (CISC 1995).

Assumptions

1. The factored gravity load is assumed to be uniformly distributed onto the base plate within a rectangle of $(h-t) \times (b-t)$ (see Fig. 10.3).
2. The base plate exerts a uniform pressure over the footing.

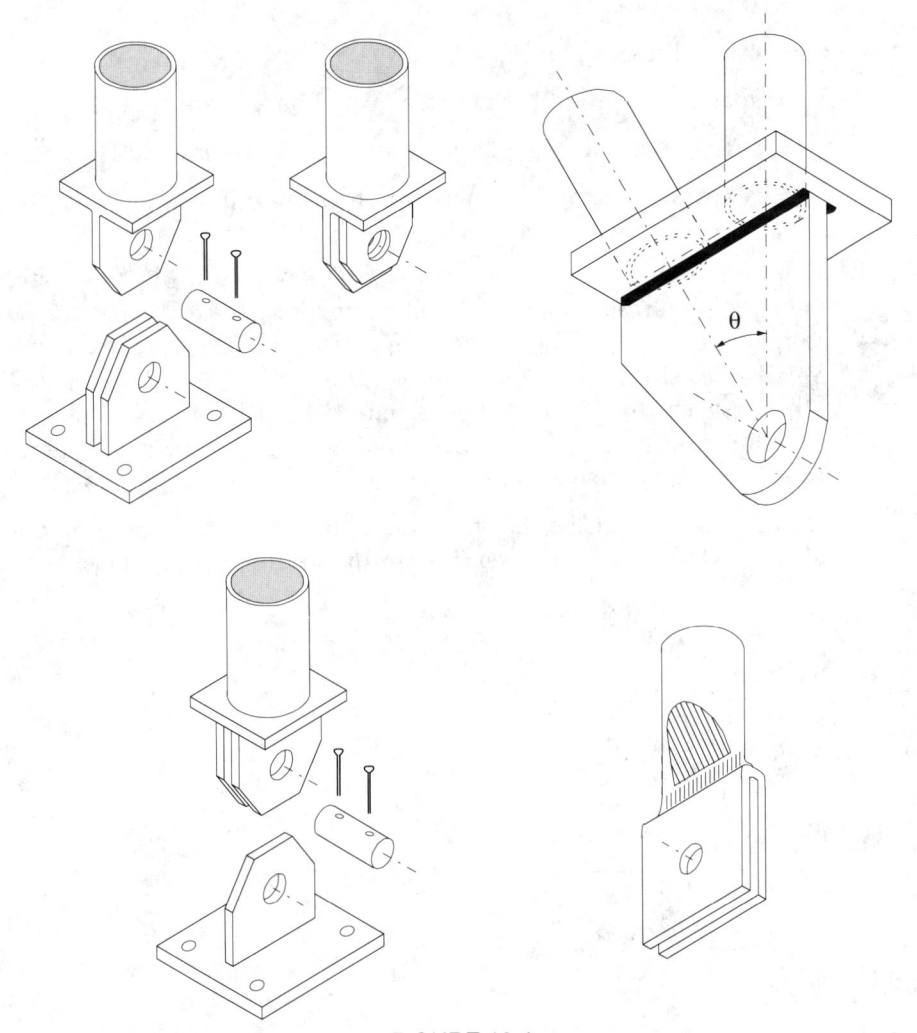

FIGURE 10.4

Examples of pin connected anchorages

3. The base plate projecting beyond the area of $(h-t) \times (b-t)$ acts as four cantilevers subjected to the uniform bearing pressure. Thus, design is based on two potential plastic collapse mechanisms for the base plate: (a) with the two overhangs of length n failing as cantilevers, or (b) with the two overhangs of length m failing as cantilevers. By the Upper Bound Theorem the larger required base plate thickness, from either of these collapse mechanisms, must be used for design.

Design

C_f = total factored column load, kN

339

$A = B \times C$ = area of plate, mm^2

t_p = plate thickness, mm

F_{yp} = specified minimum yield strength of base plate steel, MPa

f'_c = specified 28-day compressive strength of concrete, MPa

ϕ_c = resistance factor for concrete in bearing = 0.6

1. Determine the required area $A = C_f/B_r$ where B_r is the factored bearing resistance per unit of bearing area. For concrete, B_r is taken as $0.85 \phi_c f'_c$. (When the supporting concrete surface is wider on all sides than the loaded area A_1, Clause 10.8.1 of A23.3-94 (CSA 1994) states that B_r may be increased by a factor of $\sqrt{A_2/A_1}$ (but not more than 2.0) where A_2 is the dispersed bearing area defined in A23.3-94, Clause 10.0.)

2. Determine B and C so that the dimensions m and n (the projections of the plate beyond the area $(h-t) \times (b-t)$ are approximately equal.

3. Determine m and n and solve for t_p

where

$$t_p \geq \sqrt{\frac{2C_f m^2}{BC \phi F_{yp}}} \quad\quad [10.1]$$

and

$$t_p \geq \sqrt{\frac{2C_f n^2}{BC \phi F_{yp}}} \quad\quad [10.2]$$

with $\phi = 0.9$.

These formulae were derived by equating the factored moment acting on the portion of the plate taken as a cantilever to the factored moment resistance of the plate $M_r = \phi Z F_{yp}$ and solving for the plate thickness, t_p.

To minimize base plate deflection, generally the thickness should not be less than about 1/5 of the overhang, m or n.

10.2.1.2 Axial Tension

Axial tension may be a design consideration for single storey buildings subjected to uplift wind loading. Some design methods (justified by a very modest amount of testing) have been postulated for uplift of steel base plates connected to wide flange columns. However, these design models are not directly transferable to HSS column base plates because the placement of anchor bolts is quite different with HSS columns. (With the latter, they will always be located some distance away from the column perimeter, whereas with wide flange columns the anchor bolts are usually located close to the column web, thereby placing the bolts near the centre of the base plate.)

At present, there are no rational design procedures that have been validated by testing for the case of uplift on base plates connected to HSS columns. If the base plate is not rigid (very thick), prying forces will be induced in the anchor bolts and should be allowed for in the design. This is evident in the distortions shown in the HSS column base plate in Fig. 10.5, where the base plate is relatively thin.

An appropriate design procedure would be as follows:

1. Ensure that the base plate flexural resistance, as determined by possible yield line mechanisms for the anchor bolt configuration, is greater than the column factored tension load. A resistance factor (ϕ) of 1.0 is commonly applied to yield line limit strength solutions, as has been done in Tables 3.2, 3.3 and 11.1. (Since the plate yield load is well below the ultimate load, it really corresponds to a deflection-control load.)

2. Ensure that the ultimate resistance of the bolts in axial tension exceeds the maximum anticipated bolt load, with the latter allowing for bolt load amplification due to prying action. The tension load per bolt would be the column factored tension load, divided by the number of anchor bolts, then multiplied by a prying factor

FIGURE 10.5

Bolted base plate connection at failure,
with the HSS column subjected to axial tension (uplift) loading

(a) Unidirectional bending

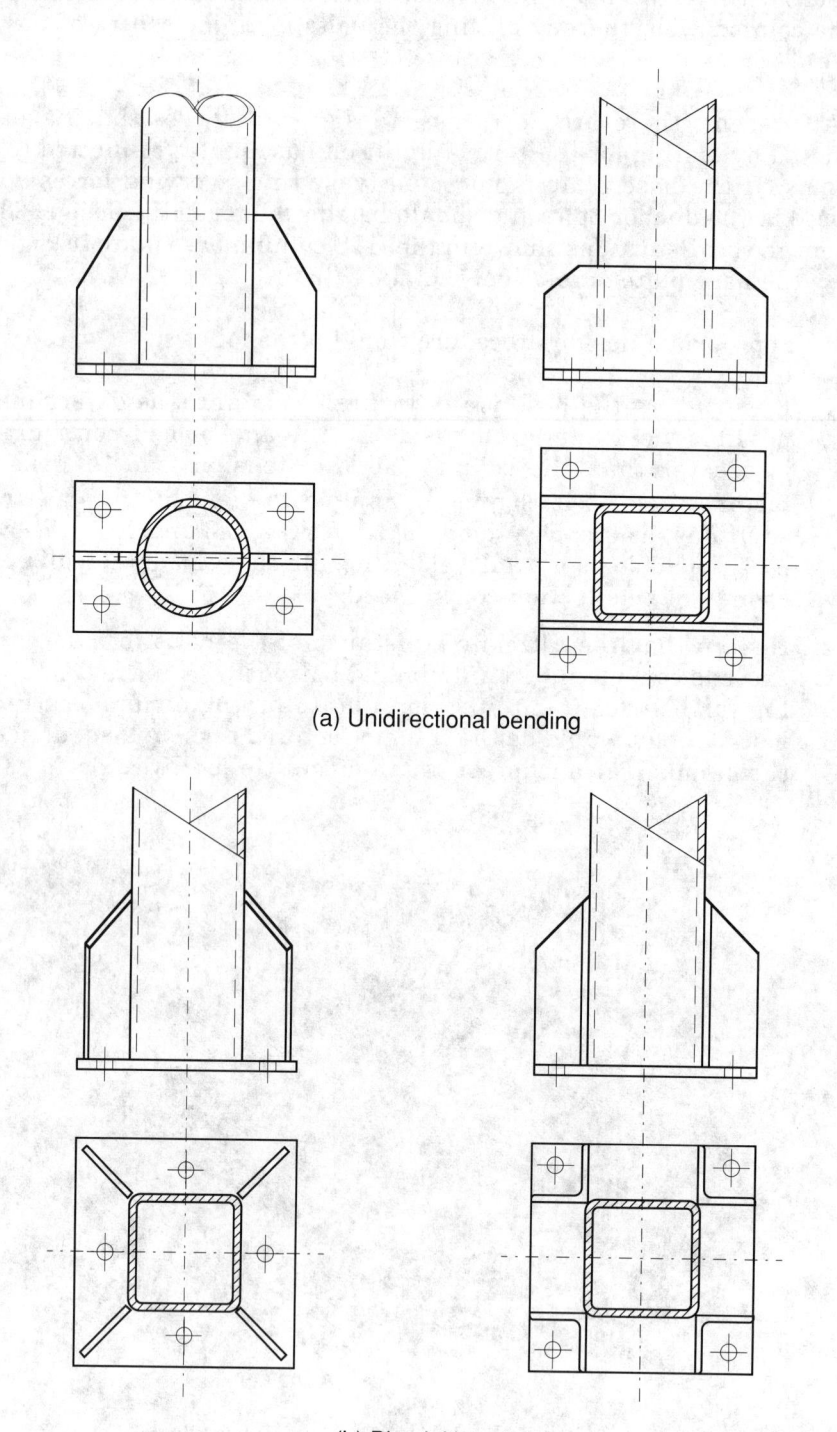

(b) Bi-axial bending

FIGURE 10.6
Examples of stiffened base plates

(typically 1.3 to 1.5). Recent research by Jaspart and Vandegans (1996) has suggested that such a prying factor is likely conservative, as the deformability of a concrete foundation under a prying force is greater than that in the case of steel-to-steel contact.

If the bolts are arranged along just two sides of a square or rectangular HSS column, one can use a design method that has been developed for bolted square and rectangular HSS flange-plate connections, and has been verified by tests (see Section 7.2.1). One can similarly use the verified design procedures for bolted circular HSS flange-plate connections for circular HSS columns subjected to uplift, provided at least four bolts are symmetrically arranged around the tube perimeter (Section 7.1).

10.2.2 Moment Resisting Base Plates

If a base plate is intended to act as a rigid base to develop the moment capacity of a connected HSS column, the base plate will invariably require reinforcement. (An exception to this might be if a very light column is welded to a very thick base plate.)

When a base plate is subjected to a bending moment that is predominantly in one direction, two possible solutions are to weld a tee section in a slot cut through the HSS column or to use two plates as shown in Fig. 10.6(a). Two methods for resisting bending moments in two directions are shown in Fig. 10.6(b).

Tests have been performed on moment-loaded, square and rectangular HSS end plate connections (the results of which could be implemented for base plate connections), but the layout was similar to Fig. 10.3. Not surprisingly, these tests verified that such connections will <u>not</u> develop the moment capacity of the connected HSS (Wheeler et al. 1995).

REFERENCES

CISC. 1995. Handbook of steel construction. Canadian Institute of Steel Construction, 6th. ed., Willowdale, Ontario, pp. 4–138.

CSA. 1994. Design of concrete structures—Structures (design), A23.3-94. Canadian Standards Association, Rexdale, Ontario.

JASPART, J.P., and VANDEGANS, D. 1996. Application of the component method to column bases. Proceedings, International Conference on Advances in Steel Structures, Hong Kong, Vol. 1, pp. 139–144.

WHEELER, A.T., CLARKE, M.J., and HANCOCK, G.J. 1995. Tests of bolted moment end plate connections in tubular members. Proceedings, 14th. Australasian Conference on the Mechanics of Structures and Materials, Hobart, Australia, pp. 331–336.

11
PLATE TO HSS CONNECTIONS

11.1 Plate to Square and Rectangular HSS Connections

Plates are sometimes welded to the faces of rectangular (or square) HSS members in order to connect components such as braces, purlins and hangers. Analytically, a pair of plates can even be used to represent the flanges of a wide flange beam moment-connected to a rectangular HSS column. In this case, the connection moment resistance can be obtained by multiplying the plate axial force resistance by the distance between the beam flange centres.

Longitudinal Plates

Plates can be welded longitudinally or transversely to the HSS member axis. A longitudinal plate will always have a very low β value and hence will be an extremely flexible connection. Figure 11.1 shows the very large out-of-plane deformations that can be produced in the HSS connection face, for a tension-loaded longitudinal plate connection, prior to connection failure. The axial resistance will thus be governed by the formation of a yield line mechanism which represents a control on the connection deformation. This condition is reflected in the low design resistance given in Table 11.1, and has been confirmed in recent laboratory tests (Hörenbaum 1996).

Limited testing to date has shown that a strength reduction factor should be included to allow for the influence of compression forces in the HSS member on the resistance of the connection. A factor $f(n')$, (Tables 11.1 and 11.2), as used for circular HSS to HSS welded connections in Chapter 3, has been found to be reasonable and conservative.

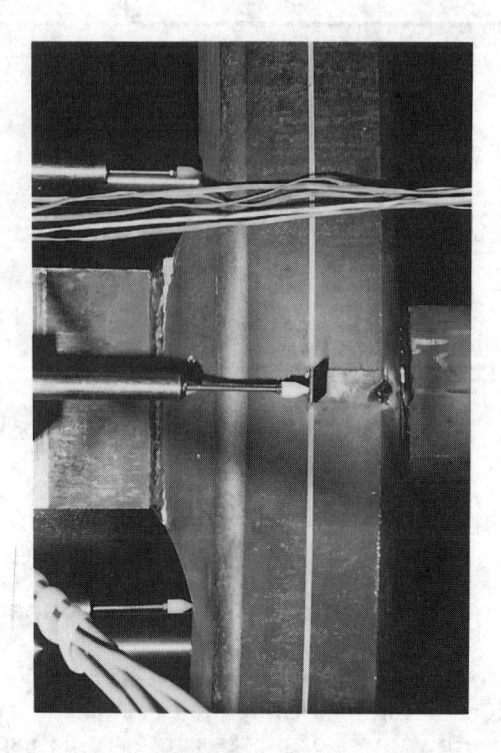

FIGURE 11.1
Longitudinal plate to HSS member connection with the plate
loaded in axial tension, at the ultimate load
(University of Toronto)

Although the longitudinal plate orientation produces a generally lower connection resistance than its transverse plate counterpart, longitudinal plates are still a popular choice for connections between brace members (i.e., in braced frames) and HSS columns. If the brace member is a diagonal, and applies an axial compression or tension load to a longitudinal plate, the connection capacity can be checked under the brace force component normal to the column axis (using Table 11.1).

If the resistance of a longitudinal plate connection is insufficient, one can use a "through-plate" connection as shown in Fig. 11.2, but this is not favoured by fabricators. The through-plate connection is believed to have twice the resistance of the single-wall longitudinal plate connection given in Table 11.1. Another alternative is to reinforce the HSS column connecting face with a tee, as shown in Fig. 11.3.

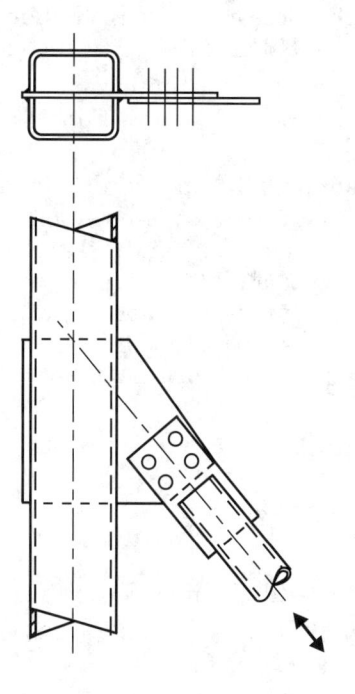

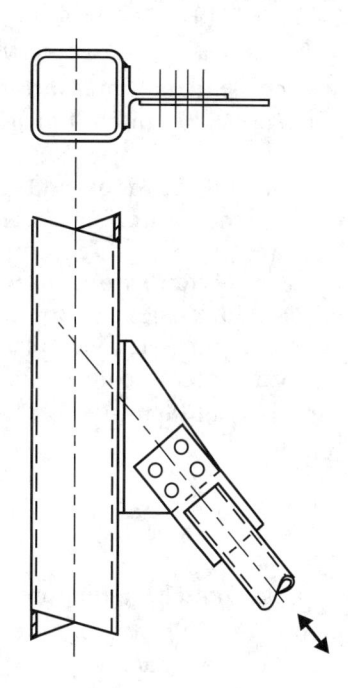

FIGURE 11.2
Brace connection to through-plate

FIGURE 11.3
Brace connection to tee

Transverse Plates

A transverse plate with a low to medium β value will also develop a yield line mechanism in the connecting face of the HSS. For large β values, but where the plate width is still less than the chord width minus twice the wall thickness (β less than $1 - 1/\gamma$), punching shear failure of the chord face is the most probable failure mode; for somewhat lesser β values a combination of flexural failure and punching shear will likely occur.

This combined failure mode has been studied by Davies and Packer (1982) but is too complicated for routine design. However, the branch load capacity can also be reduced by a non-uniform stress distribution in the branch plate, which is termed a branch "effective width" failure criterion (Wardenier *et al.* 1981, Giddings and Wardenier 1986). Moreover, it has been shown (de Koning and Wardenier 1985) that branch effective width always governs over chord face yielding as a critical failure mode for β ≤ 0.85, so the chord face yielding failure mode is omitted in Table 11.1 and a branch effective width check is included for all β values.

At $\beta = 1.0$, the plate will bear directly on the HSS side walls and so side wall failure is the pertinent failure mode for which the connection must be designed. If the branch is loaded in compression, the HSS chord side wall failure stress can still be taken as the yield stress F_{y0}, as the compression is very localized for small load lengths of t_1. (Also, note the limit of validity of $b_0/t_0 \leq 30$.) Such a local bearing failure is shown in Fig. 11.4.

Lu and Wardenier (1996) have shown that an axial load in the HSS member has an effect on the resistance of both a transverse plate connection and a beam-to-column moment connection. For HSS tension loading the connection strength and stiffness will slightly increase, which can be ignored for design purposes. For HSS compression loading, both connection strength and stiffness decrease. A regression fit to the test data was used to formulate a complex parameter $f_3(n)$ to represent the effect of the compression load, where

$$n = \frac{N_0}{A_0 F_{y0}}$$

The parameter (reduction factor) $f_3(n)$ is a function of b_0/t_0, β and the axial stress, but is presently judged to be applicable principally to the chord side wall failure mode, where $\beta \approx 1.0$.

FIGURE 11.4
Bearing failure for a full-width, transverse plate connection under compression loading
(University of Toronto)

For β = 1.0,

$$f_3(n) = 1 + 0.06\left(\frac{b_0}{t_0}\right)^{0.52} n - 0.02\left(\frac{b_0}{t_0}\right)^{0.69} n^2 \qquad [11.1]$$

for $0 > n \geq -0.6$, which is incorporated in Table 11.1, and n is negative for compression loads.

Table 11.1 provides a summary of the foregoing design criteria. Although the connections in Table 11.1 are shown on only one side of the HSS member, the same expressions can be used for connections to both sides, as all equations are functions of the connecting HSS face.

For the design of fillet welds to transverse plates, the plate effective width for all β is given by the term b_e in Table 11.1, so the effective length of the fillet welds can be taken as $2b_e$, considering that the fillet welds will be on both sides of the transverse plate.

11.2 Plate to Circular HSS Connections

Plate connections are also made to circular HSS members, with the plates oriented either longitudinal or transverse to the HSS. Both T (with connection on one side) and X (with connections on opposite sides of the member) arrangements are used. Conservative expressions by Wardenier et al. (1991) apply to both types. These are given in Table 11.2.

Sometimes a combination of plates is used to form a cruxiform detail, with plates both longitudinal and transverse to the HSS member. Since the transverse plate connection is so much stronger than the longitudinal one, the cruxiform variation is not considered to be significantly stronger than the simple transverse connection.

The plate connections must also be checked for punching shear:

$$\left(\frac{N_1}{A_1} + \frac{M_{f1}}{S_1}\right) t_1 \leq 1.16 F_{y0} t_0 \qquad ([9.19])$$

which was also used in Section 9.4.

CONNECTION TYPE	FACTORED CONNECTION RESISTANCE
Longitudinal Plate	$\beta \leq 0.85$ Basis: CHORD FACE YIELDING
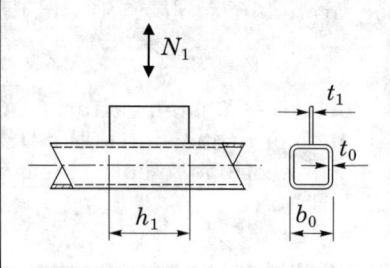 (diagram with N_1, h_1, t_1, t_0, b_0)	$N_1^* = \dfrac{F_{y0} t_0^2}{(1-\beta)} \left[2\eta + 4(1-\beta)^{0.5}\right] f(n')$ where $\beta = \dfrac{t_1}{b_0}$ and $\eta = \dfrac{h_1}{b_0}$
Transverse Plate	$\beta \approx 1.0$ Basis: CHORD SIDE WALL FAILURE
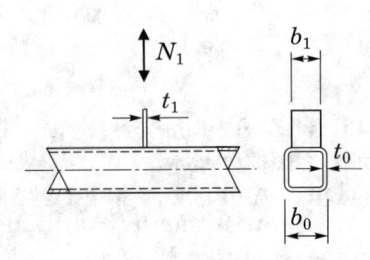 (diagram with N_1, t_1, t_0, b_1, b_0)	$N_1^* = 2 F_{y0} t_0 (t_1 + 5 t_0) f_3(n)$
	$0.85 \leq \beta \leq 1 - 1/\gamma$ Basis: PUNCHING SHEAR
	$N_1^* = \dfrac{F_{y0} t_0}{\sqrt{3}} (2t_1 + 2 b_{ep})$
	All β Basis: EFFECTIVE WIDTH
where $\beta = \dfrac{b_1}{b_0}$	$N_1^* = F_{y1} t_1 b_e$

FUNCTIONS

$$b_{ep} = \dfrac{10}{b_0/t_0} b_1 \quad \text{but} \leq b_1 \qquad b_e = \dfrac{10}{b_0/t_0} \dfrac{F_{y0} t_0}{F_{y1} t_1} b_1 \quad \text{but} \leq b_1$$

For $f(n')$ see Table 11.2
For $f_3(n)$ see Equation 11.1

RANGE OF VALIDITY

$$b_0/t_0 \leq 30$$

TABLE 11.1
Factored resistance of plate to square and rectangular HSS connections

CONNECTION TYPE	FACTORED CONNECTION RESISTANCE		
	Axial Force	Bending in Plane	Bending out of Plane
Longitudinal Plate N_1, h_1, d_0, t_0	$N_1^* = 5.0\, F_{y0}\, t_0^2\, (1 + 0.25\, \eta)\, f(n')$	$M_{r1}^* = h_1 N_1^*$	—
Transverse Plate N_1, b_1, d_0, t_0	$N_1^* = F_{y0}\, t_0^2 \left(\dfrac{5.0}{1 - 0.81\, \beta}\right) f(n')$	—	$M_{opr1} = 0.5\, b_1 N_1^*$

FUNCTIONS

$$\eta = \frac{h_1}{d_0} \qquad \beta = \frac{b_1}{d_0}$$

$f(n') = 1.0$ for $n' \geq 0$ (tension)

         $= 1 + 0.3\, n' - 0.3\, n'^2$ but ≤ 1.0 for $n' < 0$ (compression)

TABLE 11.2

Factored resistance of plate to circular HSS connections

REFERENCES

DAVIES, G., and PACKER, J.A. 1982. Predicting the strength of branch plate – RHS connections for punching shear. Canadian Journal of Civil Engineering, **9**(3): 458–467.

GIDDINGS, T.W., and WARDENIER, J. (eds.). 1986. The strength and behaviour of statically loaded welded connections in structural hollow sections. CIDECT Monograph No. 6. British Steel plc, Corby, Northants, England.

HÖRENBAUM, C. 1996. Plate connections to rectangular hollow section (RHS) columns at brace points. Diploma thesis, University of Toronto, Toronto, Ontario.

de KONING, C.H.M., and WARDENIER, J. 1985. The static strength of welded joints between structural hollow sections or between structural hollow sections and H-sections. Part 2: joints between rectangular hollow sections. Stevin Report 6-84-19, Delft University of Technology, Delft, The Netherlands.

LU, L.H., and WARDENIER, J. 1996. The influence of column pre-loading on the static strength between I-beams and RHS columns. Proceedings, International Conference on Advances in Steel Structures, Hong Kong, Vol. 1, pp. 109–115.

WARDENIER, J., DAVIES, G., and STOLLE, P. 1981. The effective width of branch plate to RHS chord connections in cross joints. Stevin Laboratory Report No. 6-81-6, Delft University of Technology, Delft, The Netherlands.

WARDENIER, J., KUROBANE, Y., PACKER, J.A., DUTTA, D., and YEOMANS, N. 1991. Design guide for circular hollow section (CHS) joints under predominantly static loading. CIDECT (ed.) and Verlag TÜV Rheinland GmbH, Köln, Federal Republic of Germany.

12

HSS WELDED CONNECTIONS SUBJECTED TO FATIGUE LOADING

12.1 Introduction

Fluctuating loads can cause local strains that are sufficient to induce localized micro structural changes in a material. This process may result in the formation of fatigue cracks, which can grow to a size sufficient to cause failure. Fatigue failure is checked at the specified (service) load level, and under the stress range produced by just live load fluctuations. Fatigue may be a governing design criterion in HSS structures such as bridges, crane girders and booms, conveyor gantries, agricultural equipment, mechanical equipment and amusement ride structures. Some examples of fatigue-critical structures are shown in Fig. 12.1.

A great deal of research has now been undertaken to determine Stress Concentration Factors (SCF) and Strain Concentration Factors ($SNCF$) for welded tubular (circular) connections because of the fatigue problems associated with connection design in offshore steel jacket-type structures. This research has taken the form of experimental measurements of strains in the vicinity of the joint, photo elastic techniques and finite element as well as other numerical models, with the results being incorporated into many national offshore structures design codes.

Unlike circular tube connections, there has been a serious lack of information available for the fatigue design of rectangular and square HSS connections because of the absence of extensive and reliable experimental data. Rectangular and square HSS *gap* connections are much more fatigue-critical than overlap connections, yet are much easier to fabricate, and hence more popular.

(a) Roller coaster

(b) Air seeder farming equipment

FIGURE 12.1
Examples of fatigue-critical HSS assemblies

The typical methods available to a design engineer for estimating fatigue life of structural connections are

— A visual classification of such connections into stress range categories
— The "hot spot stress" method.

The former approach is an approximate one, but is included in many national steel building codes. However, welded HSS connections are not generally covered, and this is the case with CAN/CSA-S16.1-94. Until relatively recently, the classification approach was recommended (Wardenier 1982) as being the more appropriate for square HSS connections (because of insufficient data), and connections were classified into groups with nearly the same fatigue resistance.

These groups were K and N overlap, plus K and N gap, all subject to a particular range of validity for various geometric parameters. In a comprehensive international state-of-the-art report by CIDECT (1982), this method was still cited as being the best available for *square* HSS connections, even though a hot spot stress method was advocated for *circular* hollow section connections.

The hot spot stress method has been found to be a more accurate and reliable method for treating fatigue in offshore circular tube connections, and this method was first recommended for all HSS connections by the International Institute of Welding Subcommission XV-E (IIW 1985). In these recommendations, *SCF* (stress concentration factor) values were given for different connection types as a function of the thickness ratio of the HSS being joined. A minimum *SCF* value of 3.0 was given for square HSS, and for circular HSS a minimum value of 2.0 is advisable. These *SCF* values could then be applied to the nominal stress range in a truss web member to determine the hot spot stress range ($S_{r-h.s.}$), and hence the fatigue life (cycles to failure) from a set of $S_{r-h.s.}$ vs. N curves, N being the number of load cycles to failure. Although revisions are now being drafted (van Wingerde et al. 1997) to these IIW recommendations, it is the IIW (1985) hot spot stress approach that is used in this chapter.

It is currently recognized that more accurate *SCF* values (based on multiple connection parameters) are necessary, particularly for square and rectangular HSS connections. This translates into obtaining more accurate *SCF* values by experimental and analytical research to produce a coherent package of *SCF* formulae and compatible $S_{r-h.s.} - N$ lines that are simple to use, while still being sufficiently accurate. This effort is presently underway in Canada and Europe (Packer et al. 1990, Puthli, et al. 1988, van Wingerde et al. 1989, 1992, 1993, 1995, 1996a, 1996b, 1996c, 1996d, Mang et al. 1989, Frater and Packer 1992, van Wingerde 1992, van Wingerde and Packer 1994, 1996, Herion 1994, Panjeh Shahi 1995, Niemi and Yrjölä 1995, and Niemi 1996).

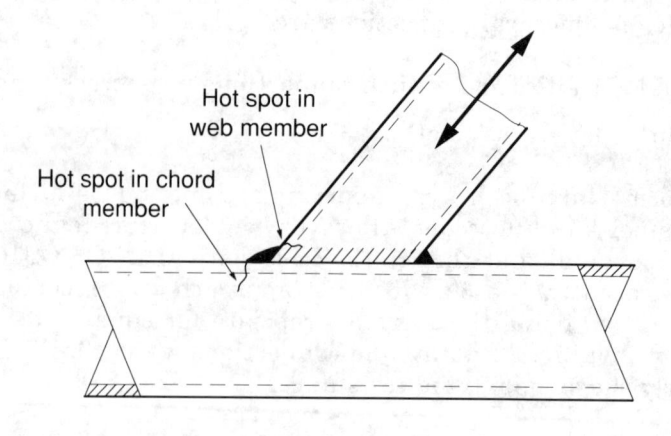

FIGURE 12.2
Potential locations of fatigue cracks in a welded HSS connection

12.2 Hot Spot Stress Approach for Fatigue Design

Fatigue failure of HSS connections occurs at the toes of the weld which connects a web member to the chord. Thus, the fatigue design procedure entails checking for cracking at two locations: at the weld toe on the web member, and at the weld toe on the chord member, as shown in Fig. 12.2. Fatigue cracking will progress through the member thickness, in either the web or chord, initiating at the most highly stressed point termed the "hot spot". It is important to note that fatigue failure through the web or chord thickness is caused by the oscillating load in the *web* member as indicated in Fig. 12.2, with the chord load having a lesser effect.

The load range in the web member produces a nominal stress range S_r which is the algebraic difference between the maximum and minimum stress in a stress cycle. In order to have the potential of producing fatigue failure, the web member stress must normally go into tension. Thus, if a web member has an alternating load entirely in the compression range, fatigue failure need not normally be considered. If the web member has a stress range which produces both tension and compression stresses, then one must consider the *total* stress range in fatigue calculations. This is due to the likely presence of residual stresses which may shift the stress ratio.

For an alternating load entirely in the compression range, a crack could possibly appear if a zone of tensile residual stress exists. Such a crack may propagate, but it usually stops upon leaving the tensile residual stress zone and entering the region of compressive only stress. The stress ratio R is defined as the ratio between the minimum and maximum stresses for constant amplitude loading, taking account of the sign of the stress. (Tension is taken as positive and compression as negative.) The stress range S_r

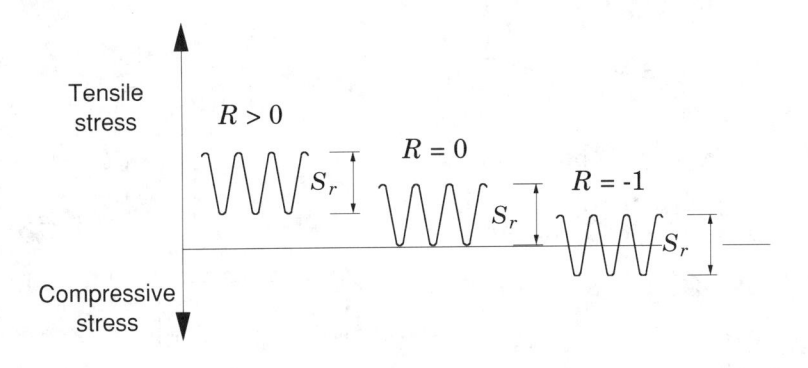

FIGURE 12.3

Stress range S_r and stress ratio R

and the stress ratio R are illustrated in Fig. 12.3. The former has the major influence on fatigue life.

For fatigue design, one must add the nominal secondary bending stress ranges produced by connection flexibility to the nominal axial stress ranges. It is usually difficult to quantify the secondary bending moments produced in a truss, but simplified amplification factors for application to the nominal axial stress ranges are given in the recommended design procedure. Then the total stress range for the *web* members must be multiplied by a Stress Concentration Factor *SCF* to determine the stress range at the "hot spot", $S_{r-h.s.}$.

In the method being advocated, this increase of stress is due to the change in member geometry at the connection, and does not include the local effects of the weld itself. The term "hot spot" is used to designate the location of the governing fatigue stress for a connection, and it is—strictly speaking—the location where the principal stress at the toe of the weld has its maximum value. An exact determination of that location and the stress magnitude could involve model studies or finite element analyses.

Numerical and experimental research has shown that the critical hot spot positions are the crown and saddle locations for circular HSS connections, and the brace (or web member) corners for square and rectangular HSS connections, as shown in Fig. 12.4 (van Wingerde 1992, van Wingerde et al. 1995, 1996a). Fortunately for the practising designer, a series of equations and coefficients has been established to estimate the maximum hot spot stress for use with circular and square HSS connections.

Fatigue life is generally specified as the number of stress cycles of a particular range that a connection sustains before failure. The relationship between the hot spot stress range ($S_{r-h.s.}$) and the number of cycles to failure N can be expressed by a set of $S_{r-h.s.}$ vs. N lines plotted on a graph with logarithmic scales. Log-log scales are used in order to produce a linear

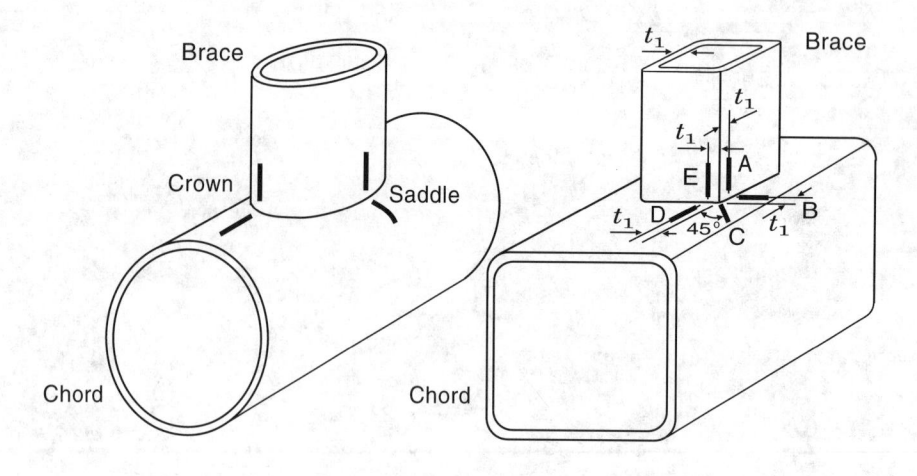

FIGURE 12.4
Critical hot spot stress positions for circular
and square (or rectangular) HSS connections

fatigue relationship, since $N(S_{r-h.s.})^m$ (where m is a constant) has been found to be a constant. The lines presented in this chapter are valid for member thicknesses up to 22 mm, which covers the current product range of North American HSS. The fatigue strengths given have been derived from a statistical analysis of experimental data, and they represent fatigue lives which are less than the mean lives by two standard deviations.

12.3 Fatigue Design Procedure for Planar Truss Connections

The following is a more detailed description of the fatigue design procedure for planar connections.

1. For the prescribed loads on a particular structure with selected members, determine the nominal axial forces and the nominal primary bending moments in the members. This can be done when the truss is loaded at the panel points by assuming all joints are pinned, in which case a set of axial forces only will be produced in all members.

   If a truss has external chord loads between panel points, it may be analysed by assuming continuous chord members which are pin connected to the web members, and this will produce in-plane bending moments in the chord members but not in the web members. (This is discussed in Section 2.2.2 of Chapter 2 and demonstrated in Sections 13.2.4 and 13.3.5 in Chapter 13 for static loading.) For circular HSS, different SCF values are available for primary in-plane bending stresses and for axial stresses, but for square HSS connections the same SCF values can be applied to both in-plane bending and axial stress ranges.

2. Having established the nominal axial forces in all members (and possibly the nominal primary bending moments if the continuous chord method of analysis was used), the designer should determine the corresponding stress ranges for axial stress ($S_{r\ ax-nom}$) and bending stress ($S_{r\ b-nom}$). The effect of member noding eccentricity on the connection can be ignored provided the ratio of the eccentricity, e to chord depth, is in the range $-0.55 \le e/h_0$ or $e/d_0 \le 0.25$. (Noding eccentricity is shown in Fig. 2.1.)

However, the connection rigidity still produces secondary bending moments in the members, and these need to be taken into account in fatigue design. To do this, the *axial* stress ranges in all members ($S_{r\ ax-nom}$) should be multiplied by the factors given in Table 12.1. Hence one obtains higher values of $S_{r\ ax-nom}$ for all truss members, termed $S'_{r\ ax}$.

TYPE OF CONNECTION		Chords	Vertical Webs	Diagonal Webs
Circular HSS Gap Connections	K-type	1.5	—	1.3
	N-type	1.5	1.8	1.4
Circular HSS Overlap Connections	K-type	1.5	—	1.2
	N-type	1.5	1.65	1.25
Square HSS Gap Connections	K-type	1.5	—	1.5
	N-type	1.5	2.2	1.6
Square HSS OverlapConnections	K-type	1.5	—	1.3
	N-type	1.5	2.0	1.4

TRUSS MEMBERS

TABLE 12.1

Amplification factors for $S_{r\ ax-nom}$ to account for secondary bending moment in HSS truss connections

3. Determine the *SCFs*. More is known about fatigue in circular HSS connections (from research for offshore structures) than about square HSS, and so the procedure is somewhat more complex for circular HSS connections.

For *square* HSS connections, a single *SCF* can be obtained from the procedure given in Section 12.4. The value covers the most critical of all the potential failure locations (see Fig. 12.4) in the web and chord member walls. This single *SCF* value can be applied to the adjusted axial stress range ($S'_{r\ ax-w}$) of web members for truss K and N connections. No other connection types are covered for square HSS connections in this chapter.

For *circular* HSS connections, *SCF* values for axial loading are determined for potential crack locations on both the web and the chord members using the procedure given in Section 12.5. The stress concentration factor for the web member cracking (SCF_{ax-w}) can then be applied to the adjusted axial stress range in the web member ($S'_{r\,ax-w}$) as shown in [12.1]. The stress concentration factor for the chord member cracking (SCF_{ax-ch}) can be applied to the same adjusted axial stress range in the web member ($S'_{r\,ax-w}$) as shown in [12.2].

If a primary in-plane bending stress range is also present in the web member ($S_{r\,b-nom-w}$), two SCF_b values must also be determined for the bending case, for web cracking (SCF_{b-w}) and chord cracking (SCF_{b-ch}) locations. These values would be implemented as shown in [12.1] and [12.2]. One should note that for both square and circular HSS truss K and N connections, the term $S_{r\,b-nom-w}$ is zero, because there is no primary bending moment in the web members, for either the fully pinned or the continuous chord/pinned web truss analysis models.

4. Determine the hot-spot stress range ($S_{r-h.s.}$) for both web cracking and chord cracking. These can be obtained from the generalized equations below, which apply to various types of planar connections under in-plane loadings.

$$(S_{r-h.s.})_{web} = (SCF_{ax-w} \cdot S'_{r\,ax-w}) + (SCF_{b-w} \cdot S_{r\,b-nom-w}) \quad [12.1]$$

$$(S_{r-h.s.})_{chord} = (SCF_{ax-ch} \cdot S'_{r\,ax-w}) + (SCF_{b-ch} \cdot S_{r\,b-nom-w}) \quad [12.2]$$

where

SCF_{ax-w} = geometrical stress concentration factor for an axial load in the web member causing cracking in the *web* member.

SCF_{b-w} = geometrical stress concentration factor for a primary in-plane bending moment in the web (or branch) member causing cracking in the *web* (or branch) member.

SCF_{ax-ch} = geometrical stress concentration factor for an axial load in the web member causing cracking in the *chord* member.

SCF_{b-ch} = geometrical stress concentration factor for a primary in-plane bending moment in the web (or branch) member causing cracking in the *chord* member.

$S'_{r\,ax-w}$ = axial stress range in the *web* member, amplified (in the case of truss K and N connections only) to take account of secondary in-plane bending moments.

$S_{r\,b-nom-w}$ = bending stress range in the *web* (or branch) member, produced by primary bending moments.

= 0 for truss K and N connections.

≠ 0 for T connections in Vierendeel trusses, for example.

SCF_{ax-w} = SCF_{ax-ch} for square HSS connections.

5. Determine the fatigue life N from the $S_{r-h.s.}$ vs. N lines given in Fig. 12.5. Separate hot spot stress ranges for the web crack location ($S_{r-h.s.})_{web}$ and the chord location ($S_{r-h.s.})_{chord}$ are needed for separate entry to the vertical axis of this figure. One then reads across to the fatigue curves A which are differentiated on the basis of the wall thickness of the member under consideration. Due to a so-called "thick-

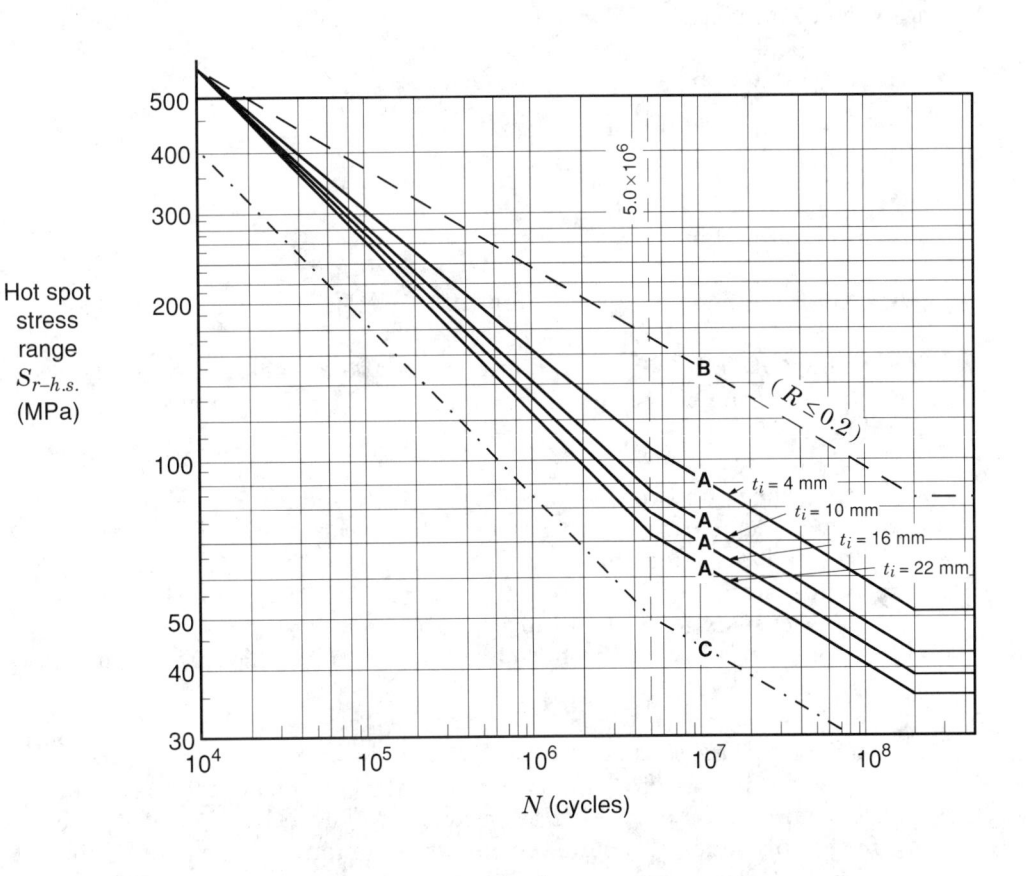

FIGURE 12.5

$S_{r-h.s.}$ vs. N curves for welded HSS connections

ness effect" a thinner member in an HSS connection has been found to have a higher fatigue strength (or for the same stress range a thinner member has a longer fatigue life). (van Wingerde 1992, van Wingerde et al. 1995, 1996a.) The web member wall t_2 will likely be smaller than the chord member wall t_0, so, for the same $S_{r-h.s.}$, the governing fatigue location would be at the weld toe in the chord.

The curves labelled A can be used for the whole range of connections covered by this chapter (viz., K and N connections of square HSS, and T, Y, X, K and N connections of circular HSS) regardless of the stress ratio R. For $t_i = 22$ mm, curve A is defined by

$$\log N = 12.271 - 3 \log S_{r-h.s.} \qquad \text{for } 10^4 \leq N \leq 5(10^6) \qquad [12.3]$$

Thus, for a particular nominal stress range in a web member, one can determine from the curves A in Fig. 12.5 the minimum number of cycles N to cause fatigue failure in either the web or chord member.

Figure 12.5 also shows another fatigue failure curve labelled B, which can be seen to have a superior fatigue performance relative to the A curves. This curve B can be used instead of the A curves for planar K and N-type connections in which

a) $t_i \leq 10$ mm

b) d_i or $b_i \leq 200$ mm

c) $R \leq 0.2$.

Curve B is defined by

$$\log N = 17.785 - 5 \log S_{r-h.s.} \qquad \text{for } 10^4 \leq N \leq 5(10^6) \qquad [12.4]$$

For R values of $0.2 < R \leq 0.8$, one can make a linear interpolation for N by using the curve B (for $R = 0.2$) and the relevant curve A (by assuming $R = 0.8$).

6. One further check on the fatigue life N determined in (5) still needs to be performed. As mentioned in Section 12.2, if the loading in the chord goes into tension, it does have some influence on the fatigue capacity of the connection, particularly on the weld toe crack in the *chord* member. In cases where the chord loading is very high, yet the web member loading is low, a further check needs to be made using curve C in Fig. 12.5.

The nominal axial stress range $S_{r\,ax-nom}$ (maximum for either side of the connection) in the *chord* member first needs to be determined. For K and N connections only, this then needs to be multiplied by 1.5 to allow for the presence of secondary bending moments in the connection (see Table 12.1) to obtain a modified axial chord stress range $S'_{r\,ax}$.

If primary bending moments are present in the chord member (e.g., Vierendeel T connections, or K and N connections in a truss analyzed on the basis of continuous chord members plus pin-ended webs) one must also calculate the bending stress range for the *chord* member $S_{r\,b-nom-ch}$. This stress is added to the modified axial stress range $(S'_{r\,ax})$ to produce the total nominal stress range in the chord.

One should then enter Fig. 12.5 at the vertical axis with this <u>nominal</u> chord stress range (S_{r-nom}) ignoring temporarily the label of "hot spot stress range" because no SCF is to be used. By reading across to curve C in this figure, one can again obtain another limiting value of N which the connection can sustain.

Curve C is defined by

$$\log N = 11.810 - 3 \log S_{r-nom} \quad \text{for} \quad 10^4 \leq N \leq 5\,(10^6) \qquad [12.5]$$

7. All of the foregoing has described a procedure for ascertaining the limiting fatigue life N of a connection when the structure is subjected to constant amplitude loading for n cycles. If a connection is subjected to n_i cycles of load at a particular amplitude of stress S_{ri} which is then varied $(i > 1)$, the acceptability of a connection under variable amplitude fatigue loading can be ascertained by applying Miner's linear cumulative damage rule. This requires that, <u>at each potential crack location</u>

$$\sum_i \frac{n_i}{N_i} \leq 1.0 \qquad [12.6]$$

where n_i = numbers of applied load cycles at stress range i for the design life of the structure

N_i = corresponding numbers of load cycles to failure at constant stress range i obtained from the $S_{r-h.s.}$ vs. N curves, Fig. 12.5.

The curves labelled A, B and C in Fig. 12.5 are actually drawn for use with variable amplitude loading with a Miner's rule calculation. If the connections are not subject to corrosion and are subjected only to constant amplitude loading, a "fatigue limit" for curves A, B and C can be adopted at $N = 5\,(10^6)$ cycles. In other words, the curves A, B and C become horizontal for $N > 5\,(10^6)$.

Tests by Pedersen and Agerskov (1991) on circular hollow section connections subjected to <u>variable amplitude</u> fatigue loading suggest that the traditional Miner's Rule summation expressed by [12.6] may be unconservative. To date, the tests indicate that the Miner sum of 1.0 should be closer to a value of 0.8, but the traditional rule will not be amended until test evidence is conclusive.

8. For inadequately protected connections in a corrosive environment (e.g., offshore structures) the curves A, B and C in Fig. 12.5 should be projected linearly onwards for $N > 5 \, (10^6)$ on their original slopes, and no "fatigue limit" can be assumed. Furthermore, the value of N so obtained from Fig. 12.5 should be divided by 2 to reflect the much more severe conditions of "stress corrosion".

These fatigue design procedures are summarized in Table 12.2.

STEP No.	SQUARE HSS PLANAR K AND N CONNECTIONS		CIRCULAR HSS PLANAR T,Y, X, K AND N CONNECTIONS	
	DESCRIPTION	REFERENCE	DESCRIPTION	REFERENCE
1	Nominal axial and nominal primary bending forces	Specified loads	Nominal axial and nominal primary bending forces	Specified loads
2a	Nominal axial and nominal primary bending stress ranges	$S_{rax-nom}$ $S_{rb-nom-w}$	Nominal axial and nominal primary bending stress ranges	$S_{rax-nom}$ $S_{rb-nom-w}$
2b	Axial stress ranges amplified to include secondary bending	S'_{rax} Table 12.1	Axial stress ranges amplified to include secondary bending	S'_{rax} Table 12.1
3	$SCF_{ax-w} = SCF_{ax-ch}$ from C and τ Validity ranges	Eqn. 12.7 Table 12.3(a) Table 12.3(b)	Find SCF^* ($\gamma = 12.5$ and $\tau = 0.5$) Validity ranges	Fig.12.6 to 12.9 Table 12.4
			Calculate SCF_{ax-ch} from SCF^* using γ & τ and $x1$ & $x2$	Eqn. 12.8 Table 12.4
			Calculate SCF_{ax-w} from SCF_{ax-ch}	Eqn. 12.8a
			$SCF_{b-ch} = SCF_{ax-ch}$ and $SCF_{b-w} = SCF_{ax-w}$	
4	Calculate $(S_{r-h.s.})_{web}$ Calculate $(S_{r-h.s.})_{chord}$	Eqn. 12.1 Eqn. 12.2	Calculate $(S_{r-h.s.})_{web}$ Calculate $(S_{r-h.s.})_{chord}$	Eqn. 12.1 Eqn. 12.2
5	Find fatigue life N from Curve B in Fig. 12.5 if validity limits in Section 12.3 (5) are met. Otherwise, find fatigue life N from Curves A in Fig. 12.5			
6	If chord goes into tension: Calculate S'_{rax} from $S_{rax-nom}$ (Table 12.1) and add $S_{rb-nom-ch}$ to get S_{r-nom} Use S_{r-nom} to find fatigue life N from Curve C in Fig. 12.5			
7	Apply Miner's linear cumulative damage rule (Equation 12.6), if applicable.			
8	Reduce value of fatigue life N as per Section 12.3 (8) if corrosion protection is inadequate.			

TABLE 12.2

Summary of steps to determine fatigue life (Section 12.3)

12.4 SCFs for Square HSS Connections

For planar K and N connections of square HSS, the stress concentration factor for axial loading is given by

$$SCF_{ax-w} = SCF_{ax-ch} = C\ f(\tau) \qquad [12.7]$$

where

$$f(\tau) = \tau, \quad \text{but} \geq \tau_{lower\ limit} \qquad [12.7a]$$

τ = ratio of web member wall thickness to chord wall thickness.

Values of the coefficient C and $\tau_{lower\ limit}$ are presented in Table 12.3(a). The ranges of validity for the relevant geometric parameters are shown in Table 12.3(b). They represent the limits of experimental verification, and are

- Angle between chord and web member must be between 40° and 90°.
- Width to wall thickness ratio of the chord must not exceed 25.
- Width of the two web members must be almost the same.
- Width of web members must be between 0.5 and 1.0 times the width of the chord.
- Gap size must be between 0.5 and 1.1 times the difference in width of the chord and the web members.
- Overlaps must be between 50% and 100%.

The stress concentration factor given in [12.7] is only quoted for axial loading because K and N connections are the only type of square HSS connections covered by this chapter, and neither method of analysis described in Section 12.2 results in primary bending moments in the truss web members. The *SCF* value is only needed for application to the *web* member stress and is not required when checking for the chord member nominal stress using curve C (see 12.3 (6)).

CONNECTION TYPE	C	$\tau_{lower\ limit}$
K and N-type with gap	6.0	0.50
K-type with overlap	3.6	0.83
N-type with overlap	4.3	0.70

TABLE 12.3(a)

Values of C and $\tau_{lower\ limit}$ for use in equations 12.7 and 12.7a for square HSS connections

CONNECTION PARAMETER	RANGE of VALIDITY
θ_i	40° to 90°
b_0 / t_0	≤ 25
b_1, b_2	$b_1 \approx b_2$
$\beta = b_i / b_0$	$0.5 \leq \beta \leq 1.0$
g	$0.5(b_0 - b_i) \leq g \leq 1.1(b_0 - b_i)$
O_v	$50\% \leq O_v \leq 100\%$

TABLE 12.3(b)
Parameter range of validity for equation 12.7

12.5 *SCFs* for Circular HSS Connections

For planar T, Y, X, K and N connections of circular HSS, the stress concentration factors for axial loading and primary in-plane bending moments are given by

$$SCF_{ax-ch} = \left(\frac{\gamma}{12.5}\right)^{x1} \left(\frac{\tau}{0.5}\right)^{x2} SCF^* \qquad \text{but} \geq 2.0 \qquad [12.8]$$

$$SCF_{ax-w} = 1 + 0.63\, SCF_{ax-ch} \qquad [12.8a]$$

where

γ = ratio of chord diameter to twice the chord wall thickness

τ = ratio of web wall thickness to chord wall thickness

$x1, x2$ = exponents depending on the connection type and loading case, as per Table 12.4

SCF^* = stress concentration factor for the considered connection type and loading case for $\gamma = 12.5$ and $\tau = 0.5$.

Four graphs of SCF^* when $\gamma = 12.5$ and $\tau = 0.5$ are presented in Figs. 12.6 to 12.9 for various connection types and load cases. These graphs are also subject to the ranges of validity for the relevant geometric parameters shown in Table 12.5.

CONNECTION TYPE	LOADING CASE	EXPONENTS $x1$	$x2$	SCF^* Figure No.
T and Y	Axial	1.00	1.00	12.6
	In-plane bending	0.60	0.80	12.7
X	Axial	1.00	1.00	12.8
	In-plane bending	0.60	0.80	12.7
K and N with gap	Axial	0.67	1.10	12.9
K and N, overlap	Axial	0.67	1.10	12.9 for $g/d_0 = 0.01$

TABLE 12.4

Values of $x1$ and $x2$ with locator of applicable SCF^* diagram for use in equation 12.8 for circular HSS connections

CONNECTION TYPE	\multicolumn{5}{c}{CONNECTION PARAMETERS}
	θ_i	γ	β	τ	g/d_0
T and Y	$30° \le \theta_i \le 90°$	$12 \le \gamma \le 32$ For $\gamma < 12$ use $\gamma = 12$ which is conservative	$0.13 \le \beta \le 1.0$		
X			$0.13 \le \beta \le 0.95$ For $\beta > 0.95$ use $\beta = 0.95$	$0.25 \le \tau \le 1.0$	
K and N (gap and overlap)			$0.30 \le \beta \le 0.80$	$0.40 \le \tau \le 0.80$	$0.01 \le g/d_0 \le 1.0$ For overlaps, use $g/d_0 = 0.01$

TABLE 12.5

Parameter range of validity for SCF^* values in Figs. 12.6 to 12.9

FIGURE 12.6

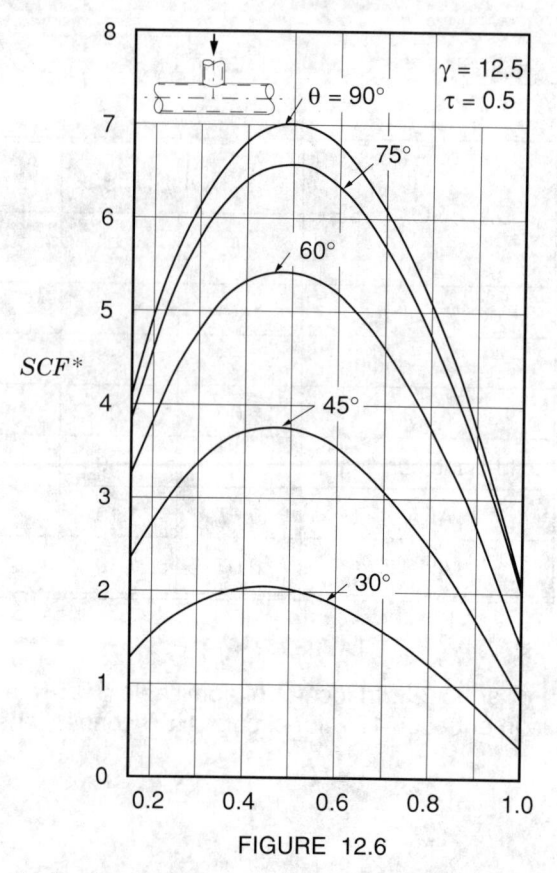

SCF* for axially loaded T and Y connections, for circular HSS

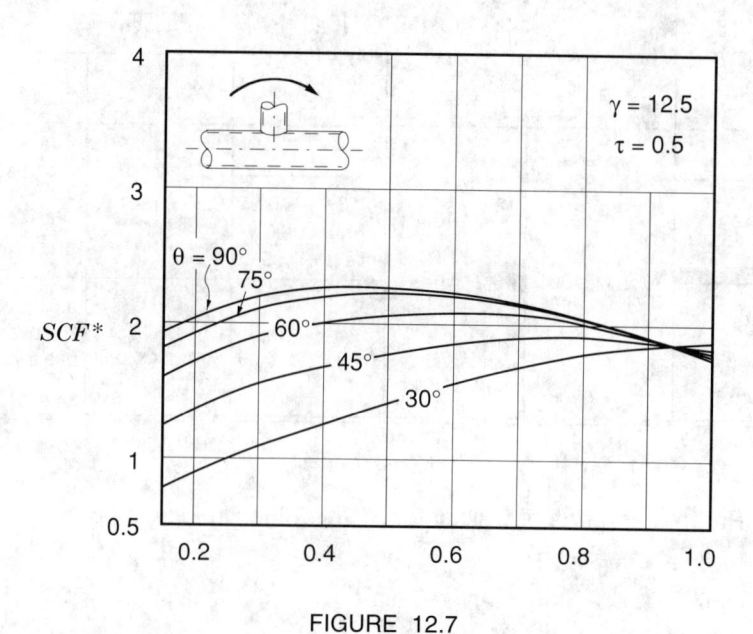

FIGURE 12.7

SCF* for T, Y and X connections loaded by in-plane bending moments, for circular HSS

FIGURE 12.8

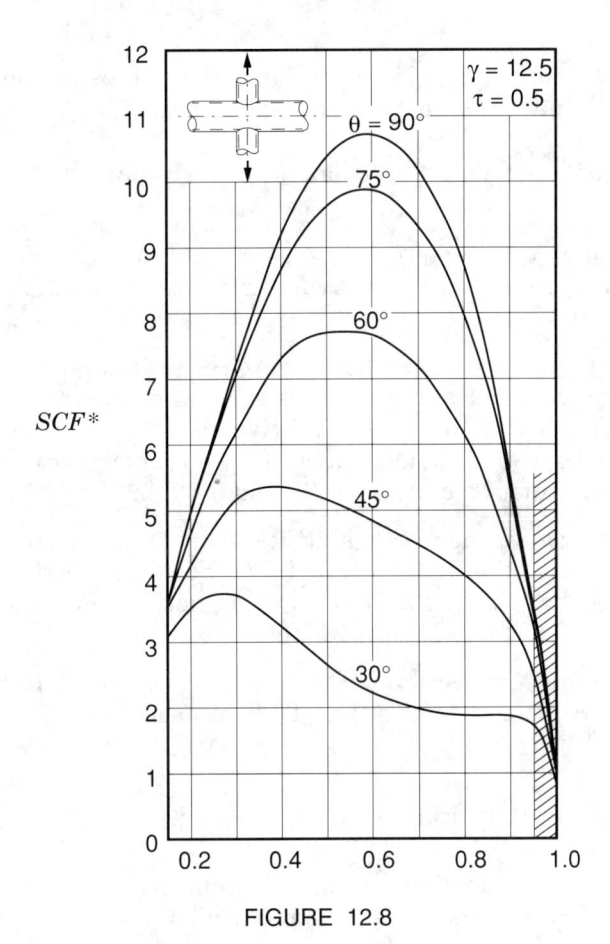

SCF^* for axially loaded X connections, for circular HSS

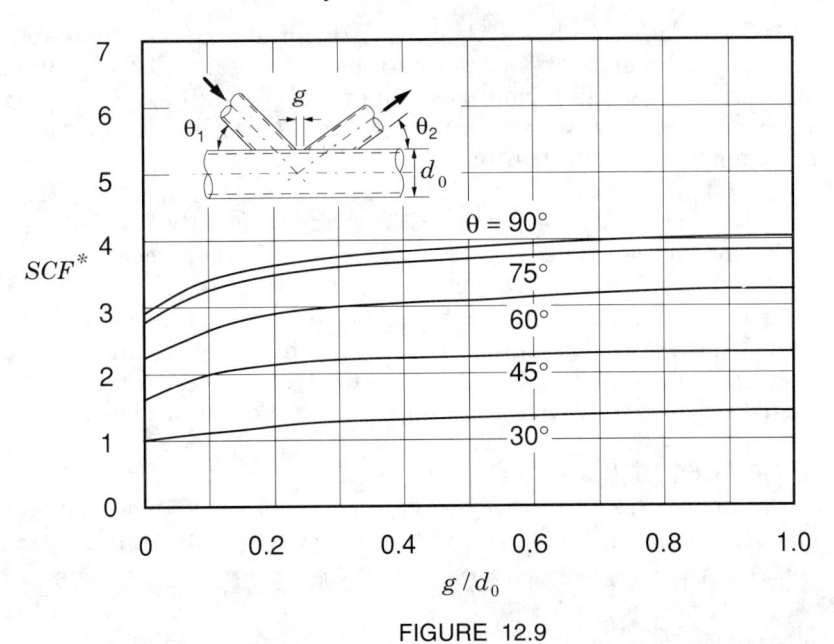

FIGURE 12.9

SCF^* for axially loaded K and N connections, for circular HSS

The parameter ranges of validity for SCF^* values to use with circular HSS T, Y, X, K and N connections are as follows:

» Angle between chord and web (or branch) must be between 30° and 90°.

» Ratio of chord diameter to twice the chord wall thickness must be between 12 and 32 (but it is conservative to use 12 when the ratio is less than 12).

» Ratio of branch member diameter to chord diameter must be between 0.13 and 1.0 for T and Y connections, or between 0.13 and 0.95 for X connections (but it is conservative to use 0.95 when the ratio is greater than 0.95), or the average web member diameter to chord diameter must be between 0.30 and 0.80 for K and N connections.

» Ratio of branch member wall thickness to chord wall thickness must be between 0.25 and 1.0 for T, Y and X connections, or the web member wall thickness to chord wall thickness must be between 0.40 and 0.80 for K and N connections.

» For K and N connections, the ratio of gap to chord diameter must be between 0.01 and 1.0 (a value of 0.01 is used for overlap connections).

12.6 Materials and Welding

The recommendations in this chapter apply equally for HSS manufactured to Class C or to Class H requirements of standard CAN/CSA-G40.20-M92.

Although CAN/CSA-G40.21-M92, Type W (weldable) steels meet specified strength requirements and are widely used for general welded construction of statically loaded structures, Types WT (weldable, notch tough) or AT (atmospheric corrosion resistant, weldable, notch tough) steels are recommended for fatigue applications.

A designer will want to specify the HSS grade (typically 350), class (C or H), type (WT or AT), and category (1 to 4). Type WT steels meet specified strength and Charpy V-Notch impact requirements and are suitable for welded construction where notch toughness at low temperature is a design requirement. Type AT steels also meet specified strength and Charpy V-Notch impact requirements; in addition, they display an atmospheric corrosion resistance of approximately four times that of plain carbon steels.

With Types WT or AT, a category of steel that establishes the Charpy V-Notch test temperature and energy level must also be specified. Category 1 corresponds to 0°C, category 2 to −20°C, category 3 to −30°C and category 4 to −45°C. A fifth category may be negotiated between customer and supplier.

Welding should be continuous around the perimeter of branch (web) members. Fillet welds, partial joint penetration groove welds and complete joint penetration groove welds, or a combination of the three, may be used. As suggested by the authors in Section 8.2.4.2, fillet weld throats should be no less than the wall thickness of the thinner member; however, fillet welds alone are not recommended for material over 8 mm thick.

One should be particularly careful not to start and stop weld sequences at the corners of square and rectangular HSS when fatigue is a design consideration, as these are the potential locations for fatigue crack initiation. A more complete discussion about welding in general is provided in Chapter 8. The potentially higher fatigue strength of an improved weld profile or an improved weld toe has not been considered in this chapter. However, a wealth of data does suggest that considerable gains in fatigue strength can be made by weld treatment.

12.7 Design Example

A circular HSS K connection is fabricated from 350WT steel, as shown in Fig. 12.10, and is contained within a truss which has been analyzed on the basis of pin-jointed members. At the specified load level (service loads) the forces on the connection are as shown in Fig. 12.10 for the static load case. Of this load, 40% is dead load and 60% is live load. The connection can be assumed to be well protected from the environment.

The aim is to determine

1) whether the static strength is adequate under the axial member loads

2) the fatigue life of the connection when the *live* load is continually fluctuating between zero and the maximum load value (Fig. 12.10)

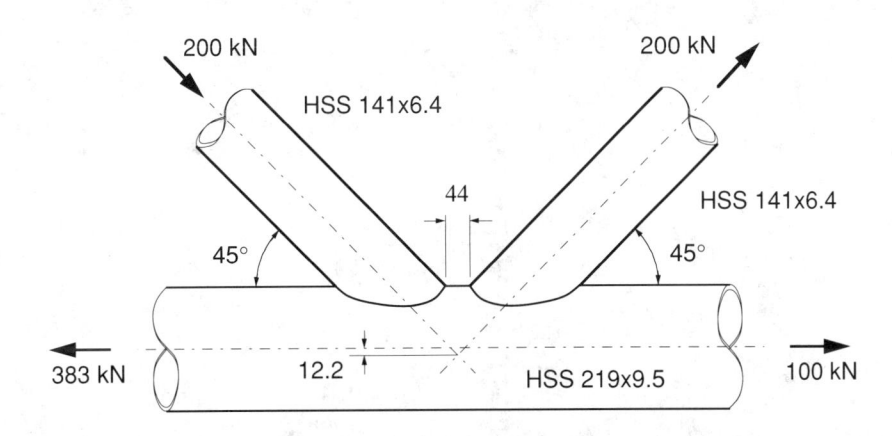

FIGURE 12.10

Circular HSS K connection used in the design example
showing total unfactored (specified) loads

for half the cycles; and between the maximum value and a minimum value which represents half the live load maximum, for the other half of the cycles. Total number of load cycles = n.

1. <u>Static Resistance</u>

$$N_1 = (1.25\,(0.40)\,200) + (1.50\,(0.60)\,200) = 280\,\text{kN}$$

$$e = \left(\frac{\sin\theta_1 \sin\theta_2}{\sin(\theta_1+\theta_2)}\right)\left[\left(\frac{d_1}{2\sin\theta_1}\right) + \left(\frac{d_2}{2\sin\theta_2}\right) + g\right] - \frac{d_0}{2} \qquad ([3.2])$$

$$= \left(\frac{0.707\,(0.707)}{1}\right)\left[\left(\frac{141}{2\,(0.707)}\right) + \left(\frac{141}{2\,(0.707)}\right) + 44\right] - \frac{219}{2}$$

$$= 12.2\,\text{mm}$$

Confirm validity of dimensional parameters (see Table 3.1(a))

$$0.2 < d_1/d_0 = d_2/d_0 = 141/219 = 0.64 \leq 1.0 \quad \therefore \text{OK}$$
$$d_1/t_1 = d_2/t_2 = 141/6.35 = 22.2 \quad \leq 50 \quad \therefore \text{OK}$$
$$\gamma = d_0/2t_0 = 219/2(9.53) = 11.5 \quad \leq 25 \quad \therefore \text{OK}$$
$$-0.55 \leq e/d_0 = 12.2/219 = 0.06 \leq 0.25 \quad \therefore \text{OK}$$
$$g = 44 \geq (t_1 + t_2) = 12.7 \quad \therefore \text{OK}$$
$$d_1/t_1 = 22.2 \quad \leq 28 \quad \therefore \text{ the connection efficiency need not be reduced in order to prevent local buckling in the compression web member}$$

Determine connection resistance of compression diagonal ($i = 1$)

$$N_1^* = \frac{F_{y0}\,t_0^2}{\sin\theta_1}(1.8 + 10.2\frac{d_1}{d_0})\;f(\gamma,g')\;f(n') \qquad \text{(Table 3.1)}$$

$$f(\gamma,g') = \gamma^{0.2}\left(1 + \frac{0.024\gamma^{1.2}}{\exp(0.5g' - 1.33) + 1}\right)$$

$$\gamma = 11.5 \qquad \text{(as above)}$$
$$g' = g/t_0 = 44/9.53 = 4.62$$

$$\therefore f(\gamma,g') = 11.5^{0.2}\left(1 + \frac{0.024\,(11.5)^{1.2}}{\exp(0.5\,(4.62) - 1.33) + 1}\right)$$

$$= 1.63\,[1 + (0.450/(2.66 + 1))] = 1.83$$

$$f(n') = 1.0 \text{ when } n' \geq 0 \quad \text{(tension)}$$

$$\therefore N_1^* = \frac{0.350\,(9.53)^2}{0.707}\left(1.8 + 10.2\,\frac{141}{219}\right)1.83\,(1.0)$$

$$= 688\,\text{kN} \quad \geq N_1 = 280 \quad \therefore \text{OK}$$

Also, $N_1^* \sin\theta_1 = N_2^* \sin\theta_2$

$$\therefore N_2^* = 688\,\text{kN} \quad \geq N_2 = 280 \quad \therefore \text{OK}$$

Check punching shear

$$N_1^* = N_2^* = \frac{F_{y0}}{\sqrt{3}}\,t_0\,\pi\,d_1\left(\frac{1+\sin\theta_1}{2\sin^2\theta_1}\right)$$

$$= \frac{0.350}{1.73}\,9.53\,(3.14)\,141\left(\frac{1+0.707}{2\,(0.50)}\right)$$

$$= 1460\,\text{kN} \quad \geq N_1 \text{ and } N_2$$

Therefore, connection static resistance is adequate.

2. **Fatigue Life**

The live load can be considered to have two constant amplitude components as shown in Fig. 12.11.

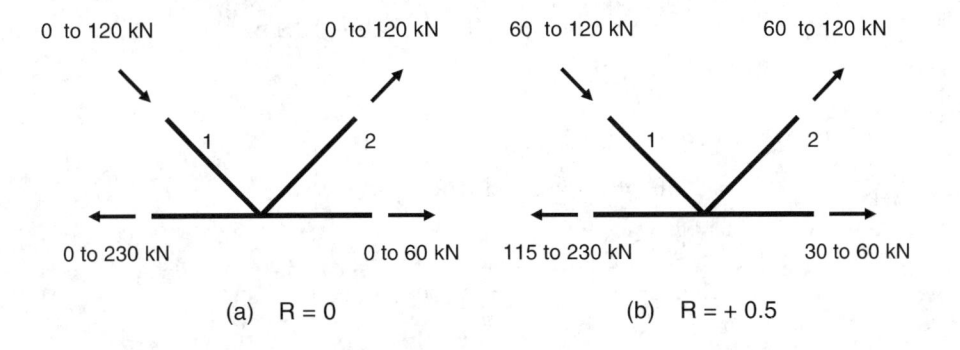

0 to 120 kN   0 to 120 kN   60 to 120 kN   60 to 120 kN

   1   2   1   2

0 to 230 kN   0 to 60 kN   115 to 230 kN   30 to 60 kN

(a) R = 0   (b) R = + 0.5

FIGURE 12.11

Constant amplitude dynamic loads acting on K connection

Confirm validity of dimensional parameters (see Table 12.5)

$30° \leq \theta_1 = \theta_2 = 45° \leq 90°$ ∴ OK

$\gamma = 11.5 \ < 12$ ∴ use $\gamma = 12$

$0.30 \leq \beta = 141/219 = 0.64 \leq 0.80$ ∴ OK

$0.40 \leq \tau = 6.35/9.53 = 0.67 \leq 0.80$ ∴ OK

$0.01 \leq g/d_0 = 44/219 = 0.20 \leq 1.0$ ∴ OK

Determine SCFs for chord and web

From Fig. 12.9, SCF^* for circular HSS K connection = 2.2 and from Table 12.4, $x1 = 0.67$ and $x2 = 1.10$.

$$\begin{aligned}
SCF_{ax-ch} &= (\gamma/12.5)^{x1} (\tau/0.5)^{x2} \, SCF^* \quad ([12.8]) \\
&= (12.0/12.5)^{0.67} (0.67/0.5)^{1.10} (2.2) \\
&= 2.95 \quad\quad \geq 2.0 \\
SCF_{ax-w} &= 1 + 0.63 \, SCF_{ax-ch} \quad ([12.8a]) \\
&= 1 + 0.63\,(2.95) = 2.86
\end{aligned}$$

(a) *Stress Ratio, R = 0*

Fatigue cracking only needs to be checked at the joint of web member 2, as this is the only web member which goes into tension.

For cracking in the web or chord, at the weld toes,

$$S_{r\,ax-nom} = 120\,000 \text{ N}/2\,690 \text{ mm}^2 = 44.6 \text{ MPa}$$

By incorporating secondary bending moments,

$$\begin{aligned}
S'_{r\,ax-w} &= 1.3\,(44.6) \quad\quad \text{(from Table 12.1)} \\
&= 58.0 \text{ MPa}
\end{aligned}$$

$$\begin{aligned}
(S_{r-h.s.})_{web} &= SCF_{ax-w} \cdot S'_{r\,ax-w} \quad ([12.1]) \\
&= 2.86\,(58.0) = 166 \text{ MPa}
\end{aligned}$$

$$\begin{aligned}
(S_{r-h.s.})_{chord} &= SCF_{ax-ch} \cdot S'_{r\,ax-w} \quad ([12.2]) \\
&= 2.95\,(58.0) = 171 \text{ MPa}
\end{aligned}$$

Curve B of Fig. 12.5 cannot be used because d_0 exceeds 200 mm.

Therefore, use curves labelled A to obtain

for web cracking, $t_i = 6.35$, so $N = 8\,(10)^5$ cycles

for chord cracking, $t_i = 9.53$, so $N = 6\,(10)^5$ cycles.

Check the effect of the nominal stress range in the chord.

$S_{r\,ax-nom}$ = 230 000 N / 6 270 mm^2 = 36.7 MPa

and $S'_{r\,ax}$ = 1.5(36.7) = 55.1 MPa   (from Table 12.1)

Curve C of Fig. 12.5 provides $N = 4\,(10)^6$ cycles for chord cracking caused by chord stress (which does not govern because $N = 6\,(10)^5$ cycles is more critical at that potential crack location).

(b) *Stress Ratio*, $R = +0.5$

For web member 2, $S_{r\,ax-nom}$ = 60 000 N / 2 690 mm^2 = 22.3 MPa

and $S'_{r\,ax-w}$ = 1.3 (22.3) = 29.0 MPa.

$\therefore (S_{r-h.s.})_{web}$ = 2.86 (29.0) = 82.9 MPa

and $(S_{r-h.s.})_{chord}$ = 2.95 (29.0) = 85.6 MPa.

Hence, for web cracking, $N = 1\,(10)^7$ cycles   (from Fig. 12.5)

and for chord cracking, $N = 5\,(10)^6$ cycles.

Once again, stress range in the chord (55.1 / 2 = 27.6 MPa) does not govern.

Therefore, one can see that the limiting number of load cycles will be governed for each constant amplitude stress range by cracking in the chord at the toe of the tension web member weld. With experience, one could see for this example that web cracking would not be critical, since $(S_{r-h.s.})_{web} < (S_{r-h.s.})_{chord}$, for both stress ratios, and $t_{web} < t_{chord}$.

Apply Miner's rule to this crack location

$$\sum_{i=1}^{2} \frac{n_i}{N_i} \leq 1.0 \qquad ([12.6])$$

$$\frac{0.5\,n}{6\,(10)^5} + \frac{0.5\,n}{5\,(10)^6} = 1.0 \text{ at the limit.}$$

$\therefore n = 1.07\,(10)^6$

The total number of live load cycles to failure is therefore $1\,(10)^6$.

12.8 HSS to Base Plate Connections

In some structures such as lighting poles and highway sign supports, HSS are fillet welded to base plates as shown in Fig. 12.12. The welded joint in such connections will be subjected to cyclic wind loading and fatigue may be a design concern.

CAN/CSA-S16.1-94, Clause 14 (1994) prescribes allowable ranges of normal stress ($S_{r\text{-}nom}$) in the HSS for such connections using a simple "classification" approach. The applicable fatigue design curve is given in Fig. 12.13, with allowable stress ranges of 109 MPa, 63 MPa and 40 MPa at 100 000, 500 000 and 2 000 000 cycles respectively. The endurance limit is reached at 18 MPa (i.e., at stress ranges of or below this magnitude, fatigue failure should not occur).

One should be particularly careful not to start and stop weld sequences at the corners of square or rectangular HSS when fatigue is a design consideration, as these are the potential locations for fatigue crack initiation.

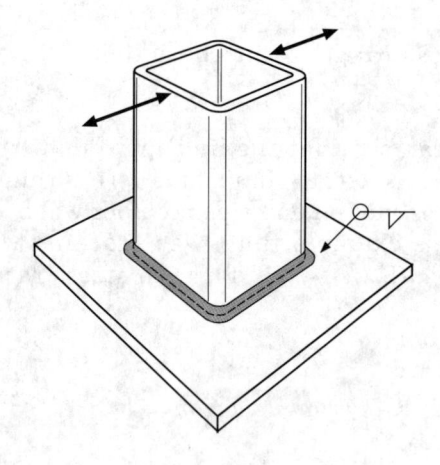

FIGURE 12.12
HSS to base plate connection subjected to cyclic loading

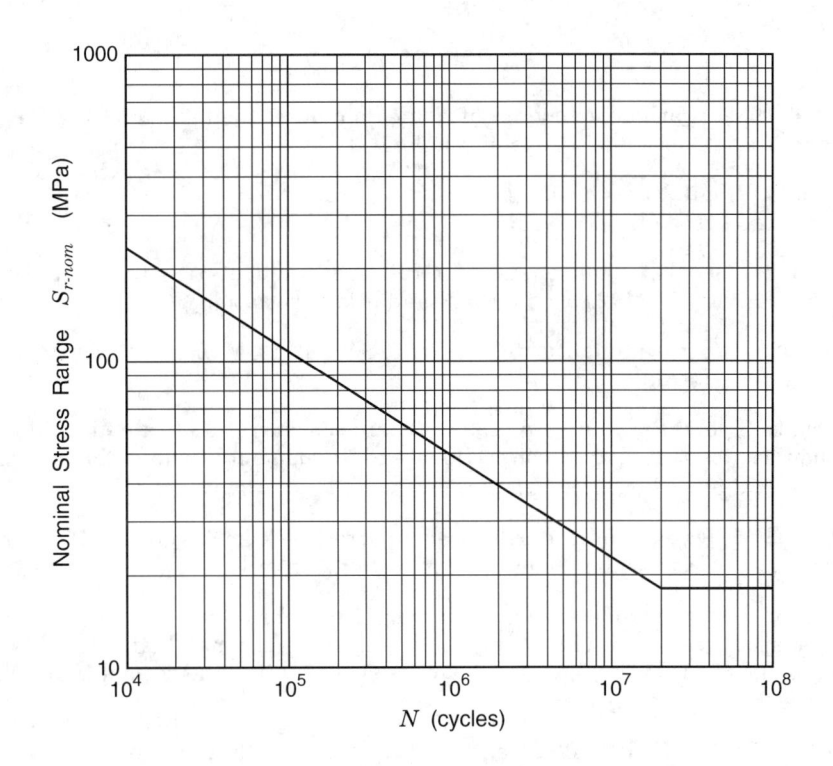

FIGURE 12.13

$S_{r\text{-}nom}$ vs. N curve for HSS to plate connection

REFERENCES

CIDECT. 1982. Fatigue behaviour of welded hollow section joints. CIDECT Monograph No. 7, Constrado, Croyden, England.

CSA. 1994. Limit states design of steel structures, CAN/CSA-S16.1-94. Canadian Standards Association, Rexdale, Ontario.

CSA. 1992. General requirements for rolled or welded structural quality steel, CAN/CSA-G40.20-M92. Canadian Standards Association, Rexdale, Ontario.

CSA. 1992. Structural quality steels, CAN/CSA-G40.21-M92. Canadian Standards Association, Rexdale, Ontario.

FRATER, G. S., and PACKER, J. A. 1993. Strain concentration factors in rectangular hollow section truss gap K-connections—an experimental study. Journal of Constructional Steel Research, **24**(2): 77–104.

HERION, S. 1994. Räumliche K-knoten aus rechteck-hohlprofilen. Ph.D. thesis, University of Karlsruhe, Federal Republic of Germany.

IIW. 1985. Recommended fatigue design procedure for hollow section joints. Part I —hot spot stress method for nodal joints. International Institute of Welding Subcommission XV-E, IIW Doc. XV-582-85, IIW Annual Assembly, Strasbourg, France.

MANG, F., HERION, S., BUCAK,Ö., and DUTTA, D. 1989. Fatigue behaviour of K joints with gap and with overlap made of rectangular hollow sections. Proceedings, 3rd. International Symposium on Tubular Structures, Lappeenranta, Finland, pp. 297–309.

NIEMI, E.J. 1996. Fatigue resistance predictions for RHS K-joints, using two alternative methods. Proceedings, 7th. International Symposium on Tubular Structures, Miskolc, Hungary, pp. 309–314.

NIEMI, E.J., and YRJÖLÄ, P. 1995. A novel fatigue analysis approach for tubular welded joints. Proceedings, Fatigue Design '95, Helsinki, Finland, pp. 189–201.

PACKER, J.A., FRATER, G.S., and ELLIOTT, K.S. 1990. Experimental determination of strain concentration factors in RHS truss gap K connections. Proceedings, Conference on Applied Stress Analysis, Nottingham, England, pp. 316–325.

PANJEHSHAHI, E. 1995. Stress concentration factors for multiplanar joints in RHS. Heron, **40**(4): 341–352.

PEDERSEN, N.T., and AGERSKOV, H. 1991. Fatigue damage accumulation in offshore tubular structures under stochastic loading. Proceedings, 4th. International Symposium on Tubular Structures, Delft University of Technology, The Netherlands, pp. 269–280.

PUTHLI, R.S., WARDENIER, J., de KONING, C.H.M., van WINGERDE, A.M., and van DOOREN, F.J. 1988. Numerical and experimental determination of strain (stress) concentration factors of welded joints between square hollow sections. Heron, **33**(2): 1–50.

WARDENIER, J. 1982. Hollow section joints. Delft University Press, Delft, The Netherlands.

van WINGERDE, A. M. 1992. The fatigue behaviour of T- and X- joints made of square hollow sections. Heron, **37**(2): 1–180.

van WINGERDE, A.M., and PACKER, J.A. 1994. Fatigue design of connections between hollow structural sections. Proceedings, International Conference on Fatigue, Toronto, Ontario, pp. 28–36.

van WINGERDE, A.M., and PACKER, J.A. 1996. Fatigue of hollow structural section welded connections. Proceedings, International Conference on Tubular Structures, Vancouver, B.C., pp. 64–73.

van WINGERDE, A.M., PACKER, J.A., and WARDENIER, J. 1995. Criteria for the fatigue assessment of hollow structural section connections. Journal of Constructional Steel Research, **35**(1): 71–115.

van WINGERDE, A.M., PACKER, J.A., and WARDENIER, J. 1996a. New guidelines for fatigue design of HSS connections. Journal of Structural Engineering, American Society of Civil Engineers, **122**(2): 125–132.

van WINGERDE, A.M., PACKER, J.A., and WARDENIER, J. 1996b. Determination of stress concentration factors for K-connections between square hollow sections. Proceedings, 6th. International Offshore and Polar Engineering Conference, Los Angeles, California, U.S.A., Vol. 4, pp. 52–59.

van WINGERDE, A.M., PACKER, J.A., and WARDENIER, J. 1996d. Stress concentration factors for K-connections between square hollow sections. Proceedings, 7th. International Symposium on Tubular Structures, Miskolc, Hungary, pp. 323–330.

van WINGERDE, A.M., PACKER, J.A., and WARDENIER, J. 1997. IIW fatigue rules for tubular joints. Proceedings, IIW International Conference on Performance of Dynamically Loaded Welded Structures, San Francisco, California, U.S.A.

van WINGERDE, A.M., PACKER, J.A., STRAUCH, L., SELVITELLA, B., and WARDENIER, J. 1996c. Fatigue behaviour of non-90° square hollow section X-connections. Proceedings, 7th. International Symposium on Tubular Structures, Miskolc, Hungary, pp. 315–322.

van WINGERDE, A.M., PACKER, J.A. WARDENIER, J., DUTTA, D., and MARSHALL, P.W. 1993. Proposed revisions for fatigue design of planar welded connections made of hollow structural sections. Proceedings, 5th. International Symposium on Tubular Structures, Nottingham, England, pp. 663–672.

van WINGERDE, A.M., PUTHLI, R.S., de KONING, C.H.M., VERHEUL, A., WARDENIER, J., and DUTTA, D. 1989. Fatigue strength of welded unstiffened RHS joints in latticed structures and vierendeel girders. CIDECT Final Report 7E+7F-89/5E, Delft University of Technology, Delft, The Netherlands.

van WINGERDE, A.M., PUTHLI, R.S., WARDENIER, J., DUTTA, D., and PACKER, J.A. 1992. Design recommendations and commentary regarding the fatigue behaviour of hollow section joints. Proceedings, 2nd. International Offshore and Polar Engineering Conference, San Francisco, California, U.S.A., Vol. 4, pp. 288–295.

13

STANDARD TRUSS EXAMPLES

13.1 General

Examples have been selected to illustrate the use of the IIW (1989) formulae and their ranges of validity, as presented in Tables 3.1, 3.2 and 3.3. A Warren truss consisting of square HSS is presented in some detail as the main example, since that configuration, for reasons mentioned in Section 2.2.1, is often the preferred solution. A similar truss using circular HSS is then shown. These trusses are followed by a number of individual connection designs which illustrate particular situations.

Calculations are performed with the HSS section "sizes" (3 significant figures) while reference to the HSS sections is by the CAN/CSA-G312.3-M92 "designations" (only 2 significant figures for some dimensions).

Material used throughout is the widely available CAN/CSA-G40.21-M92 350W, with specified minimum yield of 350 MPa, produced to the requirements of CAN/CSA-G40.20-M92, Class C. Alternative material is 350W, Class H which, because of higher resistances than Class C when used for compression members, can result in lighter sections.

Comments have been inserted throughout to draw attention to specific points.

13.2 Warren Truss with Square HSS

The length of the example truss is long enough to require chord splices, but not too long to ship the truss in one piece if circumstances are favourable. We shall assume that the chord splices are to be shop welded, and that equal width and height sections are desired at the splices. Alternative bolted chord splices will also be developed.

Table 13.1 shows the truss and factored loads along with member axial forces, determined by a pin-jointed analysis. Specified loads are 3.0 kPa live and 1.5 kPa dead, with the trusses spaced at five metres, framing into jack trusses which span between columns 10 metres apart. The top (compression) chord is considered to be laterally supported at each purlin position.

It will be noticed that this is a modified Warren truss with vertical members in the central region. Such a configuration can work well because the verticals provide support for purlins where the top chord has large axial loads. Towards the ends, where axial forces are less, bending loads from purlins can be accepted by a chord section approximately the same size as that used at the centre. This can reduce the number of different section sizes to be procured. It eliminates several verticals along with their connections and it provides a chord section more suitable for connections to the diagonal web members in the region where truss shear forces are greatest. Although there is more material in the chord, the trade-off can be advantageous.

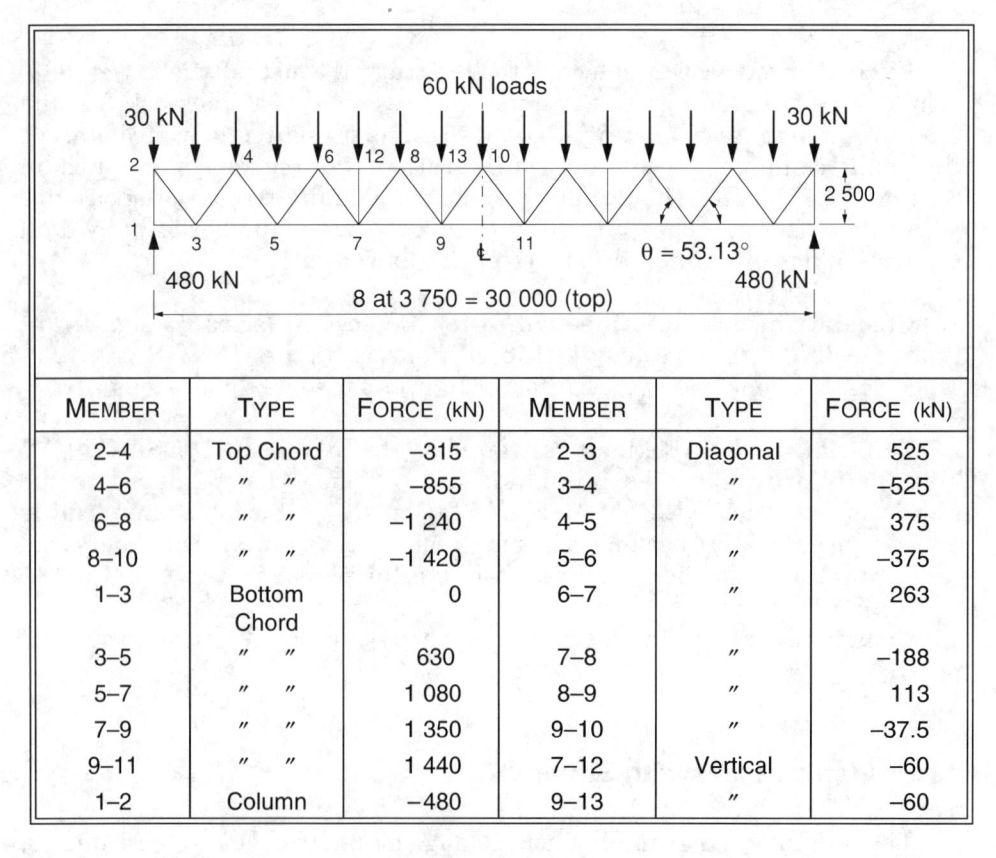

Member	Type	Force (kN)	Member	Type	Force (kN)
2–4	Top Chord	−315	2–3	Diagonal	525
4–6	" "	−855	3–4	"	−525
6–8	" "	−1 240	4–5	"	375
8–10	" "	−1 420	5–6	"	−375
1–3	Bottom Chord	0	6–7	"	263
3–5	" "	630	7–8	"	−188
5–7	" "	1 080	8–9	"	113
7–9	" "	1 350	9–10	"	−37.5
9–11	" "	1 440	7–12	Vertical	−60
1–2	Column	−480	9–13	"	−60

TABLE 13.1

Warren truss showing applied loads and resulting member forces

13.2.1 Considerations when Making Preliminary Member Selections

Time will be saved when selecting member sizes by keeping in mind the basic constraints or "limits of validity" of various dimensional parameters that must be met for connections. These were examined in Section 3.6 of the text. Also, it can be expedient to pick overcapacity sections (by as much as a third in some cases) since high efficiency connections are not always achievable when joining HSS.

The validity limits for a *gap* K or N connection with *square* chords and webs (Table 3.2(a)) are

1. Average width of the two web members must be at least 0.35 times the chord width.
2. Width of the web members, relative to the chord width, must be at least a hundredth of the chord width-to-thickness ratio, plus 0.1.
3. Compression web members must be either CAN/CSA-S16.1-94 Class 1 or Class 2 sections.
4. Maximum width to thickness ratio of tension web members is 35.
5. Width to thickness ratio of the chord must be between 15 and 35.
6. Width of the smaller web member must be at least 0.63 times the width of the other web member (otherwise resort to the more general rectangular HSS Table 3.3).
7. A limit is defined for the minimum gap as a function of chord and web member widths.
8. Absolute minimum gap is the sum of the wall thicknesses of the two web members.
9. Nodal eccentricity limits are −0.55 and +0.25 of the chord height. If the positive eccentricity allowed by this constraint is exceeded, treat the connection as a pair of T / Y connections.

For an *overlap* K or N connection with *square* chords and webs the validity limits (Table 3.2(a)) are

1. Width of the web members must be at least 0.25 times the chord width.
2. Compression web members must be CAN/CSA-S16.1-94, Class 1 sections.
3. Maximum width to thickness ratio of tension web members is 35.
4. Maximum width to thickness ratio of the chord is 40.
5. Wall thickness of the overlapping member must not exceed the wall thickness of the overlapped member (for equal yield strength materials).

6. Width of the overlapping web member must be at least 0.75 times the width of the overlapped web member.

7. Minimum overlap is 25%.

8. Nodal eccentricity limits are the same as for gap connections.

These constraints place extensive limits upon the selection of members. However, once the constraints are met, determination of the connection resistance is much simplified, and the Design Charts and Resistance Tables in Sections 3.7.1 and 3.7.2 can usually be used. For gap connections, only the plastic failure of the chord face needs to be considered; for overlapped connections, only the overlapping member needs to be examined, and then only for the "effective width" failure criterion.

Remember the final check for overlap connections. The connection efficiency, defined as the ratio of factored connection resistance to the member yield load (area times specified minimum material yield), of the overlapped member cannot be taken to be greater than the efficiency of the overlapping member. (Note later that framing with circular HSS does not have this requirement.)

13.2.2 Preliminary Member Selections

The values for compressive resistances C_r are interpolated from the CISC Handbook, pp. 4-64 to 4-67 (CISC 1995).

Central top chord

Load in length 8–10 is −1 420 kN. KL is 0.9 (1 875) = 1 690

Try HSS 178x178x9.5 $C_{r0} = -1\,870$ kN $A_0 = 6\,180$ mm^2

> It will be shown later that an HSS 178x178x8.0 ($C_{r0} = -1\,590$ kN) would have too thin a wall.

Utilization is 1 420 / 1 870 = 0.76.

Outer top chord

Load in length 4–6 is −855 kN. KL is 0.9 (3 750) = 3 380

A conservative approximation of the bending in the chord at panel point No. 4 is given by considering the portion of top chord from the end of the truss to panel point No. 6 to be a 2-span continuous member. Then the moment at panel point No. 4 would be 0.188 PL.

∴ $M_{f0} = 0.188\,(60)\,3.75 = 42.3$ kN·m

If one were to limit this moment to the yield moment of the chord, then as

$$M_{y0} = 0.9\, F_{y0}\, S_0$$

the required S_0 would be

$$42.3/0.9(0.350) = 134\,(10^3)\ \text{mm}^3.$$

Knowing that the applied forces (axial load plus bending) are to be subsequently combined, a generous member will be selected with about twice the capacity needed for the individual forces.

Try HSS 178x178x9.5

$$C_{r0} = -1\,570\ \text{kN} \qquad S_0 = 322\,(10^3)\ \text{mm}^3 \qquad A_0 = 6\,180\ \text{mm}^2$$

Again, it will be shown later that an HSS 178x178x8.0 ($C_{r0} = -1\,340$ kN) would have too thin a wall.

Central bottom chord

Load in member 9–11 is 1 440 kN.

$$A_{(min)} = T_{f0}/(\phi\, F_{y0}) = 1\,440 / 0.9\,(0.350) = 4\,570\ \text{mm}^2$$

Try HSS 152x152x9.5 $A_0 = 5\,210\ \text{mm}^2$

Utilization is $1\,440 / (0.9\,(5\,210)\,0.350) = 0.88$

Outer bottom chord

Load in member 5–7 is 1 080 kN.

$$\therefore A_{(min)} = 1\,080 / 0.9\,(0.350) = 3\,430\ \text{mm}^2$$

Try HSS 152x152x8.0 $A_0 = 4\,430\ \text{mm}^2$

Once again, it will be shown later that an HSS 152x152x6.4 ($A_0 = 3610\ \text{mm}^2$) would have too thin a wall.

Utilization is $1\,080 / (0.9\,(4\,430)\,0.350) = 0.77.$

Tension diagonal at end of truss

Load in member 2–3 is 525 kN.

$$\therefore A_{(min)} = 525 / 0.9\,(0.350) = 1\,670\ \text{mm}^2$$

Try HSS 102x102x4.8 $A_2 = 1\,790$ mm^2.

Utilization is $525 / (0.9\,(1790)\,0.350) = 0.93$.

Compression diagonal nearest to end of truss

Load in member 3–4 is -525 kN. KL is $0.75\,(3\,125) = 2\,340$

Try HSS 127x127x4.8 $C_{r1} = -595$ kN $A_1 = 2\,280$ mm^2

Utilization is $525 / 595 = 0.88$.

Compression diagonal at middle of truss

Load in member 9–10 is -37.5 kN. KL is 2 340

Try HSS 64x64x3.2 $C_{r1} = -105$ kN $A_1 = 741$ mm^2

This is just about the smallest size one would want to use in a truss this large, especially when one reflects on the member width relative to the truss depth.

Utilization is $37.5 / 105 = 0.36$

All verticals

Load is -60 kN. $KL = 0.75\,(2\,500) = 1\,880$

Try HSS 64x64x3.2 $C_{r1} = -136$ kN $A_1 = 741$ mm^2

Utilization is $60 / 136 = 0.44$.

Tension diagonal near truss centre

Load in member 8–9 is 113 kN.

$\therefore A_{(min)} = 113 / 0.9(0.350) = 359$ mm^2

Try HSS 64x64x3.2 $A_2 = 741$ mm^2

Utilization is $113 / (0.9\,(741)\,0.350) = 0.48$.

Compression diagonal 7–8

Load is -188 kN. $KL = 2\,340$ mm

Try HSS 76x76x4.8 $C_{r1} = -227$ kN $A_1 = 1\,310$ mm^2

Utilization is $188 / 227 = 0.83$.

Diagonals 4–5 and 5–6

Loads are 375 and –375 kN.

Use same sections as for members 2–3 and 3–4 to avoid a multitude of different sizes.

> It will also be shown later that diagonal 5–6 needs this size of section, even though the load is less than for diagonal 3–4.

Tension diagonal 6–7

Load in member 6–7 is 263 kN.

$\therefore A_{(min)} = 263 / 0.9(0.350) = 835$ mm^2

Try HSS 76x76x4.8 $A_2 = 1310$ mm^2

Utilization is $263 / (0.9 (1310) 0.350) = 0.64$.

> This section may appear oversize at first but the parametric requirement for a gap connection at panel point No. 6, that the width of the smaller web member (76.2 mm) be at least 0.63 times the width of the larger web member = 0.63(127 mm) = 80.0 mm, is already being stretched.

The truss with the trial members is shown in Fig. 13.1

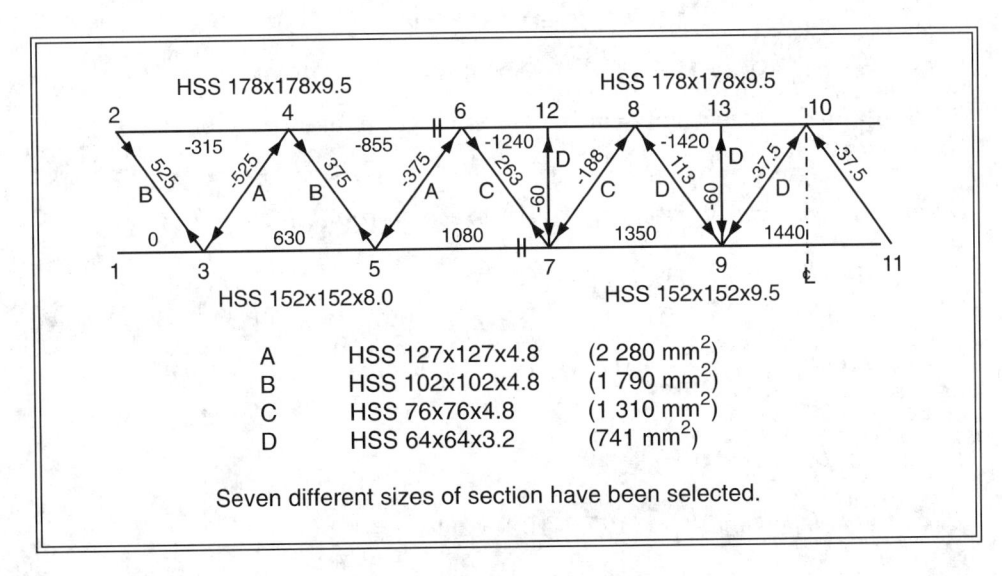

FIGURE 13.1

Preliminary selection of members—Warren truss with square HSS

13.2.3 Resistance of Gap K Connections

Panel Points No. 3 and No. 5

Confirm that a gap connection is feasible, (Fig. 3.11)

$h_0/b_0 = 1.0$ $\theta = 53.1°$ $\therefore$ max. $\beta = 0.84$

Actual $\beta = (b_1 + b_2)/(2b_0)$
$= (127 + 102)/2(152) = 0.75$ $\therefore$ OK

Confirm validity of dimensional parameters, (Section 13.2.1)

1. $\beta = 0.75$ ≥ 0.35 $\therefore$ OK
2. $b_2/b_0 = 102/152 = 0.67$
   $\geq 0.01(b_0/t_0) + 0.1 = 0.01(152/7.95) + 0.1$
   $= 0.29$ $\therefore$ OK
3. Flat width to thickness ratio of compression diagonal is
   $(127 - 4(4.78))/4.78 = 22.6$
   Requirement for Class 1 is $420/(F_{y1})^{0.5}$
   $= 420/(350)^{0.5} = 22.5$
   Although this section could fairly be considered as Class 1, only Class 2 is needed here. $\therefore$ OK
4. $b_2/t_2 = 102/4.78 = 21.3$ ≤ 35 $\therefore$ OK
5. $15 \leq b_0/t_0 = 152/7.95 = 19.1 \leq 35$ $\therefore$ OK
6. $b_2/b_1 = 102/127 = 0.80$ ≥ 0.63 $\therefore$ OK
7, 8 and 9. Effectively confirmed with gap feasibility above.

Determine connection resistance, compression diagonal 3-4 (i = 1)

Use Design Chart, Fig. 3.23.

$$N_1^* = C_{K,gap} \frac{t_0}{t_1} \frac{1}{\sin\theta_1} f(n) A_1 F_{y1}$$

$b_0/t_0 = 19.1$

$(b_1 + b_2)/2b_1 = (127 + 102)/2(127) = 0.90$

$\therefore C_{K,gap} = 0.33$

$f(n) = 1.0$ for a tension chord, (Table 3.2)

$$\therefore N_1^* = 0.33\,(7.95/4.78)\,(1/0.8)\,1.0\,(2\,280)\,0.350$$
$$= 547\,\text{kN} \quad \geq 525 \quad \therefore \text{OK}$$

Determine connection resistance of tension diagonal 2–3 ($i = 2$).

$$N_2^* = C_{K,gap}\,\frac{t_0}{t_2}\,\frac{1}{\sin\theta_2}\,f(n)\,A_2\,F_{y2}$$

$$(b_1 + b_2)/2b_2 = (127 + 102)/2(102) = 1.12$$

$$\therefore C_{K,gap} = 0.42$$

$$\therefore N_2^* = 0.42\,(7.95/4.78)\,(1/0.8)\,1.0\,(1\,790)\,0.350$$
$$= 547\,\text{kN} \quad \geq 525 \quad \therefore \text{OK}$$

Alternatively, by examining the equation for N_i^* in Table 3.2 it is apparent that $N_1^* \sin\theta_1 = N_2^* \sin\theta_2$ which, in this case, means $N_1^* = N_2^*$.

Therefore, panel points No. 3 and No. 5 are acceptable.

<small>Note that the connection resistances would have been insufficient if the chord had a 6.35 mm thickness rather than the 7.95 selected.</small>

Panel Point No. 4

Refer to Fig. 13.2. (Gap of 32 mm is minimum allowed by criterion in Table 3.2(a) for parameter 7 in Section 13.2.1.)

Confirm validity of dimensional parameters.

A review similar to that done for panel point No. 3 showed them to be adequate.

Confirm, as before, that a gap connection is feasible by using Fig. 3.11.

$$\beta = (127 + 102)/2(178) = 0.643 \quad \leq 0.84 \quad \therefore \text{OK}$$

Bending in chord from purlins

The bending moment at the connection due to purlin loads is considered to be $0.188\,PL$, as discussed when selecting preliminary member sizes in Section 13.2.2.

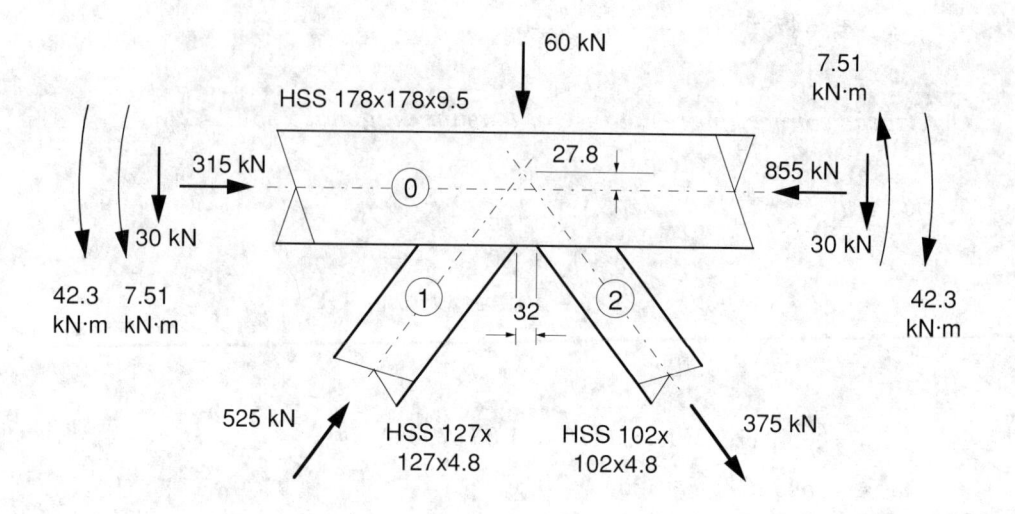

FIGURE 13.2
Panel point No. 4

$$\therefore M_{f0} = 0.188(60)3.75 = 42.3 \text{ kN·m}$$

Determine f(n) *for panel point No. 4.*

$$f(n) = 1.3 + \frac{0.4}{\beta} n \quad \text{but} \not> 1.0 \quad \text{(Table 3.2)}$$

$$\beta = 0.643$$

$$n = C_{f0}/(A_0 F_{y0}) + M_{f0}/(S_0 F_{y0}) \quad \text{(by definition)}$$
$$= (-855)/(6180(0.350)) + (-42.3)/(322(0.350))$$
$$= -0.771$$

$$\therefore f(n) = 1.3 + (0.4/0.643)(-0.771) = 0.820$$

Alternatively, use Design Chart, Fig. 3.24

to determine f(n) from n and β.

Note that only the primary moment due to transverse loading on the chord, (purlin loading), was used for M_{f0}. The primary bending moment due to nodal eccentricity was not included because the eccentricity e was within specified limits.
i.e., $-0.55 \le e/h_0 \le 0.25$.

Determine connection resistance, compression diagonal 3–4 (i = 1).

Use Design Chart, Fig. 3.23

$$N_1^* = C_{K,gap} \frac{t_0}{t_1} \frac{1}{\sin\theta_1} f(n) A_1 F_{y1}$$

$$b_0/t_0 = 178/9.53 = 18.7$$

$$(b_1 + b_2)/2b_1 = (127 + 102)/2(127) = 0.90$$

$$\therefore C_{K,gap} = 0.34$$

$$\therefore N_1^* = 0.34\,(9.53/4.78)\,(1/0.8)\,0.820\,(2\,280)\,0.350$$

$$= 554\text{ kN} \quad \geq 525 \quad \therefore \text{ OK}$$

Determine connection resistance of tension diagonal 4–5 (i = 2).

$$N_2^* = C_{K,gap} \frac{t_0}{t_2} \frac{1}{\sin\theta_2} f(n) A_2 F_{y2}$$

$$(b_1 + b_2)/2b_2 = (127 + 102)/2(102) = 1.12$$

$$\therefore C_{K,gap} = 0.43$$

$$\therefore N_2^* = 0.43\,(9.53/4.78)\,(1/0.8)\,0.820\,(1\,790)\,0.350$$

$$= 551\text{ kN} \quad \geq 375 \quad \therefore \text{ OK}$$

A couple of interesting points emerge at this stage:

First, as at panel point No. 3, Table 3.2 could have been used to show that $N_1^* = N_2^*$. The difference between them (above) of less than 1% comes from the rounding associated with the use of graphical charts.

Second, when there is an external load (such as a purlin) at a K connection, the connection resistance calculated for the more lightly loaded web member might be higher than the capacity of the member itself (as is in fact the case at panel point No. 6). This just means that the member governs, not the connection.

Therefore, panel point No. 4 is acceptable.

Note that the connection resistance of the compression diagonal would have been insufficient if the chord had a 7.95 mm thickness rather than the 9.53 selected.

Panel Point No. 6

Panel point No. 6 is similar to panel point No. 4, but has a smaller tension web member.

Validity of the dimensional parameters checked out OK.

A gap connection was confirmed feasible because

$$\beta = (127 + 76.2)/(2\,(178)) = 0.571 \quad \leq 0.84 \quad \text{(from Fig. 3.11)}.$$
$$\text{and} \geq 0.35 \quad \therefore \text{OK}$$

At panel point (node) No. 4, in order to conservatively estimate the primary bending in the chord due to the purlin load on each side of the panel point, the chord was considered to be continuous from node No. 2 to node No. 6. Now, one needs to conservatively estimate the primary bending moment in the chord at panel point No. 6 due to the purlin load between nodes No. 4 and No. 6. Consider that if the chord were fully fixed against rotation at these nodes, the bending moment at panel point No. 6 would be $PL/8$ due to the purlin load at mid span. For equilibrium, the moment to the right of node No. 6 would balance the moment to the left ($PL/8$).

Therefore, use a moment of $60(3.75)/8 = 28.1$ kN·m

Determine $f(n)$ *for panel point No. 6.*

$$\beta = 0.571$$

$$\begin{aligned} n &= C_{f0}/(A_0 F_{y0}) + M_{f0}/(S_0 F_{y0}) \\ &= (-1\,240)/(6\,180\,(0.350)) + (-28.1)/(322\,(0.350)) \\ &= -0.821 \end{aligned}$$

∴ from Design Chart, Fig. 3.24, $f(n) = 0.73$

Determine connection resistance, compression diagonal 5–6 ($i = 1$).

$$(b_1 + b_2)/2b_1 = (127 + 76.2)/2\,(127) = 0.80$$

$$b_0/t_0 = 18.7$$

$$\therefore C_{K,gap} = 0.30 \quad \text{(Fig. 3.23)}$$

$$\begin{aligned} \therefore N_1^* &= 0.30\,(9.53/4.78)\,(1/0.8)\,0.73\,(2\,280)\,0.350 \\ &= 436 \text{ kN} \quad \geq 375 \quad \therefore \text{OK} \end{aligned}$$

It is interesting to compare this value of the connection resistance with that at panel point No. 4 (554 kN) where the same size diagonal connects to the same size chord. The difference is mostly due to the smaller size tension diagonal used here, which results in lower values of $f(n)$ (because of lower β), and $C_{K,gap}$ (because of lower $(b_1 + b_2)/2b_1$).

Note again that the connection resistance of the compression diagonal requires that $t_0 = 9.53$; 7.95 would not do.

Determine connection resistance, tension diagonal 6–7 ($i = 2$).

As discussed previously, $N_2^* = N_1^*$ for this truss.

436 kN resistance ≥ 263 kN load ∴ OK

Therefore, panel point No. 6 is acceptable.

Panel point No. 8 can likewise be shown to be acceptable.

13.2.4 Top Chord as a Beam-Column between Panel Points 2 and 6

Determine nodal eccentricity at panel point No. 4.

Minimum gap allowed is $0.5(1-\beta)b_0$ (Table 3.2(a))

$= 0.5(1 - 0.643)178 = 31.8$ mm ∴ use 32 mm

$$e = \frac{\sin\theta_1 \sin\theta_2}{\sin(\theta_1+\theta_2)}\left[\frac{h_1}{2\sin\theta_1} + \frac{h_2}{2\sin\theta_2} + g\right] - \frac{h_0}{2} \quad ([3.2])$$

$= [0.8(0.8)/0.96][(127/1.6) + (102/1.6) + 32] - (178/2)$

$= 27.8$ mm

Determine bending moments in the chord at panel point No. 4 (see Fig. 13.2).

Moment from noding eccentricity $= e\,((525 + 375)\text{ kN})\cos\theta$.

$= 27.8(900)\,0.6 = 15\,000$ kN·mm

or 7.51 kN·m each side of the panel point

Moment from purlin loads is considered to be $0.188\,PL$, as discussed when selecting preliminary member sizes in Section 13.2.2.

$= 0.188(60)3.75 = 42.3$ kN·m

For design of the compression chord, *both* the moment due to transverse loading (purlin loads) and the moment due to noding eccentricity must be taken into account.

Thus, Fig. 13.3 shows the moment combinations existing at panel point No. 4:

42.3 + 7.51 = 49.8 kN·m for chord length 2–4

and 42.3 − 7.51 = 34.8 kN·m for chord length 4–6.

Determine bending moment under the purlins.

Use of the same model that was employed for the chord moment at panel point No. 4, (chord being 2 continuous spans over the panel point), gives a conservative value of 0.156 *PL*.

0.156(60)3 750 = 35.1 kN·m

Fig. 13.3 shows the moment combinations to be

35.1 − 7.51 / 2 = 31.3 kN·m for chord length 2–4

and 35.1 + 7.51 / 2 = 38.9 kN·m for chord length 4–6.

Use CAN/CSA-S16.1-94, Clause 13.8.1 for beam-columns.

Axial plus bending combination is $\dfrac{C_{f0}}{C_{r0}} + \dfrac{U_1 M_{f0}}{M_{r0}}$.

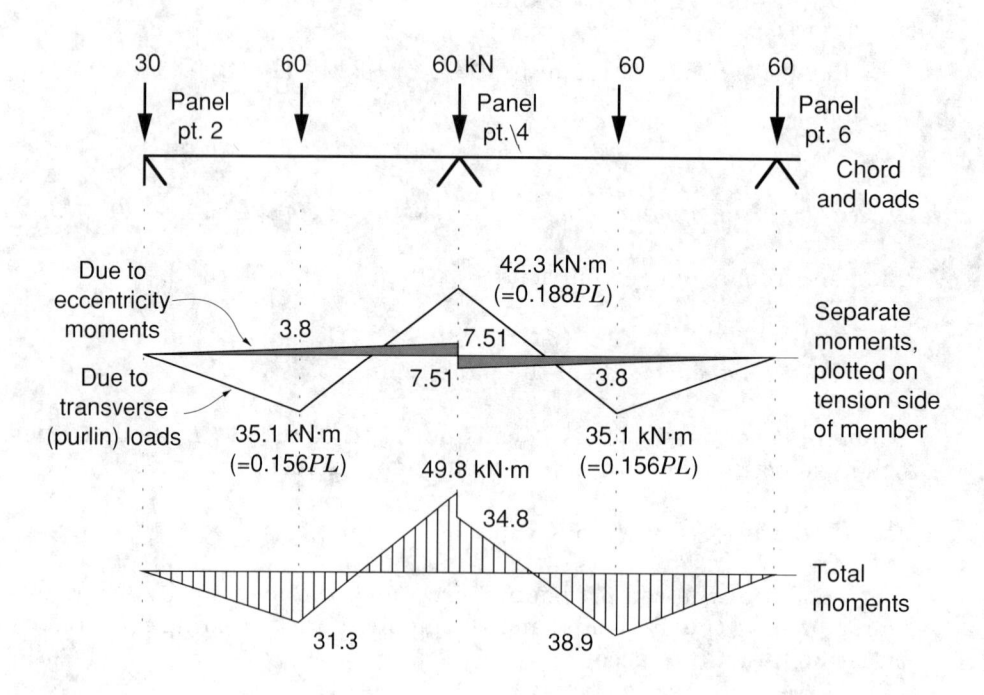

FIGURE 13.3
Simplified beam-column moments for manual design of chord

i) Overall member strength for chord length 4–6 (S16.1, Clause 13.8.1(b))

$C_{f0} = -855$ kN $\qquad KL = 1.0(3\,750) = 3\,750$

$C_{r0} = -1\,490 \qquad (K = 1.0$ for S16.1, Clause 13.8.1(b))

$$U_1 = \frac{\omega_1}{1 - \dfrac{C_{f0}}{C_e}} \qquad \text{(S16.1, Clause 13.8.3)}$$

$$C_e = \frac{\pi^2 EI}{L^2}$$

$\qquad = 3.14^2\,(200)\,28.6(10)^6/3\,750^2 = 4\,010$ kN

$\omega_1 = 0.85 \qquad$ (S16.1, Clause 13.8.4(c))

$\therefore U_1 = 0.85 / (1 - (855 / 4\,010)) = 1.08$

M_{f0} = maximum moment = 38.9 kN·m (see Fig. 13.3)

(chord length 4–6 considered as two segments as per S16.1, Clause 13.8.4(c))

$M_{r0} = \phi Z_0 F_{y0} = 0.9(385)0.350 = 121$ kN·m

$\therefore$ combination $= 855 / 1\,490 + 1.08\,(38.9) / 121$

$\qquad\qquad = 0.92 \quad \leq 1.0 \quad \therefore$ OK

ii) Overall member strength for chord length 2–4

$C_{f0} = -315$ kN

$C_{r0} = -1490$ kN

$U_1 = 0.85/(1 - (315/4\,010)) = 0.922$

$M_{f0} = 49.8$ kN·m

$M_{r0} = 121$ kN·m

$\therefore$ combination $= 315/1490 + 0.922\,(49.8)/121$

$\qquad\qquad = 0.59 \quad \leq 1.0 \quad \therefore$ OK

iii) Cross-sectional strength for chord length 4–6 (S16.1, Clause 13.8.1(a))

$C_{f0} = -855$ kN

$$C_{r0} = -\phi A_0 F_{y0} = -0.9\,(6\,180)\,0.350 = -1\,950 \text{ kN}$$
$$U_1 = 1.0$$
$$M_{f0} = 38.9 \text{ kN·m}$$
$$M_{r0} = 121 \text{ kN·m}$$

which, by comparison to case i), will not govern.

iv) Cross-sectional strength for chord length 2–4

$$C_{f0} = -315 \text{ kN}$$
$$C_{r0} = -1\,950 \text{ kN}$$
$$U_1 = 1.0$$
$$M_{f0} = 49.8 \text{ kN·m}$$
$$M_{r0} = 121 \text{ kN·m}$$

$\therefore$ combination $= 315/1950 + 1.0\,(49.8)/121$
$$= 0.57 \quad \leq 1.0 \quad \therefore \text{ OK}$$

Therefore, the chord is acceptable as a beam-column.

13.2.5 Resistance of an X Connection

Panel point No. 13 is an X connection because the transverse compression load is transferred through the chord member. It is assumed that the width of the load application point on the chord is no less than the width of the web member on the underside of the chord (63.5 mm).

Refer to Design Chart, Fig. 3.21

$$\beta = 63.5/178 = 0.357$$

$$N_1^* = C_X \frac{t_0}{t_1} \frac{1}{\sin\theta_1} f(n)\, A_1 F_{y1}$$

$$b_0/t_0 = 178/9.53 = 18.7$$

$\therefore C_X = 0.24$

$$n = (-1\,420)/(6\,180\,(0.350)) = -0.66$$

$\therefore f(n) = 1.3 + (0.4/0.36)(-0.66) = 0.57$

$\therefore N_1^* = 0.24\,(9.53/3.18)\,(1.0)\,0.57\,(741)\,0.350$
$$= 106 \text{ kN} \quad \geq 60 \quad \therefore \text{ OK}$$

Panel point No. 10 is also an X connection, transferring the same force through the chord as in the example above (60kN). The two web members framing into the chord at panel point No. 10 are both HSS 64x64x3.2 (as above), but the load will be dispersed over a greater length along the underside of the chord, effectively increasing the dimension h_1 of the web member on one side of the chord, and also the connection resistance. This connection will hence be adequate.

13.2.6 Resistance of KT Overlap Connections

Panel Point No. 7

Panel point No. 7 is a KT connection; refer to Fig. 13.4.

It can be quickly seen that a gap connection is not feasible within the nodal eccentricity limits; therefore, an overlap connection will be necessary. Overlap the vertical member onto the diagonals, 25% onto each.

Confirm validity of parameters.

1. $b_3/b_0 = 63.5/152 = 0.42 \geq 0.25$ ∴ OK
2. Requirement for Class 1 is $420/(F_{y1})^{0.5}$

    $= 420/(350)^{0.5} = 22.5$

    Flat width to thickness ratio of HSS 64x64x3.2 is

    $(63.5 - 4(3.18))/3.18 = 16.0 \leq 22.5$ ∴ OK

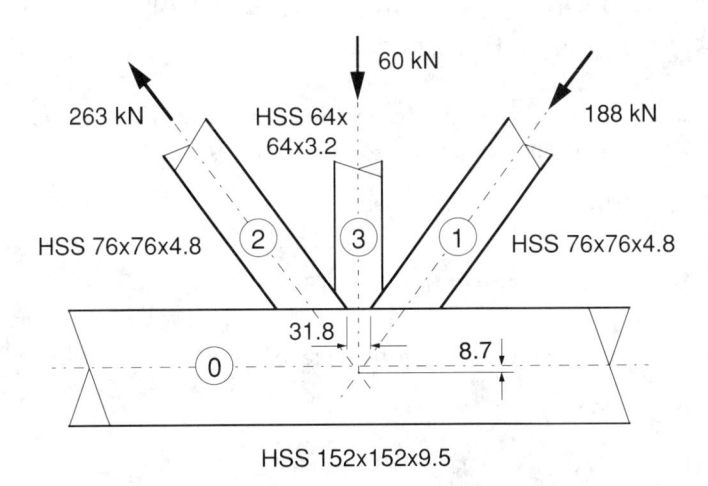

FIGURE 13.4
Panel point No. 7

Flat width to thickness ratio of HSS 76x76x4.8 is

$$(76.2 - 4(4.78))/4.78 = 11.9 \quad \leq 22.5 \quad \therefore \text{OK}$$

3. $b_2/t_2 = 76.2/4.78 = 15.9 \quad \leq 35 \quad \therefore \text{OK}$
4. $b_0/t_0 = 152/9.53 = 16.0 \quad \leq 40 \quad \therefore \text{OK}$
5. $t_3/t_j = 3.18/4.78 = 0.67 \quad \leq 1.0 \quad \therefore \text{OK}$
6. $b_3/b_j = 63.5/76.2 = 0.83 \quad \geq 0.75 \quad \therefore \text{OK}$
7. $25 \leq \%\text{ overlap} = (15.85/63.5)100 = 25.0\% \leq 100 \quad \therefore \text{OK}$
8. $-0.55 \leq e/h_0 = 8.7/152 = 0.06 \leq 0.25 \quad \therefore \text{OK}$

Consider Design Chart, Fig. 3.25.

We cannot use it because overlap is less than 50%.

Therefore, we will use the equation from Table 3.2.

$$N_3^* = F_{y3} t_3 \left(\frac{O_v}{50} (2h_3 - 4t_3) + b_e + b_{e(ov)} \right)$$

$b_e = 0$ because neither "flange" of the vertical lands on the chord, but there will be two $b_{e(ov)}$ terms.

$$b_{e(ov)} = \frac{10}{b_j/t_j} \frac{t_j}{t_3} b_3$$

$$= (10/(76.2/4.78))(4.78/3.18)\,63.5 = 59.9 \text{ mm}$$

$\therefore N_3^* = 0.350(3.18)\,[(25/50)(127 - 12.7) + 2(59.9)]$

$\quad = 197 \text{ kN} \quad \geq 60 \quad \therefore \text{OK}$

Confirm that the connection efficiency of the overlapped members does not exceed that of the overlapping member.

Efficiency of the vertical is

$N_3^*/(A_3 F_{y3}) = 197/(741(0.350)) = 0.76$

$\therefore$ efficiency of diagonals cannot exceed 0.76

so $N_2^* = 0.76\,(1310 \text{ mm}^2)\,0.350$

$\quad = 348 \text{ kN} \quad \geq 263 \quad \therefore \text{OK}$

Alternatively, one may use the convenient relationship:

$$N_j^* \leq N_i^* \frac{F_{yj} A_j}{F_{yi} A_i}$$

i.e., $N_2^* \leq \dfrac{197 \times 1310}{741} = 348$ kN

Therefore, panel point No.7 is acceptable.

Panel Point No. 9

Panel point No. 9 is another KT connection, similar to panel point No. 7 (see Fig. 13.5). Repeat the previous calculations for the members of this panel point.

Confirm validity of the new parameters.

3. $b_2/t_2 = 63.5/3.18 = 20.0 \quad \leq 35 \quad \therefore$ OK
5. $t_3/t_j = 1.0 \quad \leq 1.0 \quad \therefore$ OK
6. $b_3/b_j = 1.0 \quad \geq 0.75 \quad \therefore$ OK
8. $-0.55 \leq e/h_0 = -1.9/152 = -0.01 \leq 0.25 \quad \therefore$ OK

Determine resistance of the overlapping member.

$$N_3^* = F_{y3} \, t_3 \left(\dfrac{O_v}{50} (2h_3 - 4t_3) + b_e + b_{e(ov)} \right)$$

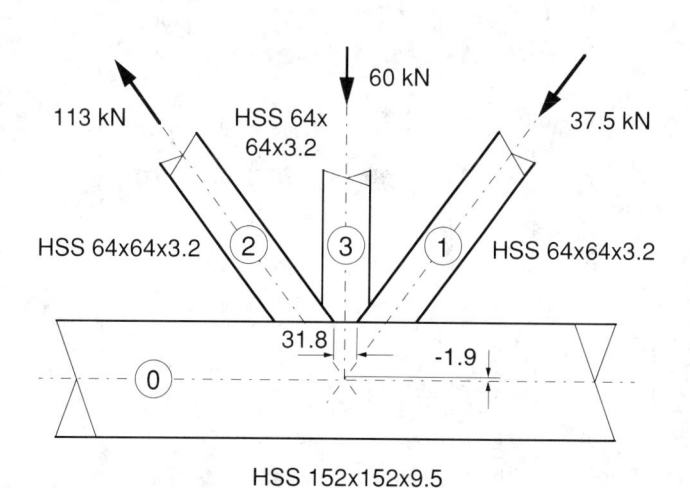

FIGURE 13.5
Panel point No. 9

$$b_{e(ov)} = \frac{10}{b_j/t_j} \frac{t_j}{t_3} b_3$$

$$= (10/(63.5/3.18))(3.18/3.18)63.5 = 31.8 \text{ mm}$$

$$\therefore N_3^* = 0.350\,(3.18)\,[(25/50)\,(127-12.7)+2(31.8)]$$
$$= 134 \text{ kN} \quad \geq 60 \quad \therefore \text{OK}$$

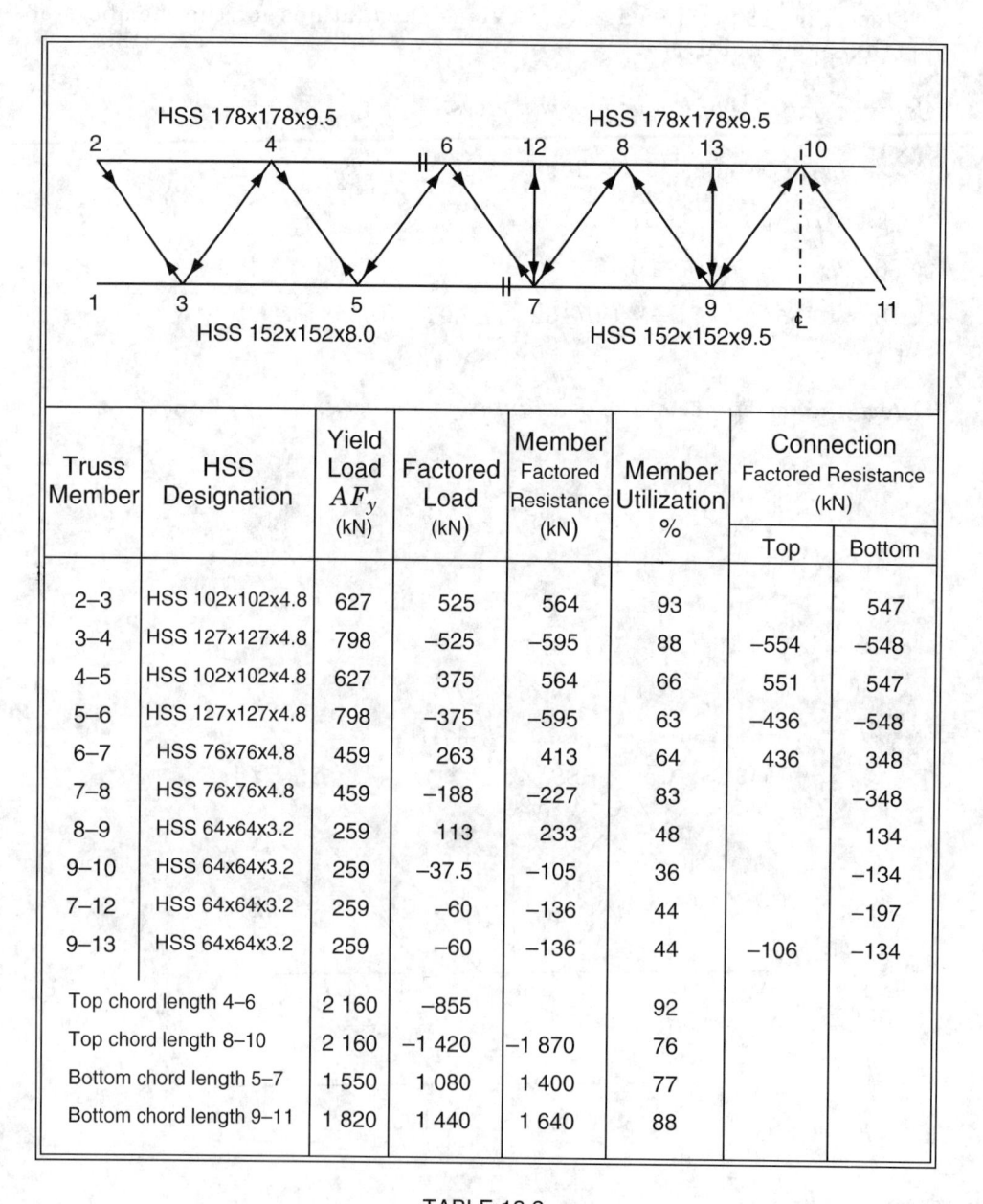

```
                HSS 178x178x9.5              HSS 178x178x9.5
  2          4              6    12       8         13      10
  1     3          5              7             9           11
         HSS 152x152x8.0                   HSS 152x152x9.5
```

Truss Member	HSS Designation	Yield Load AF_y (kN)	Factored Load (kN)	Member Factored Resistance (kN)	Member Utilization %	Connection Factored Resistance (kN) Top	Connection Factored Resistance (kN) Bottom
2–3	HSS 102x102x4.8	627	525	564	93		547
3–4	HSS 127x127x4.8	798	−525	−595	88	−554	−548
4–5	HSS 102x102x4.8	627	375	564	66	551	547
5–6	HSS 127x127x4.8	798	−375	−595	63	−436	−548
6–7	HSS 76x76x4.8	459	263	413	64	436	348
7–8	HSS 76x76x4.8	459	−188	−227	83		−348
8–9	HSS 64x64x3.2	259	113	233	48		134
9–10	HSS 64x64x3.2	259	−37.5	−105	36		−134
7–12	HSS 64x64x3.2	259	−60	−136	44		−197
9–13	HSS 64x64x3.2	259	−60	−136	44	−106	−134
Top chord length 4–6		2 160	−855		92		
Top chord length 8–10		2 160	−1 420	−1 870	76		
Bottom chord length 5–7		1 550	1 080	1 400	77		
Bottom chord length 9–11		1 820	1 440	1 640	88		

TABLE 13.2

Truss of square HSS members: members, loads and resistances

Efficiencies

The members are all sections of the same size, and since the efficiencies of the overlapped members are not allowed to exceed that of the overlapping member, this results in

$$N_2^* = N_3^* \qquad \text{because} \quad N_j^* \leq \left(\frac{F_{yj} A_j}{F_{yi} A_i}\right) N_i^*$$
$$= 134 \text{ kN} \quad \geq 113 \quad \therefore \text{OK}$$

Therefore, Panel point No. 9 is acceptable.

See Table 13.2 for the final truss members, loads, and member and connection resistances.

13.2.7 Design of Bolted Flange-Plate Splice Connection

The bottom chord splice to the left of panel point No. 7 can be made readily with flange plates bolted along two sides as shown in Fig. 13.6.

Bolting examples in this Design Guide have been worked generally with metric bolts, as they are expected to become the norm. However, this example is presented with imperial bolts, partly because they are still in general use, but more because their values result in particular points being better illustrated for this load.

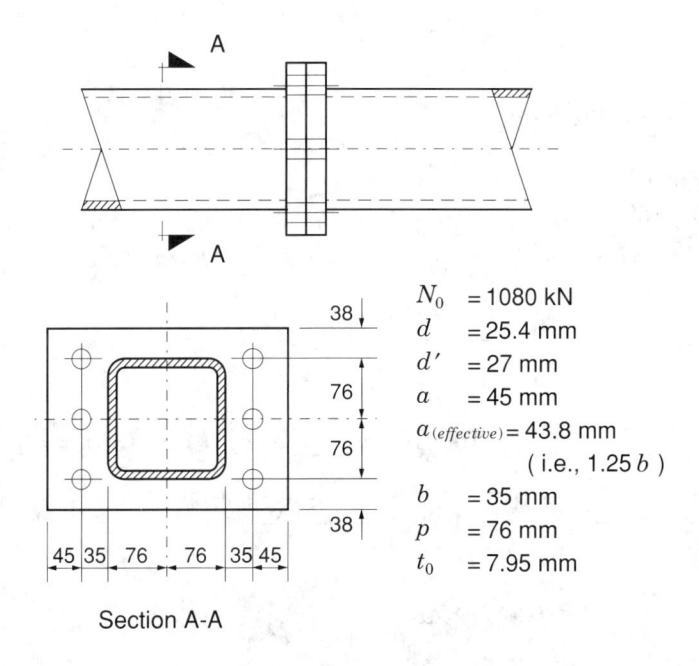

$N_0 = 1080$ kN
$d = 25.4$ mm
$d' = 27$ mm
$a = 45$ mm
$a_{(effective)} = 43.8$ mm
(i.e., $1.25\,b$)
$b = 35$ mm
$p = 76$ mm
$t_0 = 7.95$ mm

Section A-A

FIGURE 13.6
Tension chord splice

Refer to Section 7.2.1, and make a trial arrangement.

Load is 1 080 kN.

T_r of 1 inch diameter A490 bolts is 264 kN.

If 6 bolts are used, P_f = 180 kN and P_f/T_r = 0.68

$$\delta = 1 - d'/p \quad ([7.5])$$
$$= 1 - 27/76 = 0.645$$

$$b' = b - d/2 + t_0 \quad ([7.4])$$
$$= 35 - (25.4/2) + 7.95 = 30.3 \text{ mm}$$

$$K = 4 b'/(\phi_p F_{yp} p) \quad ([7.7])$$
$$= 4 (30.3)/(0.9 (0.300) 76) = 5.91$$

CSA G40.21, Grade 300W plate has been assumed since that is the structural material most commonly used by Canadian fabricators.

$$t_{min} = (KP_f/(1+\delta))^{0.5} \quad \text{(from [7.6])}$$
$$= [5.91 (180)/(1 + 0.645)]^{0.5} = 25.4 \text{ mm}$$

$$t_{max} = (K P_f)^{0.5} \quad \text{(from [7.6])}$$
$$= (5.91 (180))^{0.5} = 32.6 \text{ mm}$$

∴ try 25.4 mm (1 inch) plate

Determine, the ratio of the "sagging" plate moment at the edge of the bolts to the "hogging" plate moment within the tube.

$$\alpha = \left(\frac{K T_r}{t_p^2} - 1\right) \left(\frac{a + (d/2)}{\delta (a + b + t_0)}\right) \quad ([7.8])$$

Maximum effective a is 1.25 b = 1.25 (35) = 43.8.

$$\therefore \alpha = \left(\frac{5.91 (264)}{25.4^2} - 1\right) \left(\frac{43.8 + (25.4/2)}{0.645 (43.8 + 35 + 7.95)}\right)$$

$$= 1.42 (1.01) = 1.43$$

Calculate the splice tensile resistance.

$$N_0^* = \frac{t_p^2 (1 + \delta \alpha) n}{K} \quad ([7.9])$$

$$= \frac{25.4^2 (1 + 0.645 (1.43)) \, 6}{5.91}$$

$$= 1\,260 \quad \geq 1\,080 \quad \therefore \text{ OK}$$

Examine the actual total bolt tension, including prying.

$$T_f \approx P_f \left(1 + \frac{b'}{a'} \left(\frac{\delta \alpha}{1 + \delta \alpha} \right) \right) \quad ([7.10])$$

$$a' = a + d/2 = 43.8 \text{ (effective)} + 25.4/2 = 56.5 \text{ mm}$$

$$\alpha = \left(\frac{K P_f}{t_p^2} - 1 \right) \frac{1}{\delta} \quad \text{(becomes [7.6] for } \alpha = 1 \text{ or } 0\text{)}$$

$$= \left(\frac{5.91 \, (180)}{25.4^2} - 1 \right) \frac{1}{0.645}$$

$$= 1.01$$

$$\therefore T_f \approx 180 \left(1 + \frac{30.3}{56.5} \left(\frac{0.645 \, (1.01)}{1 + 0.645 \, (1.01)} \right) \right)$$

$$\approx 180 \, (1.21) \approx 218 \text{ kN}$$

Therefore, prying tension is about $218 - 180 = 38$ kN per bolt.

Note that this arrangement, which uses the minimum thickness plate, causes major flexure of the plate and hence considerable prying forces in the A490 bolts, as is indicated by the fact that $\alpha > 1.0$, the realm where the plate governs. An alternative approach would be to use a 30 mm plate with A325 bolts, bearing in mind that such a plate thickness is beyond the range of plates used for experimental validation at the time of writing.

Try 30 mm plates with six 1 inch diameter A325 bolts

T_r of 1 inch diameter A325 bolts is 210 kN, and $P_f / T_r = 0.86$.

$$\alpha = \left(\frac{K T_r}{t_p^2} - 1 \right) \left(\frac{a + (d/2)}{\delta (a + b + t_0)} \right) \quad ([7.8])$$

$$= \left(\frac{5.91\,(210)}{30^2} - 1\right)\left(\frac{43.8 + (25.4/2)}{0.645(43.8 + 35 + 7.95)}\right)$$

$$= 0.379\,(1.01) = 0.382$$

Calculate the splice tensile resistance.

$$N_0^* = \frac{t_p^2\,(1 + \delta\alpha)\,n}{K}$$

$$= \frac{30^2\,(1 + 0.645\,(0.382))\,6}{5.91}$$

$$= 1140 \geq 1080 \quad \therefore \text{ OK}$$

Examine the actual total bolt tension, including prying.

$$T_f \approx P_f\left(1 + \frac{b'}{a'}\left(\frac{\delta\alpha}{1 + \delta\alpha}\right)\right) \quad ([7.10])$$

$$\alpha = \left(\frac{KP_f}{t_p^2} - 1\right)\frac{1}{\delta} \quad \text{(related to [7.6])}$$

$$= \left(\frac{5.91\,(180)}{30^2} - 1\right)\frac{1}{0.645} = 0.282$$

$$\therefore T_f \approx 180\left(1 + \frac{30.3}{56.5}\left(\frac{0.645\,(0.282)}{1 + 0.645\,(0.282)}\right)\right)$$

$$\approx 180\,(1.08) \approx 194 \text{ kN}$$

Therefore, prying tension is about 194 − 180 = 14 kN per bolt.

This would appear to be a superior arrangement in that it employs readily available A325 bolts rather than the more demanding and less frequently used A490 type. Furthermore, the introduction of a few connections with A490 bolts on a project that uses A325 bolts elsewhere may present logistical and quality control problems which are best avoided when possible.

This particular connection, as laid out, works for M24 A490M bolts, but not for M24 A325M bolts, whereas both types of 1 inch diameter bolts were strong enough. The slightly smaller 24 mm bolts have less capacity than the 1 inch bolts.

13.2.8 Design of Welded Joints

Consider panel point No. 4 (Fig. 13.2), a conventional gap K connection.

Tension Web Member

HSS 102x102x4.8 $t_2 = 4.78$ mm $\theta_2 = 53.1°$

Load = 375 kN

Effective weld lengths

Refer to Section 8.2.4.2.2 for effective weld lengths.

Weld along the heel of a web member is
fully effective when $\theta_2 \leq 50°$ (see [8.12])
not effective when $\theta_2 \geq 60°$ (see [8.13])

Interpolate to get 69% effective for 53.1°.

Therefore, total effective weld length is

$(2h_2 / \sin\theta_2) + 1.69 b_2$
$= 2(102)/ 0.80 + 1.69 (102) = 427$ mm

If a more precise determination of the weld length is required, the expressions at the bottom of Table 8.5 may be used. The difference in weld length given by these expressions compared with those above can become significant for small thick-walled HSS.

Weld length is $K_a (4\pi t_2 + 2(b_2 - 4t_2) + 2(h_2 - 4t_2))$
where $K_a = ((h_2/\sin\theta_2) + b_2)/(h_2 + b_2)$.

Thus for $\theta_2 = 90°$, total weld length would be 392 mm,
or 98.0 mm along each of the four faces.

For $\theta_2 = 53.1°$, total weld length = 441 mm,
or 98 mm along the heel and toe
plus 245 mm along the two side of the web member.

Therefore, interpreting [8.12] and [8.13] for a square hollow section with round corners, effective weld length (with $\theta_2 = 53.1°$) is

245 + 1.69(98) = 411 mm.

Default Method for sizing the welds

As discussed in Section 8.2.4.2.1, 90° fillet welds with a *throat* 1.10 times the wall thickness of a web member may be considered to develop the resistance of that member, providing the angle between the axis of the weld and the direction of the force ≥ 49°. (That is the angle below which the CAN/CSA-S16.1-94, Clause 13.13.2.2 resistance of the weld throat [8.3] is less than the resistance of the fusion face [8.4], and also less than the tensile resistance of the HSS material.) For the connection at panel point No. 4, the angle of the load to the weld axis is greater than 49° (either 53.1° or 90°) for all four faces of the web member.

Thus the default 90° fillet weld *size* (leg) for a 4.78 mm wall is

4.78 (1.10) 1.41 = 7.4 mm   say, 8 mm.

Therefore, 8 mm fillet welds (or welds with at least the same resistance, considering the heel weld where the cross section is less than 90°) all around the web member would be more than adequate.

For confirmation that a 7.4 mm 90° fillet weld develops the tensile resistance of an HSS102x102x4.78 (if the full length of the weld is effective), consider:

T_r of the HSS is $0.9 \, (1\,790 \text{ mm}^2) \, 0.350 = 564 \text{ kN}$

V_r of a 7.4 mm fillet weld with an effective length of 392 mm (the full perimeter of the HSS as calculated on the previous page) is

0.67 (0.67) 7.4 (0.450) 392 = 586 kN   (from [8.4])

Since the weld develops the resistance of the HSS wall, this default method for sizing the weld does not require consideration of effective weld lengths, and is an upper limit on the required weld size.

Simple Method for sizing the welds

Welds may be sized on the basis of load, rather than on wall thickness. This can result in smaller welds (greater economy), but effective weld lengths must be used.

The simple method is to use resistances for fillet welds without the enhancement introduced in CAN/CSA-S16.1-94 for welds loaded at angles other than parallel to the weld axis, i.e., without the $(1.00 + 0.50 \sin^{1.5} \theta)$ expression in [8.3]. Then, the resistance of fillet welds (cross sections are from 60° to 120°) to HSS is governed only by the weld throat, not by the weld fusion face, which is the case when the enhancement is added and the angle of loading θ ≥ 49°.

Without the enhancement, the required *size* (leg) of a fillet weld with the previously calculated effective weld length of 411 mm, for an applied axial load on the member of 375 kN is

$$375 \,/\, (0.67(0.67)0.707(0.480)411) = 5.99 \,\text{mm} \qquad \text{(from [8.3])}$$

Therefore, 6 mm fillet welds (or welds with at least the same resistance) all around the web member would be adequate. Note that the throat of a 6 mm fillet weld is $6(0.707) = 4.24$ mm, which will be referred to again.

For confirmation that the resistance of a 6 mm 90° fillet weld with an effective length of 411 mm exceeds 375 kN, examine Table 3-24 in the CISC Handbook (CISC 1995). Therein the resistance is given as 0.914 kN per mm of effective weld length, which gives 376 kN.

Precise Method for sizing the welds

The least amount of welding results from the use of all the provisions of CAN/CSA-S16.1-94 in conjunction with effective weld lengths.

Then the required *size* (leg) of a 90° fillet weld with the previously calculated effective weld length of 411 mm, for an applied axial load on the member of 375 kN can be determined from [8.4]. Equations [8.4a] and [8.4b] for base metal failure will govern instead of [8.3] for weld metal failure because all four welds have their axes at either 53.1° or 90° to the direction of the load, which exceeds the 49° limit. Thus,

$$375 \,/\, (0.67(0.67)0.450(411)) = 4.5 \,\text{mm}$$

Therefore, 5 mm fillet welds (or welds with at least the same resistance) all around the web member would be adequate.

For confirmation that the resistance of a 5 mm 90° fillet weld with an effective length of 411 mm exceeds 375 kN, examine Table 3-25 in the CISC Handbook, (CISC 1995). Therein, the resistance is given at 1.01 kN per mm of effective weld length when the angle between the weld axis and the direction of force exceeds 49°, which gives 415 kN.

Differences in the weld sizes from the above three methods can be quite significant, costwise, for larger HSS, particularly if more than one pass is required to make a weld.

Even though the welds along the heel and toe (53.1° and 126.9°) are actually beyond the CSA W59-M1989 range of angles for fillet welds ($60° \leq \theta \leq 120°$) it is common practice for the designer to identify the required weld resistance in terms of the size of a 90° fillet weld. In such an instance, the onus is on the fabricator to develop a practical weld of the same resistance when drawing the connection details.

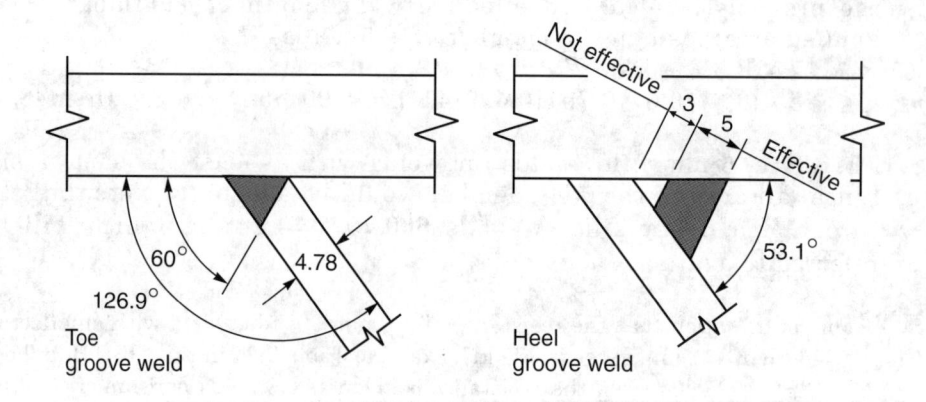

(a) Toe and heel welded for actual load
(sized by the Simple Method)

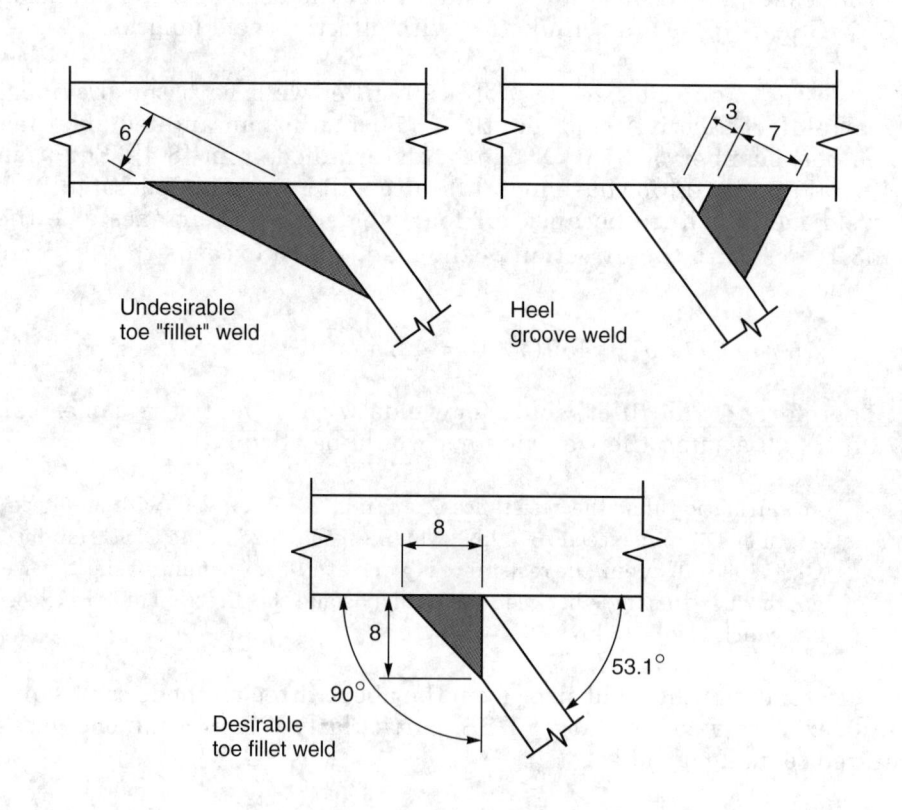

(b) Toe and heel welded for capacity of the member
(sized by the Default Method)

FIGURE 13.7
Welds for web member 4–5

Details for the welds sized by the Simple Method

Use the calculated 6 mm fillet welds along the sides of the web member. Options for the heel and toe are shown in Fig. 13.7(a).

The weld at the heel is a partial penetration weld, so [8.5] and [8.6] are used when calculating the weld shear resistance. Since [8.5] and [8.3] are identical except for the enhancement available for fillet welds, and that enhancement is not used by the Simple Method, the required throat for this PJPG weld is the same as for the fillet weld calculated earlier, i.e., 4.24 mm. The weld illustrated has an effective throat of 5 mm, while discounting 3 mm at the root (when $\theta < 60°$) as per W59-M1989, Fig. M-3.

The weld at the toe is a partial penetration weld loaded in tension with a nominal throat of 4.78 mm, when the included weld angle is $\geq 60°$. Since the configuration does not precisely conform to that for pre-qualified PJPG welds in Section 10 of CSA W59-M1989, it may be imprudent to use [8.7] to determine its resistance. However, the required resistance is provided by even the lower shear resistance from [8.5], when applied to an effective throat of 4.24 mm, which is reasonable.

One would blend the various weld types at the corners of the HSS web member to produce smooth full strength transitions. Also, one should apply the heel weld to the full length of the heel, even though it was sized for a lower effective length of that weld.

Weld details for welds sized by the Precise Method would be similar, but smaller.

Details for the welds sized by the Default Method

Use the calculated 8 mm fillet welds along the sides of the web member. Options for the heel and toe are shown in Fig. 13.7(b).

Since the resistance of the HSS wall is 0.9 (4.78) 0.350 = 1.51 kN per mm of length, the required throat of the partial penetration weld at the heel is

$$1.51/(0.67 (0.67) 0.480) = 7.00 \text{ mm} \qquad \text{(from [8.5])}$$

Therefore, the 7 mm throat shown.

At the toe, a "fillet" weld with the "default" throat of 1.10 times the wall thickness (1.10(4.78) = 5.26, say 6 mm) has a large volume of weld metal relative to more desirable alternatives (Fig. 13.7(b)).

One might consider the partial penetration weld in Appendix L of W59-M1989 that is regarded as equivalent in strength to a full penetration weld, but it has exacting details and is not prequalified. A better option is to use an 8 mm fillet weld as on the side walls, which can be done by preparing the toe as shown. The necessary resistance is obtained, provided the weld has a leg size of at least 7.4 mm (see earlier calculations).

Compression Web Member

HSS 127x127x4.8 $t_1 = 4.78$ mm $\theta_1 = 53.1°$

Load = –525 kN

Length of effective weld ≈ 2 (127) / 0.80 + 1.69(127) = 532 mm

> The approximate required size of fillet weld (using the Simple Method and the unit factored shear resistance of weld metal from the CISC Handbook Table 3-22 (CISC 1995)) is
>
> 525 kN / (532 mm (0.152 kN/mm^2)) = 6.5 mm.

Therefore, weld the sides of this web member with 8 mm fillets (the next standard size of fillet weld).

The throat of a 6.5 mm fillet would be 4.6 mm. Therefore, the same welds at the heel and the toe that were used for the tension web member by the Simple Method in Fig. 13.7(a) will suffice for this member.

To connect for the member capacity (by using the Default Method), use the same welds as for the tension member (same wall thickness).

13.2.9 Truss to Column Connection

One of several possible arrangements for connecting the truss to a column (illustrated in Fig. 13.8) is to weld a tee across the ends of the truss chord and diagonal, and to field bolt the web of that tee to a gusset slotted into the column.

Connection details are usually chosen to satisfy practical considerations. Adequacy of the details to transmit the loads can then be checked based on simple but conservative procedures as used below.

The assumed column is an HSS 305x305x9.5 with a 16 mm plate slotted into the top. Such a plate is conservatively stressed, even with the eccentricity of the single lap joint to the truss.

A 30 mm plate is suggested for the flange of the tee to promote effective load distribution from the web of the tee to the side walls of the HSS members. (This assumes a stress distribution ratio of about 2.5 to 1 from the face of the web of the tee.)

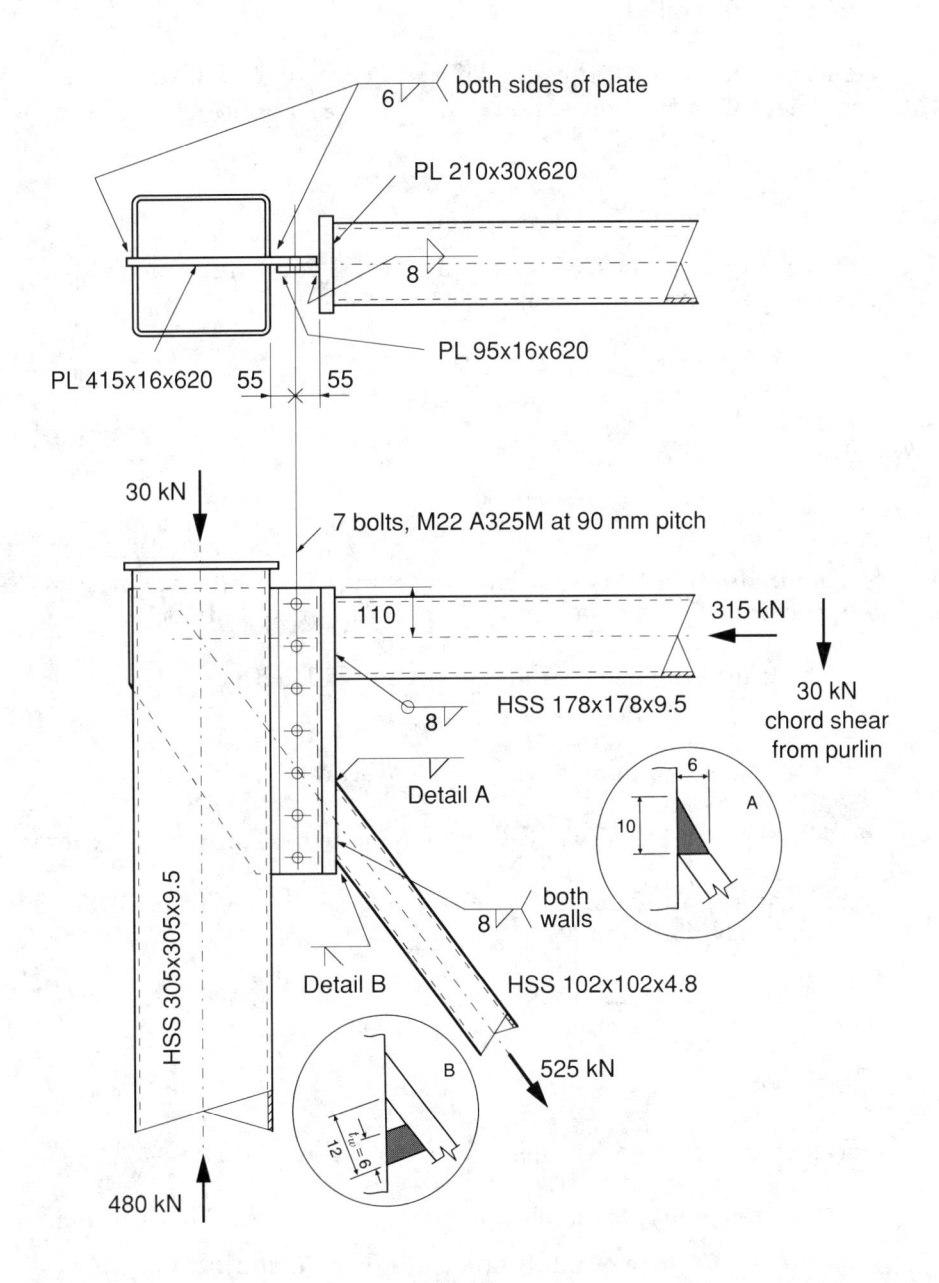

FIGURE 13.8
Truss to column connection

Bolts

The M22 A325M bolts are selected based on an eccentricity that reaches to the centre of the column.

Factored shear resistance per bolt is 127 kN (CISC Handbook, Table 3–4, threads excluded).

Using a reaction at the end of the truss of 450 kN, the required coefficient C in the bolt eccentricity Table 3–14 of the CISC Handbook is

$$450/127 = 3.54.$$

Eccentricity $L = 153 + 55 = 208$ mm. Interpolating from Table 3–14, 7 bolts at 90 mm pitch gives

$$C = 4.25 \quad \geq 3.54 \quad \therefore \text{OK}$$

Welds

Tee

Try double 8 mm fillets for the web/flange joint of the fabricated tee and confirm that they are adequate at the tension diagonal.

The load is applied over an effective weld length of at least

$$102/\cos 53.1° + 2.5\,(30) = 245 \text{ mm}$$

(assuming not much tee material below the HSS)

Therefore, a conservative weld load is $525 / 245 = 2.14$ kN/mm

Resistance of double 8 mm fillets (as confirmed from CISC Handbook Table 3-24) is

$$16\,(0.152) = 2.43 \text{ kN/mm} \quad > 2.14 \quad \therefore \text{OK}$$

Column plate

Check welding of the "through plate" to the column.

Required length of double 6 mm fillets at each wall of the column is

$$450/(4(6)0.152) = 123 \text{ mm} \quad \text{OK by examination}$$

HSS diagonal to the 30 mm tee flange plate

All the weld lengths are fully effective, as the base is a "rigid" plate.

Toe weld

Since the 30 mm plate requires heat input equivalent to that from either a PJPG weld with an 8 mm deep preparation or an 8 mm fillet weld (CSA W59-M1989, Tables 4-3 and 4-4), the PJPG weld used in Fig. 13.7(a) will not do for the toe. However, there is room for a 6x10 mm fillet weld if the edge preparation is increased to 90° as shown in Detail A of Fig. 13.8, which should provide the heat input needed (to avoid an unsound weld due to quick cooling).

Use the resistance of a 6 mm fillet weld (0.914 kN/mm, CISC Handbook, Table 3-24) to obtain

102 (0.914) = 93 kN

Heel weld

For resistance based on effective throat, an equivalent partial penetration weld for the heel (of the type shown in Detail B of Fig. 13.8) would have an effective throat of 6(0.707) = 4.24 mm. Since the fusion face is almost the same area as the throat in this case, the requirement for the effective throat is more like 4.24 (0.480/0.450) = 4.52 mm. Five mm would do, but use 6 in this instance.

Since θ is between 45° and 30°, there will be a further 6 mm in the root that is not effective for a total of 12, as per Fig. 3–10 in the CISC Handbook (CISC 1995). The volume of the resulting effective weld is similar to that of an 8 mm fillet weld, so the heat input requirement is satisfied.

Side welds

The length of the 8 mm fillet welds on the sides of the HSS can be calculated using the K_a formula below Table 8.5. One obtains

K_a = 1.333, and hence a contact perimeter of 522 mm.

Therefore, the length of the two fillet welds on the sides of the HSS is

522 − 2(102) = 318 or 159 mm on each side.

The resistance of these two welds is then given by

2 (159) 1.22 (CISC, Table 3-24) = 388 kN

Then, total resistance of the four welds is

2 (93) + 388 = 574 kN   > 525   ∴ OK

(without using the enhanced value for fillet welds in [8.3]).

Check crippling adequacy of the tee's web at the top chord.

    Check with CAN/CSA-S16.1-94, Clause 21.3(a).

    Assume the plate is 300W steel.

$$B_r = \phi\, w_c\, (t_b + 5k)\, F_{yc}$$

    where $t_b + 5k$ is the bearing length

    (here, for the HSS chord, it is $110 + 178/2 + 2.5(30) = 274$ mm

    $\therefore B_r = 0.9\,(16)\,274\,(0.300)$

           $= 1180$ kN $\gg 315$ $\therefore$ OK

Check adequacy of the tee's flange at the tension diagonal.

    Check with CAN/CSA-S16.1-94, Clause 21.3(b).

$$T_r = 7\,\phi\, t_c^{\,2}\, F_{yc}$$

       $= 7\,(0.9)\,30^2\,(0.300)$

       $= 1700$ kN $\gg 525\,(\cos 53.1°) = 315$ $\therefore$ OK

Therefore, the connection is adequate.

13.2.10 Truss Deflection

    The maximum truss deflection under the specified live load of 3.0 kPa (see Section 13.2) should also be checked. This maximum occurs at the centre of the truss (at panel point 10, see Table 13.1). Under the unfactored live load only, the forces in the truss members will be simply 3.0/(1.5(3.0)+1.25(1.5)) or 0.47 of those under the factored live and dead loads, as shown in Table 13.1. By using the Principle of Virtual Work (by applying a unit load to the truss at panel point 10) the deflection of panel point 10 can then be easily computed. The resulting calculations are tabulated in Table 13.3, and a truss deflection of 59.6 mm is obtained. One should note that this is derived from a pin-jointed analysis of the truss.

    Apart from the KT connections at panel points 7, 9, 11 and 17 on the bottom chord (which are overlapped), all other K connections are gapped. It was noted in Section 2.4 that for gap-connected trusses a more conservative approach was preferable for estimating truss deflections, by using 1.15 times the deflection calculated from a pin-jointed analysis. On this basis, a predicted service load deflection of 68.5 mm is expected under live loads. The maximum value for deflection recommended in Appendix I of CAN/CSA-S16.1-94 (CSA 1994), whether for an industrial or non-industrial building and with any material covering, is 1/360 of the truss span, or 83.3

mm. The estimated service load deflection (68.5 mm) is therefore acceptable.

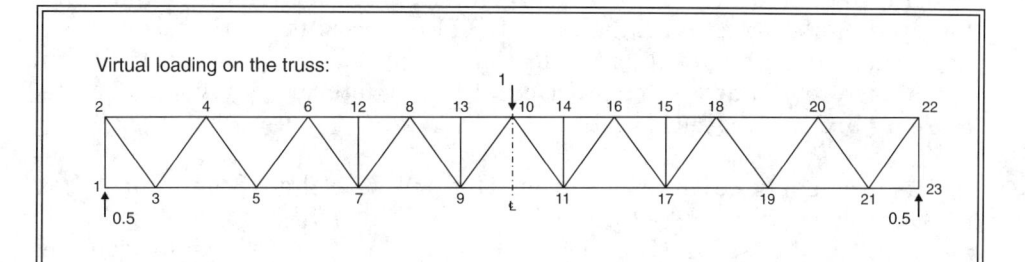

MEMBERS	REAL FORCE [P] (kN)	LENGTH [L] (mm)	AREA [A] (mm^2)	VIRTUAL [VL]	$\delta = [VL]\left(\dfrac{PL}{AE}\right)$ (mm)
1-2 and 22-23	−225.0	2500	11 000	−0.500	2 x 0.13
2-4, 20-22	−147.7	3750	6 180	−0.375	2 x 0.17
4-6, 18-20	−400.8	3750	6 180	−1.125	2 x 1.37
6-8, 16-18	−580.2	3750	6 180	−1.875	2 x 3.30
8-10, 10-16	−664.5	3750	6 180	−2.625	2 x 5.29
1-3, 21-23	0	1875	4 430	0	0
3-5, 19-21	+295.3	3750	4 430	+0.750	2 x 0.94
5-7, 17-19	+506.3	3750	4 430	+1.500	2 x 3.21
7-9, 11-17	+632.9	3750	5 210	+2.250	2 x 5.12
9-11	+675.1	3750	5 210	+3.000	1 x 7.29
2-3, 21-22	+246.1	3125	1 790	+0.625	2 x 1.34
3-4, 20-21	−246.1	3125	2 280	−0.625	2 x 1.05
4-5, 19-20	+175.8	3125	1 790	+0.625	2 x 0.96
5-6, 18-19	−175.8	3125	2 280	−0.625	2 x 0.75
6-7, 17-18	+123.1	3125	1 310	+0.625	2 x 0.92
7-8, 16-17	−87.9	3125	1 310	−0.625	2 x 0.66
8-9, 11-16	+52.7	3125	741	+0.625	2 x 0.69
9-10, 10-11	−17.6	3125	741	−0.625	2 x 0.23
7-12, 15-17	−28.1	2500	741	0	0
9-13, 11-14	−28.1	2500	741	0	0

∴ Maximum truss deflection = $\Delta_{\text{panel point 10}}$ = $\Sigma(\delta)$ = 59.6 mm ↓

TABLE 13.3

Tabulation of virtual work calculations for truss deflection

13.3 Warren Truss with Circular HSS

The truss used in Section 13.2 will now be designed for circular HSS. Design formulae for T, Y, X, K and N connections were presented in Table 3.1. Extensive use will be made of the design charts illustrated in Figs. 3.12 to 3.18. In general, the design of circular HSS connections is simpler than for rectangular (including square) HSS connections as there are fewer potential failure modes, but fabrication is appreciably more difficult. The truss, factored loads and member axial forces determined by a pin-jointed analysis are illustrated in Table 13.1.

As previously, calculations are performed with the HSS section "sizes" (3 significant figures) while reference to the HSS sections is by the CAN/CSA-G312.3-M92 (CSA 1992) "designations" (only 2 significant figures for some dimensions).

13.3.1 Considerations when Making Preliminary Member Selections

Once again it is prudent to bear in mind the "limits of validity" that must be met for various dimensional parameters, and to select chords conservatively because minimum mass members often lead to undesirable reinforcing at the connections.

The validity limits (discussed in Section 3.6) are

1. Diameter of web members must be between 0.2 and 1.0 times the chord diameter.

2. Diameter to wall thickness ratio of a member must not exceed 50 (40 for chord members in X connections).

3. Wall thickness of the overlapping member must not exceed the wall thickness of the overlapped member (for equal yield strength materials).

4. Nodal eccentricity limits are −0.55 and +0.25 of the chord diameter (same as for square HSS, if one considers the diameter as the overall height).

5. Minimum overlap is 25% (as for square HSS).

6. Minimum gap is the sum of the wall thicknesses of the members on either side of the gap.

7. With increasing diameter to wall thickness ratio of a web compression member (beyond 28 for 350 MPa yield material), there is an increasing reduction in the permitted connection resistance relative to the yield capacity of the member. These reduction factors

are: 0.98 for a web diameter to thickness ratio of 30, 0.88 for 35, 0.82 for 40, 0.78 for 45, and 0.76 for 50.

It will be noticed that these limits are much less onerous than those for square HSS. Relative sizes of web members are less critical, and there are no gap restrictions as a function of chord and web member widths.

13.3.2 Preliminary Member Selections

The values for compressive resistances C_r are interpolated from the CISC Handbook, pp. 4-72 to 4-76 (CISC 1995).

Central top chord

Load in length 8–10 is −1420 kN. KL is $0.9(1875) = 1690$

Try HSS 219x9.5 $C_{r0} = -1910$ kN $A = 6\,270$ mm^2

> It will be shown later that an HSS 219x8.0 ($C_{r0} = -1610$ kN) would have too thin a wall.

Utilization is $1420/1910 = 0.74$.

Outer top chord

Load in length 4–6 is −855 kN. KL is $0.9(3\,750) = 3\,380$

As in the previous example, the purlin loads between panel points cause a bending moment in the chord at panel point No. 4 which can be conservatively approximated by $0.188\,PL$ (2 equal spans, continuous).

Then M_{f0} would be $0.188\,(60)\,3.75 = 42.3$ kN·m which would require an

$$S_0 \text{ of } M_{f0}/\phi F_{y0} = 42.3/(0.9\,(0.350))$$
$$= 134\,(10^3) \text{ mm}^3$$

As previously, knowing that the axial and bending loads in the chord will be combined, a generous member will be selected with more than twice the capacity for the individual forces.

Try HSS 219x9.5

$C_{r0} = -1660$ kN $S_0 = 315\,(10^3)$ mm^3 $A_0 = 6\,270$ mm^2

It will be shown later in the check of connection No. 4 that an HSS 219x8.0 ($C_{r0} = -1\,400$) would be too light, and that the 9.5 wall is just adequate.

Central bottom chord

Load in member 9–11 is 1440 kN.

$$A_{(min)} = T_{f0} / (\phi F_{y0}) = 1440/(0.9\,(0.350)) = 4\,570 \text{ mm}^2$$

Try HSS 168x9.5 $A_0 = 4\,750 \text{ mm}^2$

Utilization is $1440/(0.9\,(4\,750)\,0.350) = 0.96$.

Outer bottom chord

Load in member 5–7 is 1080 kN.

$$\therefore A_{(min)} = 1080 / (0.9\,(350)) = 3\,430 \text{ mm}^2$$

Try HSS 168x8.0 $A_0 = 4\,000 \text{ mm}^2$

Utilization is $1080/(0.9\,(4\,000)\,0.350) = 0.86$.

Tension diagonal at end of truss

Load in member 2–3 is 525 kN.

$$\therefore A_{(min)} = 525 / (0.9\,(350)) = 1670 \text{ mm}^2$$

Try 141x4.8 $A_2 = 2\,050 \text{ mm}^2$

Utilization is $525 / (0.9\,(2\,050)\,0.350) = 0.81$.

Compression diagonal nearest to end of truss

Load in member 3–4 is −525 kN. KL is $0.75(3\,125) = 2\,340$

Try HSS 141x4.8 $C_{r1} = -528$ kN $A_1 = 2\,050 \text{ mm}^2$

Utilization is $525/528 = 0.99$.

Compression diagonal at middle of truss

Load in member 9–10 is −37.5 kN. KL is $2\,340$ mm

Try HSS 60x3.2 $C_{r1} = -62$ kN $A_1 = 571$ mm^2

Utilization is $37.5/62 = 0.60$.

This is about the smallest size one would want to use in a truss this large, especially when one reflects on the member width relative to the truss depth.

All verticals

Load is -60 kN. KL is $0.75(2\,500) = 1\,880$ mm

Try HSS 60x3.2 $C_{r1} = -84$ kN $A_1 = 571$ mm^2

Utilization is $60/84 = 0.71$.

Tension diagonal near truss centre

Load in member 8–9 is 113 kN.

$\therefore A_{(min)} = 113/(0.9\,(350)) = 359$ mm^2

Try HSS 60x3.2 $A_2 = 571$ mm^2

Utilization is $113/(0.9\,(571)\,0.350) = 0.63$.

Compression diagonal 7–8

Load is -188 kN. $KL = 2\,340$ mm

Try HSS 89x4.8 $C_{r1} = -226$ kN $A_1 = 1260$ mm^2

Utilization is $188/226 = 0.83$.

Tension diagonals 6–7

Load is 263 kN. $\therefore A_{(min)} = 263/(0.9\,(350)) = 835$ mm^2

Try HSS 89x4.8 $A_2 = 1260$ mm^2 (standardize with 7–8)

Utilization is $263/(0.9\,(1260)\,0.350) = 0.66$.

Compression diagonal 5–6

Load is -375 kN. $KL = 2\,340$ mm

Try HSS 114x4.8 C_{r1} = –372 kN   < 375   ∴ no good

Use HSS 141x4.8 as for diagonal 3–4

C_{r1} = –528 kN A_1 = 2 050 mm^2

Utilization is 375/528 = 0.71.

Tension diagonal 4–5

Load is 375 kN.   ∴ $A_{(min)}$ = 375 / (0.9 (350)) = 1 190 mm^2

Try HSS 89x4.8 A_2 = 1260 mm^2

Utilization is 375 / (0.9 (1260) 0.350) = 0.94.

The truss with the trial members is shown in Fig. 13.9.

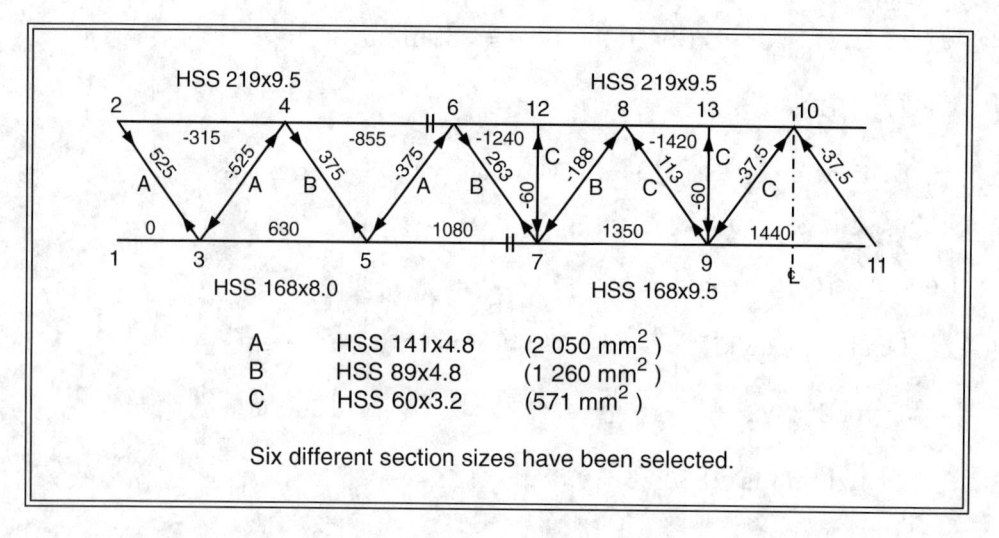

FIGURE 13.9
Preliminary selection of members—Warren truss with circular HSS

13.3.3 Resistance of Gap K Connections

Panel Point No. 3

See Fig. 13.10 for layout of panel point No. 3.

Maximum nodal eccentricity of 0.25 d_0 = 168/4 = 42.0 mm produces a gap of 12.7 mm which exceeds the minimum of 2(4.78) = 9.6 mm. Therefore, a gap connection is feasible.

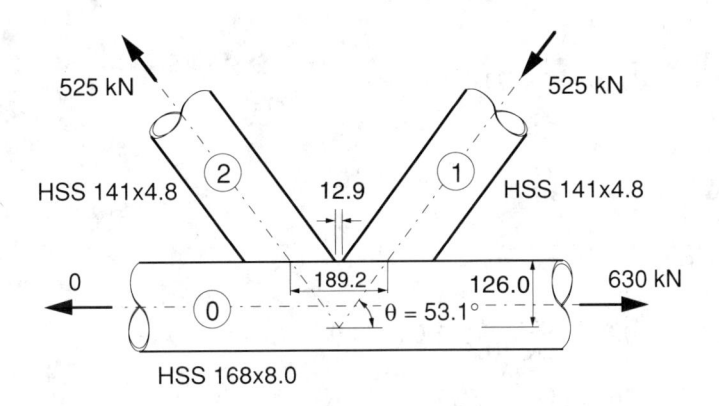

FIGURE 13.10
Panel point No. 3

Calculation of the gap

Height from intersection point of webs to top of chord is
0.75(168) = 126 mm

Distance along top of chord between web centrelines is
2(126)/tan 53.1° = 189.2 mm

Gap is
189.2 − 2 (141 / (2 sin 53.1°)) = 12.9 mm

Confirm validity of dimensional parameters, (Section 13.3.1).

1. $0.2 < d_1/d_0$, d_2/d_0 = 141/168 = 0.84 ≤ 1.0   ∴ OK
2. d_1/t_1, d_2/t_2, = 141/4.78 = 29.5 ≤ 50   ∴ OK
   d_0/t_0, = 168/7.95 = 21.1 ≤ 50   ∴ OK
3. and 5.   Not applicable
4. and 6.   Nodal eccentricity and gap are already confirmed.
7. No effect in this case

Determine connection resistance of compression diagonal 3–4.

$$g' = g/t_0 = 12.9/7.95 = 1.6$$

$$\therefore \text{ use Design Chart, Fig. 3.14} \quad (\text{for } g' = 2.0)$$

$$N_1^* = C_K \frac{t_0}{t_1} \frac{1}{\sin\theta_1} f(n') A_1 F_{y1}$$

$$d_1/d_0 = 141/168 = 0.84$$
$$d_0/t_0 = 168/7.95 = 21.1$$

$$\therefore C_K = 0.38$$

$$f(n') = 1.0 \text{ for a tension chord}$$

$$\therefore N_1^* = 0.38 \frac{7.95}{4.78} \frac{1}{0.8} 1.0 \,(2\,050)\,0.350$$

$$= 567 \text{ kN} \quad \geq 525 \quad \therefore \text{ OK}$$

Determine connection resistance of tension diagonal 2–3.

$$N_2^* = N_1^* \frac{\sin\theta_1}{\sin\theta_2} = 567 \text{ kN also} \quad (\text{Table 3.1})$$

$$\geq 525 \quad \therefore \text{ OK}$$

Therefore, panel point No. 3 is acceptable.

Panel Point No. 5

A sketch and calculations similar to those at panel point No. 3 show that a layout with the same maximum positive eccentricity of 42 mm gives a gap of 45.4 mm. A comparison with panel point No. 3 confirms that the dimensional parameters are within limits of validity.

Determine connection resistance of compression diagonal 5–6.

$$N_1^* = C_K \frac{t_0}{t_1} \frac{1}{\sin\theta_1} f(n') A_1 F_{y1}$$

$$g' = 45.4/7.95 = 5.7$$

Use Design Chart Fig. 3.15 ($g' = 6$) for conservative results.

$d_1/d_0 = 0.84$, and $d_0/t_0 = 21.1$, as previously

$\therefore C_K = 0.33$

$f(n') = 1.0$ for a tension chord

$\therefore N_1^* = 0.33 \; \dfrac{7.95}{4.78} \; \dfrac{1}{0.8} \; 1.0 \,(2050)\,0.350$

$ = 492 \text{ kN} \quad \geq 375 \quad \therefore \text{ OK}$

Determine connection resistance of tension diagonal 4–5.

$N_2^* = N_1^* \; \dfrac{\sin\theta_1}{\sin\theta_2} = 492 \text{ kN also}$

$ \geq 375 \quad \therefore \text{ OK}$

Therefore, panel point No. 5 is acceptable.

Panel Point No. 4

Try a gap connection; refer to Fig. 13.11(a). A near-minimum gap of 10 mm gives a nodal eccentricity of −7.1 mm.

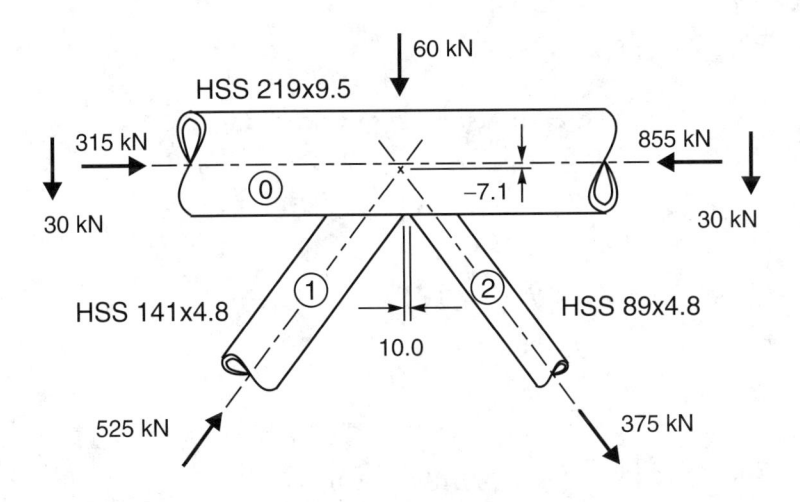

FIGURE 13.11(a)
Panel point No.4—gap connection

Validity of other dimensional parameters checked out, as it did at panel point No. 3.

The bending moment at the connection due to purlin loads is considered to be 0.188 PL, as discussed when selecting preliminary member sizes,

$$= 0.188\,(60)\,3.75 = 42.3 \text{ kN·m}$$

Determine f(n') for panel point No. 4.

$$f(n') = 1 + 0.3\,n' - 0.3\,n'^2 \quad \text{but} \not> 1.0 \quad \text{(Table 3.1)}$$

$$n' = N_{0p}/(A_0\,F_{y0}) + M_{f0}/(S_0\,F_{y0}) \quad \text{(by definition)}$$
$$= (-315)/(6\,270\,(0.350)) + (-42.3)/(315\,(0.350))$$
$$= -0.527$$

$$\therefore f(n') = 1 + 0.3\,(-0.527) - 0.3\,(-0.527)^2 = 0.759$$

Or alternatively use Design Chart, Fig. 3.18.

Note that it was not necessary to include the primary bending moment from the nodal eccentricity because the eccentricity e was within specified limits.

$$\text{i.e.,} \quad -0.55 \le e/d_0 \le 0.25$$

Determine connection resistance of compression diagonal 3–4.

$$g' = 10.0/9.53 = 1.05$$

$\therefore$ conservatively use Design Chart, Fig. 3.14.

$$N_1^* = C_K\,\frac{t_0}{t_1}\,\frac{1}{\sin\theta_1}\,f(n')\,A_1\,F_{y1}$$

$$d_1/d_0 = 141/219 = 0.64$$
$$d_0/t_0 = 219/9.53 = 23.0$$

$$\therefore C_K = 0.38$$

$$\therefore N_1^* = 0.38\,\frac{9.53}{4.78}\,\frac{1}{0.8}\,0.759\,(2\,050)\,0.350$$

$$= 516 \text{ kN} \quad < 525 \quad \therefore \text{ no good}$$

A diagonal with a thicker wall would not be any better here because the reduced t_0/t_1 ratio offsets the increased area and the connection resistance is much the same. A thicker chord wall would work, but that is rather inefficient.

The resistance above (derived from the Design Charts) is 98% of the factored load. When results from the charts are within ±5% of the requirement, it is usually advisable to work out a more precise value by using the actual equations. From Table 3.1:

$$N_1^* = \frac{F_{y0} t_0^2}{\sin\theta_1}\left(1.8 + 10.2 \frac{d_1}{d_0}\right) f(\gamma, g') \, f(n')$$

$$\gamma = d_0/2t_0 = 219/(2(9.53)) = 11.5$$

$$g' = g/t_0 = 10/9.53 = 1.05$$

$$f(\gamma, g') = \gamma^{0.2}\left(1 + \frac{0.024\,\gamma^{1.2}}{\exp(0.5g' - 1.33) + 1}\right)$$

$$= 1.63\left(1 + \frac{0.024(18.7)}{\exp(0.5(1.05) - 1.33) + 1}\right)$$

$$= 1.63\,[1 + (0.449/(0.447 + 1))] = 2.14$$

$$\therefore N_1^* = \frac{0.350(9.53)^2}{0.8}\left(1.8 + 10.2\,\frac{141}{219}\right) 2.14\,(0.751)$$

$$= 534\text{ kN} \geq 525 \quad \therefore \text{ OK}$$

The equations provide greater accuracy and give a resistance 2% more than the requirement.

If the connection still had not been adequate, an overlap connection could have been used since overlap connections have higher resistances than do gap connections. See Section 13.3.4.

Note that the connection resistance of the compression diagonal would have been insufficient if the chord had a 7.95 mm thickness rather than the 9.53 selected.

Punching shear resistance (Table 3.1) must also be checked:

$$N_i^* = \frac{F_{y0}}{\sqrt{3}}\, t_0\, \pi\, d_i\, \frac{1 + \sin\theta_i}{2\sin^2\theta_i}$$

This need not be calculated when the Design Charts are used because the upper bound of the charts represents this value, but it should be checked when using the equations. Here:

$$N_1^* = \frac{0.350}{\sqrt{3}} \, 9.53 \, \pi \, 141 \, \frac{1 + 0.8}{2\,(0.8)^2}$$

$$= 1200 \, \text{kN} \quad \geq 525 \quad \therefore \text{OK}$$

Determine connection resistance of tension diagonal 4–5.

From Table 3.1, it can be seen that $N_2^* = N_1^*$ for the chord plastification check and since $N_2 = 375$ kN $< N_2^* = 534$ kN, this will be adequate.

For the punching shear check:

$$N_2^* = \frac{0.350}{\sqrt{3}} \, 9.53 \, \pi \, 89 \, \frac{1 + 0.8}{2\,(0.8)^2}$$

$$= 757 \, \text{kN} \quad \geq 375 \quad \therefore \text{OK}$$

Therefore, panel point No. 4 is acceptable.

Panel Point No. 6

Dimensional parameters check out for a gap connection, with zero nodal eccentricity giving a 20.6 mm gap.

At panel point (node) No. 4, in order to conservatively estimate the primary bending in the chord due to the purlin load on each side of the panel point, the chord was considered to be continuous from node No. 2 to node No. 6. Now, the primary bending moment in the chord at panel point No. 6 due to the purlin load between nodes No. 4 and No. 6 needs to be conservatively estimated. Consider that, if the chord were fully fixed against rotation at these nodes, the bending moment at panel point No. 6 would be *PL/8* due to the purlin load at mid span. For equilibrium, the moment to the right of node No. 6 would balance the moment to the left (*PL/8*).

Therefore, use a moment of $60\,(3.75)/8 = 28.1$ kN·m.

Determine f(n') *for panel point No. 6.*

$$n' = N_{0p}/(A_0 F_{y0}) + M_{f0}/(S_0 F_{y0})$$

$$= (-855)/(6\,270\,(0.350)) + (-28.1)/(315\,(0.350))$$

$$= -0.645$$

$\therefore f(n') = 0.67$ (Fig. 3.18)

Determine connection resistance of compression diagonal 5–6.

$$N_1^* = C_K \frac{t_0}{t_1} \frac{1}{\sin\theta_1} f(n') A_1 F_{y1}$$

$$g' = 20.6/9.53 = 2.2$$

$\therefore$ use Design Chart, Fig. 3.14 for $g' = 2$.

$C_K = 0.38$, as for diagonal 3–4 at panel point No. 4

$$\therefore N_1^* = 0.38 \frac{9.53}{4.78} \frac{1}{0.8} 0.67 (2050) 0.350$$

$$= 455 \text{ kN} \quad \geq 375 \quad \therefore \text{ OK}$$

Check punching shear:

$$N_1^* = \frac{0.350}{\sqrt{3}} 9.53 \pi \, 141 \frac{1 + 0.8}{2 (0.8)^2}$$

$$= 1200 \text{ kN} \quad \geq 375 \quad \therefore \text{ OK}$$

Determine connection resistance of tension diagonal 6–7.

As previously, $N_2^* = N_1^* = 455 \text{ kN} \quad \geq 263 \quad \therefore \text{ OK}$

For punching shear check, $N_2^* = 757 \text{ kN} \quad \geq 263 \quad \therefore \text{ OK}$

Therefore, panel point No. 6 is acceptable.

This panel point is an example of a K connection (with an external load) that has a tension member with a connection resistance ($N_2^* = 455$ kN) that is larger than the tensile resistance of the member itself. The member resistance is

$\phi A_2 F_{y2} = 0.9 (1260) 0.350 = 397$ kN, which will govern.

13.3.4 Resistance of an Overlap K Connection

This discussion continues from Section 13.3.3, where it briefly appeared that an overlap connection might be necessary for panel point No. 4. Had that been the case, the following would have been used.

Confirm validity of dimensional parameters (Section 13.3.1).

See Fig. 13.11(b).

Parameters not previously confirmed are

3. $t_i/t_j = 4.78/4.78 \leq 1.0$ ∴ OK
4. $-0.55 \leq e/d_0 = -32.3/219 = -0.15 \leq 0.25$ ∴ OK
5. Minimum overlap of 25% is satisfied.

Determine connection resistance of compression diagonal 3–4.

Use Design Chart, Fig. 3.17.

$$N_1^* = C_K \frac{t_0}{t_1} \frac{1}{\sin\theta_1} f(n') A_1 F_{y1}$$

$d_1/d_0 = 0.64$, as before
$d_0/t_0 = 23.0$, as before

$C_K = 0.43$

$$\therefore N_1^* = 0.43 \frac{9.53}{4.78} \frac{1}{0.8} 0.759 (2\,050) 0.350$$

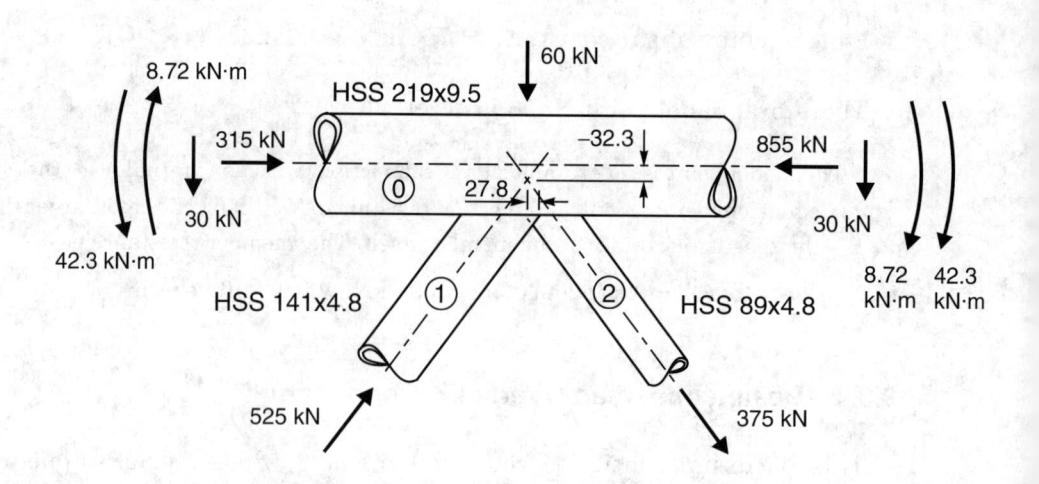

$= 584 \text{ kN} \geq 525$ ∴ OK

FIGURE 13.11(b)
Panel point No. 4—overlap connection

It is worth keeping in mind that connection resistances for circular HSS increase only marginally with increasing overlap, unlike the case for rectangular HSS.

No punching shear check is necessary for an overlap connection.

Determine connection resistance of tension diagonal 4–5.

$$N_2^* = N_1^* \frac{\sin\theta_1}{\sin\theta_2} = 584 \text{ kN also}$$

$$\geq 375 \quad \therefore \text{ OK}$$

Therefore, panel point No. 4 is acceptable.

Once again, there is a value for N_2^* that exceeds the resistance of the tension member itself,

$$\phi A_2 F_{y2} = 0.9(1260)0.350 = 397 \text{ kN}$$

which means that the member governs.

13.3.5 Top Chord as a Beam-Column between Panel Points 2 and 6

For illustrative purposes, consider the chord as though it had the overlap connection at panel point No. 4 from Section 13.3.4, rather than the gap connection actually used (Section 13.3.3). Nodal eccentricity of the overlap connection was larger, 32.3 mm.

Determine bending moments in the chord at panel point 4.

See Fig.13.11(b).

Moment from noding eccentricity is $e \, ((525 + 375) \text{ kN}) \cos\theta$

$$= 32.3 \,(900)\, 0.6 = 17\,400 \text{ kN·mm}$$

or 8.72 kN·m each side of the panel point.

Moment from purlin loads is considered to be $0.188 \, PL$, as discussed when selecting preliminary member sizes.

$$= 0.188 \,(60)\, 3.75 = 42.3 \text{ kN·m}$$

For design of the compression chord, *both* the moment due to transverse loading (purlin loads) and the moment due to noding eccentricity must be taken into account.

Fig. 13.12 shows the moment combinations existing at panel point No. 4:

42.3 − 7.25 = 35.0 kN·m for chord length 2–4

and 42.3 + 7.25 = 49.6 kN·m for chord length 4–6

Determine bending moment under the purlins.

Considering the chord as 2 continuous spans over panel point No. 4 (as was done for the chord moment at that panel point) gives a conservative value for the moment of 0.156 PL.

0.156(60)3750 = 35.1 kN·m

Fig. 13.12 shows the moment combinations at the purlins which are

35.1 + 8.72/2 = 39.5 kN·m for chord length 2–4

and 35.1 − 8.72/2 = 30.7 kN·m for chord length 4–6.

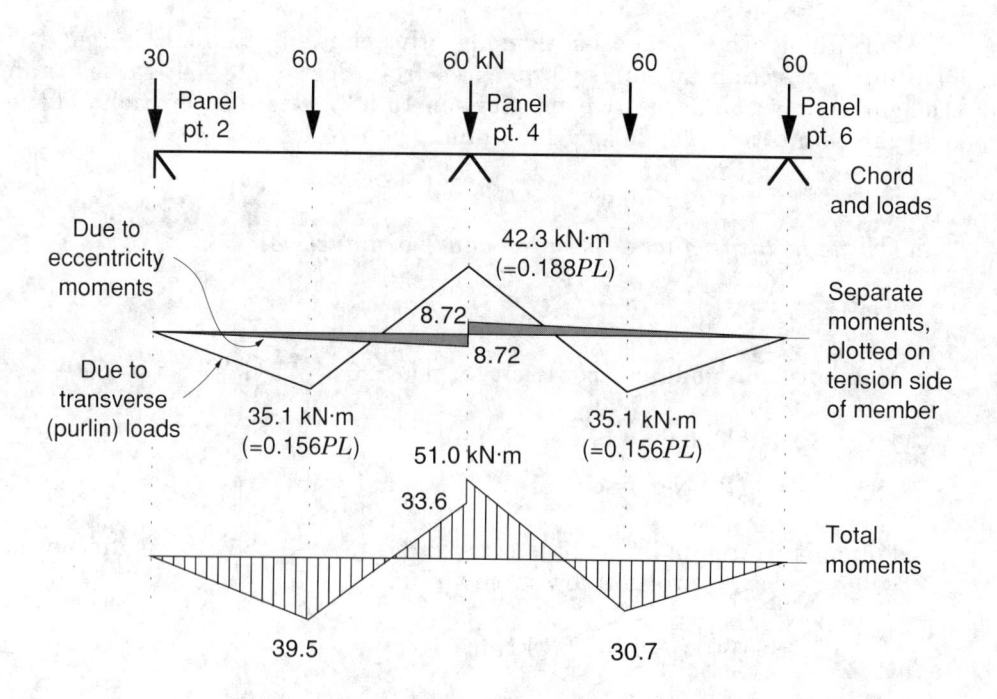

FIGURE 13.12
Simplified beam-column moments for manual design of chord

Use CAN/CSA-S16.1-94, Clause 13.8.1 for beam-columns.

Axial plus bending combination is $\dfrac{C_{f0}}{C_{r0}} + \dfrac{U_1 M_{f0}}{M_{r0}}$.

i) Overall member strength for chord length 4–6 (S16.1, Clause 13.8.1(b))

$C_{f0} = -855 \text{ kN} \qquad KL = 1.0\,(3\,750) = 3\,750$

$C_{r0} = -1\,580 \qquad (K = 1.0 \text{ for S16.1, Clause 13.8})$

$U_1 = \dfrac{\omega_1}{1 - \dfrac{C_{f0}}{C_e}} \qquad$ (S16.1, Clause 13.8.3)

$C_e = \dfrac{\pi^2 EI}{L^2}$

$\qquad = 3.14^2\,(200)\,34.5\,(10)^6 / 3\,750^2 = 4\,840 \text{ kN}$

$\omega_1 = 0.85 \qquad$ (S16.1, Clause 13.8.4(c))

$\therefore U_1 = 0.85 / (1 - (855 / 4\,840)) = 1.03$

M_{f0} = maximum moment = 51.0 kN·m (see Fig. 13.12)

(chord length 4–6 considered as two segments as per S16.1, Clause 13.8.4(c))

$M_{r0} = \phi Z_0 F_{y0} = 0.9\,(419)\,0.350 = 132 \text{ kN·m}$

$\therefore$ combination $\dfrac{C_{f0}}{C_{r0}} + \dfrac{U_1 M_{f0}}{M_{r0}}$

$\qquad = 855 / 1580 + 1.03\,(51.0) / 132$

$\qquad = 0.94 \quad \leq 1.0 \quad \therefore \text{ OK}$

ii) Overall member strength for chord length 2–4

$C_{f0} = -315 \text{ kN}$

$C_{r0} = -1\,580 \text{ kN}$

$U_1 = 0.85 / (1 - (315 / 4\,840)) = 0.909$

$M_{f0} = 39.5 \text{ kN·m}$

$M_{r0} = 132 \text{ kN·m}$

$$\therefore \text{combination} = 315/1580 + 0.909\,(39.5)/132$$
$$= 0.47 \quad \leq 1.0 \quad \therefore \text{OK}$$

iii) Cross-sectional strength for chord length 4–6 (S16.1, Clause 13.8.1(a))

$C_{f0} = -855\,\text{kN}$

$C_{r0} = -\phi A_0 F_{y0} = -0.9\,(6\,270)\,0.350 = -1\,980\,\text{kN}$

$U_1 = 1.0$

$M_{f0} = 51.0\,\text{kN·m}$

$M_{r0} = 132\,\text{kN·m}$

$$\therefore \text{combination} = 855/1980 + 1.0\,(51.0)/132$$
$$= 0.82 \quad \leq 1.0 \quad \therefore \text{OK}$$

iv) Cross-sectional strength for chord length 2–4

$C_{f0} = -315\,\text{kN}$

$C_{r0} = -1\,980\,\text{kN}$

$U_1 = 1.0$

$M_{f0} = 39.5\,\text{kN·m}$

$M_{r0} = 132\,\text{kN·m}$

$$\therefore \text{combination} = 315/1980 + 1.0\,(39.5)/132$$
$$= 0.46 \quad \leq 1.0 \quad \therefore \text{OK}$$

Therefore, the chord is acceptable as a beam-column.

13.3.6 Resistance of an X Connection

Panel point No. 13 is an X connection with the purlin on top and the vertical member underneath the chord.

Refer to Design Chart, Fig. 3.13.

$$N_1^* = C_X \frac{t_0}{t_1} \frac{1}{\sin\theta_1} f(n')\, A_1 F_{y1}$$

$$d_1/d_0 = 60.3/219 = 0.275$$
$$d_0/t_0 = 219/9.53 = 23.0, \text{ as before}$$

$\therefore C_X = 0.36$

$$n' = -1420/(6\,270\,(0.350)) = -0.65 \text{ and } f(n') = 0.67$$

$$1/\sin\theta = 1.0$$

$$\therefore N_1^* = 0.36\,\frac{9.53}{3.18}\,\frac{1}{1.0}\,0.67\,(571)\,0.350$$

$$= 144\,\text{kN} \quad \geq 60 \quad \therefore \text{ OK}$$

Therefore, panel point No. 13 is acceptable.

Panel point No. 10 is also an X connection, transferring the same force through the chord as in the example above (60 kN). The two web members framing into the chord at panel point No. 10 are both HSS 60x3.2 (as at panel point No. 13), but the load will be dispersed over a greater length along the underside of the chord, effectively increasing the connection resistance. This connection will hence be adequate.

13.3.7 Resistance of a KT Gap Connection

Panel point No. 7 is a KT connection. Unlike rectangular HSS, circular HSS do not have large minimum gaps as a function of relative member widths; this makes it possible to employ small gaps as a function of web member wall thicknesses. Consequently, gap connections are possible here, whereas overlap connections were necessary for the previous example that used square HSS.

Refer to Fig. 13.13.

Maximum permitted positive eccentricity of 168/4 = 42.0 mm gives gaps of 8.9 mm between adjacent members, which exceed the minimum required gap of 4.78 + 3.18 = 7.96 mm.

For KT connections the "gap" is taken as the largest gap between two web members having significant forces acting in the opposite sense. Here this is the distance between web members 2 and 3.

Determine connection resistance of compression diagonal 7–8.

For KT connections, the force components (perpendicular to the chord) of the two members acting in the same sense are added together to represent the load. The connection resistance component (perpendicular to the chord) of the remaining diagonal is then required to exceed that load.

$$N_1 \sin\theta_1 + N_3 \sin\theta_3 = 188\,(0.8) + 60\,(1.0)$$

$$= 210\,\text{kN}$$

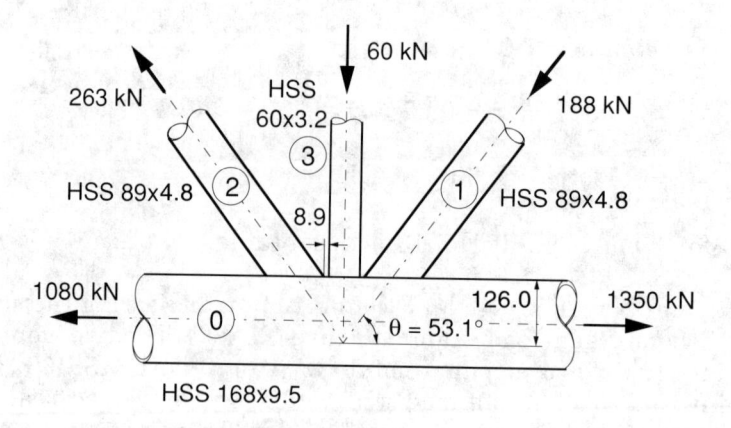

FIGURE 13.13
Panel point No. 7

Also, for KT connections, d_1/d_0 is replaced by $(d_1 + d_2 + d_3)/3d_0$
$$= (88.9 + 88.9 + 60.3) / 3 (168) = 0.472$$

$$N_1^* = C_K \frac{t_0}{t_1} \frac{1}{\sin\theta_1} f(n') A_1 F_{y1}$$

$$d_0/t_0 = 168/9.53 = 17.7$$

$$g' = 8.9/9.53 = 0.93$$

$\therefore C_K = 0.47$ from Design Chart, Fig. 3.14

$$\therefore N_1^* = 0.47 \frac{9.53}{4.78} \frac{1}{0.8} 1.0 (1260) 0.350 = 517 \text{ kN}$$

Since $N_2^* = N_1^* \frac{\sin\theta_1}{\sin\theta_2}$, and $\theta_1 = \theta_2$ in this example,

$$N_2^* = N_1^* = 517 \text{ kN}$$

$$\therefore N_2^* \sin\theta_2 = 517(0.8) = 414 \text{ kN} \quad \geq 210 \quad \therefore \text{ OK}$$

Another situation for HSS gap KT connections exists when there is an external cross-chord load (not the case in this example). Such a load could result in a controlling load for the diagonal which is acting alone in compression or tension. (See Section 3.2.3 and Figs. 3.10(c) and 3.10(d).)

Therefore, panel point No. 7 is acceptable, *with member resistances governing rather than connection resistances.*

The remaining panel points can be confirmed acceptable by comparing them with the ones already examined.

Table 13.4 shows this example truss with members, loads, and member and connection resistances.

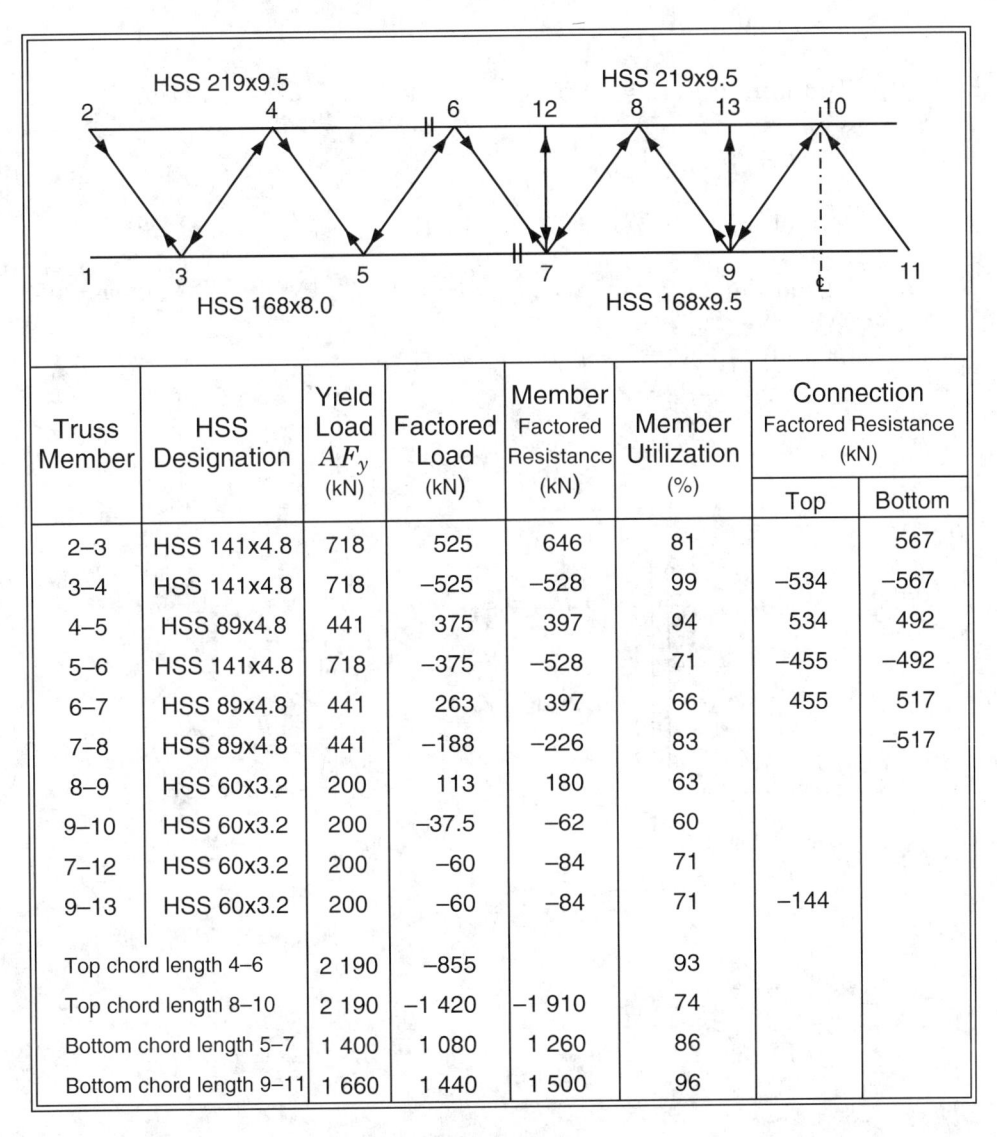

Truss Member	HSS Designation	Yield Load AF_y (kN)	Factored Load (kN)	Member Factored Resistance (kN)	Member Utilization (%)	Connection Factored Resistance (kN) Top	Connection Factored Resistance (kN) Bottom
2–3	HSS 141x4.8	718	525	646	81		567
3–4	HSS 141x4.8	718	−525	−528	99	−534	−567
4–5	HSS 89x4.8	441	375	397	94	534	492
5–6	HSS 141x4.8	718	−375	−528	71	−455	−492
6–7	HSS 89x4.8	441	263	397	66	455	517
7–8	HSS 89x4.8	441	−188	−226	83		−517
8–9	HSS 60x3.2	200	113	180	63		
9–10	HSS 60x3.2	200	−37.5	−62	60		
7–12	HSS 60x3.2	200	−60	−84	71		
9–13	HSS 60x3.2	200	−60	−84	71	−144	
Top chord length 4–6		2 190	−855		93		
Top chord length 8–10		2 190	−1 420	−1 910	74		
Bottom chord length 5–7		1 400	1 080	1 260	86		
Bottom chord length 9–11		1 660	1 440	1 500	96		

TABLE 13.4

Truss of circular HSS members: members, loads and resistances

13.3.8 Design of Bolted Flange-Plate Splice Connection

The splice in the bottom chord to the left of panel point No.7 can be made readily with bolted flange plates as shown in Fig. 13.14.

Refer to Section 7.1 and make a trial arrangement.

Load is 1080 kN

Try M24 A325M bolts with $T_r = 188$ kN

Use $a = b = 40$ mm

$d_0 = 168$ mm and $t_0 = 7.95$ mm (weaker side of connection)

$(d_0 - t_0)/(d_0 + 2b)$

$= (168 - 7.95)/(168 + 80) = 0.645$

∴ from Fig. 7.2, $f_3 = 5.1$

Flange plate thickness is given by [7.1] as

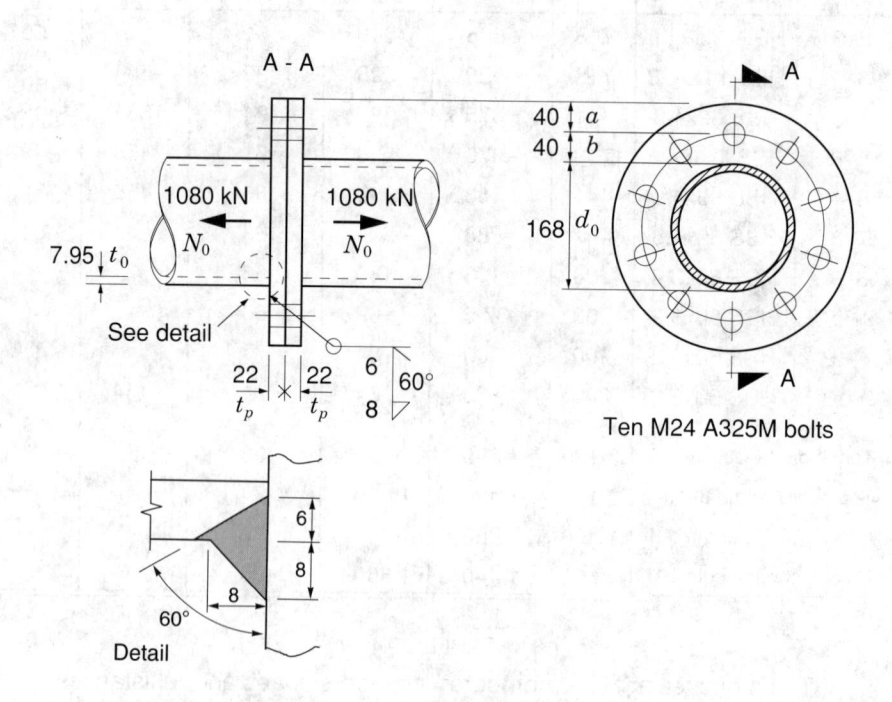

FIGURE 13.14
Flange-plate splice for bottom chord

$$t_p \geq \sqrt{\frac{2N_0}{\phi F_{yp} \pi f_3}}$$

$$\geq \sqrt{\frac{2(1\,080)}{0.9\,(0.300)\,\pi\,(5.1)}} = 22.3$$

∴ use 22 mm thick plate (accepting the 1.4% underdesign)

The required number of bolts is given by [7.3] as

$$n \geq \frac{N_0}{\phi T_r}\left(1 - \frac{1}{f_3} + \frac{1}{f_3 \ln(r_1/r_2)}\right)$$

$$r_1 = \frac{d_0}{2} + 2b = \frac{168}{2} + 2(40) = 164$$

$$r_2 = \frac{d_0}{2} + b = \frac{168}{2} + 40 = 124$$

$$\ln(r_1/r_2) = \ln(164/124) = 0.280$$

$$\therefore n \geq \frac{1\,080}{0.9\,(188)}\left(1 - \frac{1}{5.1} + \frac{1}{5.1\,(0.280)}\right)$$

$$\geq 9.6 \quad \therefore \text{ use 10 bolts}$$

Circumferential bolt spacing at the bolt line is

$$\pi\,(168 + 80)\,/\,10 = 78 \text{ mm} = 3.25d$$

Minimum recommended bolt spacing is $3d$
 but not less than $2.7d$ (CAN/CSA-S16.1-94, Clause 22.5)

Therefore, spacing is acceptable.

Weld

Use a reinforced partial joint penetration groove weld to attach the flange plate for the tensile resistance of the HSS wall (see Fig. 13.14).

Wall resistance is $\phi t_0 F_{y0}$

$= 0.9\,(7.95)\,0.350$

$= 2.50$ kN per mm of circumference

PJPG weld resistance T_r from CAN/CSA-S16.1-94, Clause 13.13.3.2 is

$\phi_w A_n F_u$ per mm of circumference.

If A_n is 6.0 mm (requires a 60° angle for the 6 mm preparation),

$T_r = 0.67 \, (6.0) \, 0.450 = 1.81$ kN < 2.50 ∴ not enough

Try reinforcing the PJPG weld with an 8 mm fillet weld:

$$T_r = \phi_\omega \sqrt{(A_n F_u)^2 + (A_w X_u)^2} \qquad \text{(S16.1-94, Clause 13.13.3.3)}$$

$$= 0.67 \sqrt{(6.0 \, (0.450))^2 + (0.707 \, (8) \, 0.480)^2}$$

$$= 2.57 \text{ kN} \qquad > 2.50 \qquad \therefore \text{ OK}$$

Use 8 mm fillet weld reinforcement on 6 mm PJPG weld.

An alternative would be a fillet weld with a throat equal to 1.10 times the wall thickness of the HSS (see Section 8.2.4.2.1).

i.e., $1.10 \, (7.95) \, 1.414 = 12.4$ ∴ a 14 mm fillet weld

13.3.9 Truss to Column Connection

A common truss to column connection, shown in Fig. 13.15, uses a gusset plate which is slotted into the end members of the truss. The gusset is field bolted to the web of a tee welded onto the face of the column.

The depth of the slot into an HSS member for the gusset is influenced by the length of welds needed to develop the member load after shear lag considerations (discussed in Section 7.5.1).

Weld for the diagonal

Determine required effectiveness of the diagonal.

Load = 525 kN   Area = 2 050 mm^2

Effective net area (18 mm slots) = $2\,050 - (2 \, (18) \, 4.78) = 1\,880$ mm^2

∴ required effectiveness is $525 / (0.9 \, (1\,880) \, 0.350) = 0.887$

Hence, the shear lag factor should exceed 0.887.

The shear lag factor is 0.87 when the weld lengths are at least 1.5 times the distance around the HSS wall between the welds (CAN/CSA-S16.1-94, Clause 12.3.3.3(b)(ii)). If this 2% deficiency is not acceptable, increase the 1.5 factor above to 2.0 as in Clause 12.3.3.3(b)(i) of S16.1, and thereby avoid any loss of section due to shear lag.

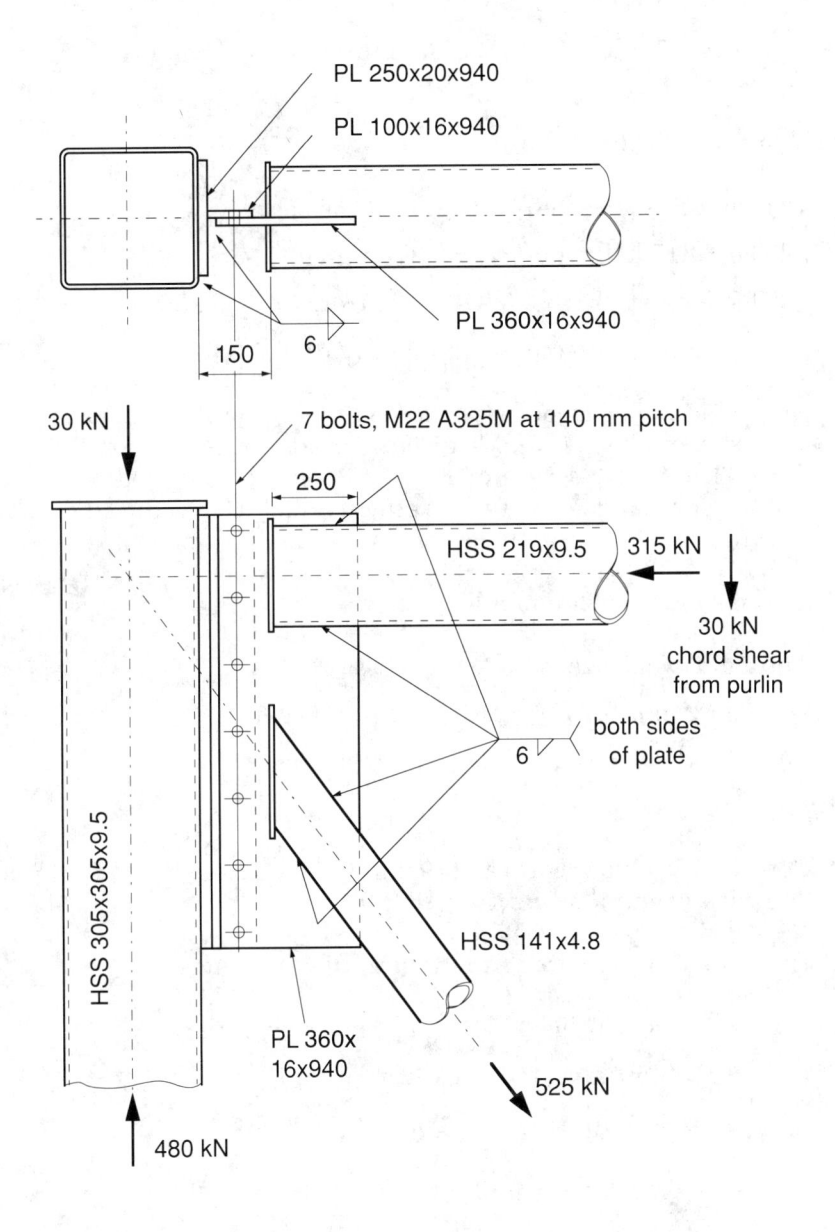

FIGURE 13.15
Truss to column connection

Distance between welds is

$(\pi d_2 / 2) - t_p = \pi(141)/2 - 16 = 205$ mm

so the minimum weld length is $2.0(205) = 410$ mm

and horizontal length of slot is $410(\cos 53.1°) = 246$, say 250 mm.

439

Minimum fillet weld size is

$525 / (410 (4) 0.152) = 2.1$ mm

∴ use 6 mm   (W59-M1989 minimum for 16 mm plate)

Weld for the chord

Determine required effectiveness of the chord.

Load = 315 kN   Area = 6 270 mm^2

Effective net area is $6\,270 - (2\,(18)\,9.53) = 5\,930$ mm^2

∴ required effectiveness is $315 / (0.9\,(5\,930)\,0.350) = 0.17$

Shear lag is not a consideration when required member effectiveness is less than about 0.5. See discussion at end of Section 7.7.2 and Fig. 7.26. (The vertical axis of Fig. 7.26 can be regarded as the member effectiveness in the context of effective net area reduced for shear lag. A value of 0.17 is way below the shear lag realm.)

Chord welds are 250 mm long

Minimum weld size is

$315/(250(4)0.152) = 2.1$ mm   ∴ use 6

Design of the tee

Seven M22 A325M bolts at 140 mm pitch will be sufficient (by comparison with example in Section 13.2.9).

Examine the weld of the tee web to the tee flange:

Using CISC Table 3–34 with weld size = 6 mm:

$P = 450$ kN $t = 16$ mm

$l = 940$ mm $al = 305/2 + 20 = 173$ mm

∴ $a = 0.18$ and $C' = 1.84$

Weld resistance is $1.84\,(940) = 1730$ kN   ≥ 450   ∴ OK

Weld of the tee flange to the column is similar   ∴ OK

The width of the tee flange was selected to be almost the full width of the flat portion of the column face, leaving just enough for a pair of fillet welds. A width to thickness ratio of 13 is recommended for tee flanges when used to connect beams (Section 9.2.1), but 12.5 is not expected to be a problem with this application.

REFERENCES

CISC. 1995. Handbook of steel construction, 6th ed. Canadian Institute of Steel Construction, Willowdale, Ontario.

CSA. 1994. Limit states design of steel structures, CAN/CSA-S16.1-94. Canadian Standards Association, Rexdale, Ontario.

CSA. 1989. Welded steel construction (metal arc welding), W59-M1989. Canadian Standards Association, Rexdale, Ontario.

CSA. 1992. General requirements for rolled or welded structural quality steel, CAN/CSA-G40.20-M92. Canadian Standards Association, Rexdale, Ontario.

CSA. 1992. Structural quality steels, CAN/CSA-G40.21-M92. Canadian Standards Association, Rexdale, Ontario.

CSA. 1992. Metric dimensions of structural steel shapes and hollow structural sections, CAN/CSA-G312.3-M92. Canadian Standards Association, Rexdale, Ontario.

IIW. 1989. Design recommendations for hollow section joints—predominantly statically loaded, 2nd. ed. International Institute of Welding Subcommission XV-E, IIW Doc. XV-701-89, IIW Annual Assembly, Helsinki, Finland.

INDEX

Abbreviations, *38*
Advantages of HSS, *1, 4, 45-46*
Architectural Institute of Japan (AIJ), *13*
American Inst. of Steel Construction (AISC), *12, 13, 18, 193*
American Society for Testing and Materials (ASTM), *12-16, 19*
Analysis of trusses, *50, 60, 358*
American Welding Society (AWS), *10, 12, 173, 272*
Area,
    Gross, *236, 250*
    HSS sections, *20-31*
    Net, *236, 248*
    Effective net, *236, 245, 250*
    Effective net, reduced for shear lag, *238, 247, 251*
Atlas Tube, *14*
Australia, *14*
Base plates,
    Bi-axial bending, *340*
    Fatigue, *376, 377*
    Pinned anchorages, *337*
    Simple, *335*
    Tension, *338*
    Unidirectional bending, *340*
Beam-columns, *199, 393, 429*
Bending HSS,
    Cold bending, *260-266*
    Hot bending, *266*

Bending moments,
    Primary, *50-52*
    Secondary, *51, 52*
    With axial forces, *92, 191, 210*
Block shear, *236*
Bolted connections,
    Beam to column, *301, 309*
    Blind, *227*
    Flow-drilled, *229-231*
    Gusset plates, *233, 239, 244, 248*
    Holes, *237*
    Huck, *227, 228*
    Pretensioned, *224*
    Prying, *222, 339, 403, 436*
    Purlins, *242*
    Shear tabs, *309-313, 316, 328*
    Tensile, *215, 220, 224, 241, 401, 436*
    With HSS ends flattened, *243*
    With welded attachments, *241*
Bridges, *2, 55, 56*
Canadian Inst. of Steel Construction (CISC), *vi, 7*
Canadian Standards Association (CSA), *4, 14-19, 271, 272*
Canadian Welding Bureau (CWB), *272, 279, 296*
Charpy V-notch, *16, 370*
China, *13*
CIDECT, *7, 8, 10, 50, 53-56, 67, 148, 355*
Class (C, H), *4, 14, 16, 370, 381*
Class (1, 2), *18, 89-91, 195, 201*

443

Colour codes, *257*
Compression members,
   Beam-columns, *199, 393, 429*
   Chords, *50, 55, 59*
   Effective length, *53-55*
   Laterally unsupported, *55*
Computer programs, *114, 148*
Concrete-filled HSS, *4-6, 10, 18, 127, 328*
Connections,
   Bird-mouth, *155*
   Concrete reinforced, *127, 328*
   Cranked chord, *135*
   Flange-plate, *215, 220, 224, 401, 436*
   Gap, *47, 61, 68, 133, 134, 388, 420*
   Gusset plate, *244, 249*
   KK, *169, 171, 177*
   KT, *77, 397, 433*
   Knee, *201*
   Moment, (see **Moment connections**)
   Multiplanar, *163*
   Nailed, *267-270*
   Overlap, *47, 54, 61, 68, 397, 427*
   Plate (longitudinal) to rectangular HSS, *345*
   Plate (transverse) to rectangular HSS, *347*
   Plate to circular HSS, *349*
   Reinforced, *121-127, 306, 308, 320, 325*
   Simple shear, *302, 309, 328*
   TT, *169, 174*
   TX, *167, 174*
   Tensile, *215, 220, 224, 241*
   Truss to HSS columns, *333, 410, 438*
   Welded, (see **Welds**)
   Wide flange to HSS columns, *309-331*
   X, *396, 432*
   XX, *165, 174*
Connection resistance tables, *80-91, 110, 115-120, 265, 266, 350, 351*
Continuity from beam to beam,
   Flange diaphragms, *317*
   Strap angles, *316*
   Through-plates, *316*
Continuous beams, *314*
Corners of HSS, *14, 16, 270, 280, 281*
Corrosion, *198, 270, 364*
Costs, *4, 45, 46, 121, 143, 312*
Cracks, *258, 270, 356*
Cropped web members, *148*
Deflection (trusses), *60, 152, 414*

Design examples,
   Bolted gusset plates, *244*
   Fatigue, *371*
   Gap K connection, *133, 388, 420*
     concrete-filled, *134*
   KT gap connection, *433*
   KT overlap connection, *397*
   Overlap K connection, *427*
   Plate slotted into HSS, *249*
   Rectangular flange-plate splice, *401*
   Round flange-plate splice, *436*
   TX connection, *167*
   Truss chord as beam-column, *199, 393, 429*
   Truss to column connection, *410, 438*
   Vierendeel truss, *192*
   Warren truss with circular HSS, *416*
   Warren truss with square HSS, *381*
   Welded joints, *405, 412*
   X connection, *396, 432*
   XX connection, *166*
Designation of HSS sections, *19-31*
   (see also **Size of HSS sections**)
Deutsches Institut für Normung (DIN), *201*
Eccentricity, *50, 67, 174, 312*
Effective,
   gusset width, *239*
   length, *53, 54*
   net area, *236, 245, 250*
   net area reduced for shear lag, *238, 247, 251*
   welds, *288, 290-292*
   width failure, *72, 75*
Efficiency, *99*
Efficiency charts, *100-114*
Elastic modulus, E, *32*
Electrodes, *272*
Elongation, *15*
Eurocode 3, *11, 52, 83, 288*
Examples, (see **Design examples**)
Fabrication, *46, 257-270*
Failure modes, *70, 186, 231, 236, 269, 320*
Fatigue,
   Amplification factors, *359*
   Base plate, *376, 377*
   Chapter, *353-379*
   Classification approach, *355, 376*
   Fatigue life curves, *361, 377*
   Hot spot stress, *355, 360*

Material, *370*
Miner's rule, *363*
Stress concentration factors (SCF), *353, 360, 365, 366*
Stress range, *356*
Stress ratio, *356*
Strain concentration factors (SNCF), *353*
Summary, *364*
Weld throat, *293*
Fillet welds, (see **Welds**)
Fire protection, *4, 6, 10*
Flattened HSS ends, *148, 258*
Flexibility, *192, 301*
Flowdrilling, *229*
Germany, *13*
Grades (of steel), *12, 15, 16*
Groove welds, (see **Welds**)
Gusset plates, *233, 239, 244, 248*
Haunches, *308*
Holes, *229, 237*
Home study, *297*
Inspection,
  Liquid penetrant, *294, 296*
  Magnetic particle, *294, 296*
  Radiographic, *295, 296*
  Ultrasonic, *295, 296*
  Visual, *294, 296*
Interaction equations,
  Axial plus in-plane moments, *92*
  Axial plus in-plane and out-of-plane moments, circular HSS, *210*
  Beam-columns, *199, 393, 429*
  Vierendeel trusses, *191*
International Institute of Welding (IIW), *8-10*
Ipsco Inc., *14*
Japan, *13*
Joists, *158*
K-factors (see **Effective length**)
Lateral-torsional buckling, *193*
Mass of HSS sections, *12, 17, 20-31*
Modulus of elasticity, E, *32*
Moment connections, *186*
  Behaviour, *186*
  Interaction with axial load, *191*
  Reinforced, *306*
  Rigid, *192, 206, 211*
  Semi-rigid, *192, 206, 211*
  T and X (in-plane), *205, 206*
  T and X (out-of-plane), *207, 209*

  Wide flange to HSS columns, *313-331*
Moments of inertia of HSS, *20-31*
Nailing HSS, *267-270*
Notch tough steel, *16*
Parameters, validity ranges, *88-91, 92-98, 149, 151, 154, 176, 191, 202, 365, 370*
Plastic design, *52, 193*
Properties of HSS sections, *20-31*
Prying action on bolts, *222, 339, 403, 436*
Purlin connections, *242*
Radius of gyration of HSS sections, *20-31*
Reinforcing (by),
  Angles, *325*
  Haunches, *308*
  Plates, *121-127, 306, 320*
  Tees, *327*
Researchers,
  Agerskov, *363*
  Ales, *309*
  Astaneh, *302, 313*
  de Back, *221*
  Banks, *230*
  Bauer, *178*
  Bellerini, *230*
  Bergmann, *10, 13*
  Birkemoe, *221, 223*
  Brockenbough, *185*
  Brodka, *185*
  Caravaggio, *224*
  Cassidy, *292*
  Chidiac, *145*
  Coutie, *60, 178*
  Cran, *8*
  Crockett, *176, 177*
  Cute, *183*
  Czechowski, *60*
  Davies, *174, 176, 177, 185, 189, 309, 347*
  Dawe, *146, 302, 313, 320, 323, 329*
  Dekkers, *230*
  Duff, *183*
  Dutta, *10, 13, 123, 201, 215*
  Eastwood and Wood, *7*
  Fang, *183, 302, 309, 311, 312*
  Fear, *128*
  Frater, *60, 292, 355*
  Galambos, *53*
  Giddings, *185, 347*
  Giroux, *317*
  Goel, *18*

Govil, *185*
Griffiths, *215*
Grondin, *320, 324*
Guravich, *323, 324*
Haleem, *7*
Harrison, *185*
Henderson, *227*
Herion, *355*
Horne, *196*
Igarashi, *216*
Jaspart, *341*
Kamba, *313, 317, 326*
Kanatani, *185, 317*
Kato, *220, 224, 317*
Kennedy, *260*
Kitipornchai, *241*
de Koning, *189, 347*
Korol, *127, 145, 147, 148, 185, 186, 190, 231, 252, 306-308*
Kosteski, *270*
Krutzler, *267, 268*
Kurobane, *70, 73, 163-170, 172-174*
Lau, *146, 154*
Lazar, *183*
Liu, *176-177*
Lu, *313, 348*
Luff, *143, 148*
Lui, *18*
Makino, *125*
Mang, *185, 201, 220, 224, 355*
Mansour, *185*
Mehendale, *302, 313*
Mehrotra, *183, 185*
Mirza, *185*
Mitri, *147*
Morita, *174*
Morris, *150, 154, 196*
Mouty, *53, 54*
Mukai, *220, 224*
Nader, *302, 313*
Niemi, *10, 355*
Ono, *156*
Owen, *156*
Packer, *7, 9-11, 13, 60, 73, 98, 110, 128, 130, 137, 154, 163, 178, 189, 197, 215, 220, 221, 223, 267, 268, 292, 309, 328, 347, 355*
Panjeh Shahi, *185, 355*
Parik, *11*
Paul, *170*

Pedersen, *363*
Philiastides, *60*
Picard, *317*
Post, *297*
Puthli, *355*
Redwood, *178, 183*
Reusink, *99*
Riviezzi, *266*
Rockey, *215*
Rondal, *10, 13, 53, 54*
Sadri, *227*
Shanmugam, *326*
Sedlacek, *210*
Sherman, *18, 230, 301, 309, 312*
Shinouda, *123*
Staples, *185*
Stark, *70*
Struik, *221*
Szlendak, *185*
Tanaka, *232, 313*
Tabuchi, *313*
Thornton, *239*
Ting, *326, 327*
Traves, *241*
Twilt, *10, 13*
Vandegans, *341*
van der Vegte, *166, 167*
Wardenier, *8-11, 13, 70, 72, 73, 99, 123, 148-150, 156, 163, 174, 186-191, 201, 206, 209, 210, 241, 259, 260, 329-331, 347, 348, 355*
Watson, *47, 48, 121*
White, *302, 309, 311, 313*
Whitmore, *239*
van Wingerde, *11, 12, 294, 355, 357, 362*
de Winkel, *331*
Würker, *10, 13, 123, 201, 215*
Yamamoto, *328*
Yeomans, *178, 230, 231*
Yrjölä, *355*
Residual stress, *14*
Resistance factor, *17, 139, 191*
Section modulus of HSS sections, *20-31*
Seismic use, *18, 313*
Shear,
   Block, *236*
   HSS beams, *306*
   Lag, *238, 247, 251, 438, 440*
   Tabs, *309-313, 316, 328*

Shear modulus, G, *33*
Size of HSS sections, *15, 19-31*
  (See also **Designation of HSS sections**)
Sonco Steel Tube, *14*
Sources of HSS, *14*
South Africa, *14*
Staggered holes, *236*
Standards, *12, 14, 53, 61, 271, 272*
Standard Tube Canada Inc., *14*
Stelco, *7, 8, 270*
Steels,
  Grades, *12, 15, 16*
  Mechanical properties, *15*
STI, *12, 13, 17*
Stress range (see **Fatigue**)
Summaries,
  Fatigue life, *364*
  Moments for truss design, *52*
  Truss design, *62*
Symbols, *32*
Tensile strength, *15*
Tension members,
  Effective net area, *236, 245, 250*
  Effective net area reduced for shear lag, *238, 247, 251*
  Failure modes, *236*
  Gross area, *236, 250*
  Holes, *237*
  Net area, *236*
  Resistance, *236*
Terminology, *68*
Trusses,
  Analysis, *50, 60, 358*
  Chord walls, *60*
  Column connections, *333, 410, 438*
  Compression chord, *50, 60, 199, 393, 429*
  Computer modeling, *50*
  Cranked chord, *135*
  Cropped web members, *148*
  Deflection, *60, 152, 414*
  Design examples, *192, 381, 416*
  Design summary, *62*
  Double chord, *141*

Fatigue, *358*
Fink, *49*
Flattened web member ends, *148, 258*
Pony, *55*
Pratt, *49, 69*
Preliminary members, *383, 384, 416, 417*
Static strength, *60*
Triangular, *171, 177*
Vierendeel, *183*
Warren, *49, 69, 381, 416*
Web members, *53, 60, 66*
Webs framing on to chord corners, *155*
Wide flange chords, *79, 86*
U.S.A., *12*
Uses of HSS, *1*
Vierendeel trusses, *183*
  Elastic design, *199*
  Plastic design, *193*
Welds,
  Complete penetration, *278, 284, 298*
  Effectiveness, *288, 290-292*
  Fatigue, *293, 370*
  Fillet, *48, 250, 274, 280, 288, 297, 371, 406*
  Flare bevel, *48, 281-283*
  Inspection, (see **Inspection**)
  Length, *287, 290*
  Overlapped members, *79*
  Partial penetration, *276, 281, 282, 297, 437*
  Prequalified, *279, 282, 284, 285*
  Shear lag, *239, 251, 438, 440*
Welded Tube of Canada Ltd., *14*
Welding,
  Electrodes, *272*
  Problems, *297*
  Processes, *273, 274*
  Symbols, *297*
Wide flange members,
  Truss chords, *79, 86*
  Beams, *309-331*
Yield strength, *15*

NOTES